*Mayo Ethnobotany*

*The publisher gratefully acknowledges the generous contribution to this book provided by the General Endowment Fund of the Associates of the University of California Press.*

# Mayo Ethnobotany

*Land, History, and Traditional Knowledge in Northwest Mexico*

DAVID YETMAN AND THOMAS R. VAN DEVENDER

*University of California Press*
BERKELEY   LOS ANGELES   LONDON

University of California Press
Berkeley and Los Angeles, California

University of California Press, Ltd.
London, England

© 2002 by the Regents of the University of California

Library of Congress Cataloging-in-Publication Data

Yetman, David, 1941–
    Mayo ethnobotany : land, history, and traditional knowledge in northwest Mexico / David Yetman and Thomas R. Van Devender.
      p.    cm.
    Includes bibliographical references and index.
    ISBN 0-520-22721-2 (cloth : alk. paper)
    1. Mayo Indians—Ethnobotany.  2. Ethnobotany—Mexico—Sonora (State).  3. Ethnobotany—Mexico—Sinaloa (State).
I. Van Devender, Thomas R.  II. Title.
F1221.M3 Y46 2002
581.6'0972'17—dc21                       2001003101

Manufactured in the United States of America

11  10  09  08  07  06  05  04  03  02

10  9  8  7  6  5  4  3  2  1

The paper used in this publication meets the minimum requirements of ANSI/NISO Z39.48-1992 (R 1997) (*Permanence of Paper*).♾

# Contents

*List of Illustrations* — vii
*Preface* — xi
*Acknowledgments* — xiii

PART ONE

1. The People and the Land — 3
2. A Brief Ethnography of the Mayos — 16
3. Historical and Contemporary Mayos — 30
4. Plant and Animal Life — 62
5. Eight Plants That Make Mayos Mayos — 79

PART TWO

6. Plant Uses — 109
7. An Annotated List of Plants — 127

*Appendix A. Mayo Region Place Names and Their Meanings* — 273
*Appendix B. Yoreme Consultants* — 276
*Appendix C. Gazetteer of the Mayo Region* — 281
*Appendix D. Mayo Plants Listed by Spanish Name* — 293
*Appendix E. Mayo Plants Listed by Mayo Name* — 303
*Appendix F. Glossary of Mayo and Spanish Terms* — 313
*Notes* — 319
*Works Cited* — 331
*Index* — 337

# Illustrations

Unless otherwise credited in the caption, all photographs are by David Yetman. All drawings are by Paul Mirocha.

| | |
|---|---|
| The land of the Mayos (map) | 2 |
| Coastal thornscrub near Jambiolobampo | 5 |
| *Pascola* and deer dancers, Masiaca | 6 |
| Nelo López, Yavaros | 7 |
| View of coastal flats and thornscrub | 9 |
| The upper Río Mayo north of San Bernardo | 14 |
| Norma Humo and her son, José Luis Baisegua, Sirebampo | 20 |
| Holy Saturday in Masiaca | 21 |
| *Fariseos* (Pharisees), Benito Juárez | 21 |
| *Fariseo*, Etchojoa | 22 |
| Men watching *pascolas*, Masiaca plaza, Holy Week | 23 |
| Don Reyes Baisegua, Sirebampo | 24 |
| Doña Amelia Verdugo, Los Muertos | 25 |
| Plaza, Nahuibampo | 27 |
| Mayo transportation, Masiaca | 28 |
| Burros, Masiaca | 28 |
| House of woven *pitahaya* arms, Bachomojaqui | 31 |
| House of adobe walls and dirt roof, Teachive | 32 |
| The Catholic church in Masiaca | 37 |
| Mayo family, Yópori | 58 |

Clearing for buffelgrass — 59
Vicente Tajia contemplates the *pitahayal* — 60
Mangrove estuary near Topolobampo, Sinaloa — 63
Coastal thornscrub near Camahuiroa — 64
Coastal thornscrub near Jambiolobampo — 66
Foothills thornscrub, Cerro Terúcuchi — 68
Foothills thornscrub, Mesa Masiaca — 69
Tropical deciduous forest near Alamos, September — 70
Mayo home, tropical deciduous forest, Los Capomos, Sinaloa — 71
Transitional tropical deciduous forest near Yocogigua, March — 72
Tropical deciduous forest in drought near Alamos — 73
Tropical deciduous forest with *sabinos* (*Taxodium distichum*), Río Cuchujaqui — 75
Río Cuchujaqui gorge near Alamos — 75
*Ébano* (*Caesalpinia sclerocarpa*) on Arroyo Camahuiroa — 76
*Pascola* mask carved by Francisco Gámez — 80
Filemón Navarete, Ejido Fernando Solís — 82
Vicente Tajia gathering *pitahaya* fruits near Coteco — 83
*Aaqui nábera* (crestate *pitahaya*) — 84
*Pitahaya* fence, near Bachoco — 85
Pulp of the *pitahaya* fruit — 86
*Aca'ari* (*pitahaya* gathering basket) made by Serapio Gámez — 88
*Pitahaya zarca* (white *pitahaya* fruit) — 89
*Cu'u* (*mezcal*; *Agave vivipara*) — 91
Don Reyes Baisegua with a mass of *ixtle* (agave fiber) — 93
*Morral* (*ixtle* handbag) woven by Don Reyes Baisegua — 94
Vicente Tajia with *etcho* cactus (*Pachycereus pecten-aboriginum*) near Camahuiroa — 96
*Etcho* fruit — 97
Living fence of *etchos*, *sahuiras*, and *pitahayas* near Güirocoba — 98
Overgrazed pasture with *etchos* — 99
*Sato'oro* (*Jatropha cordata*) in thornscrub near Teachive — 101
*Sato'oro*'s shiny bark — 102

| | |
|---|---:|
| *Saya* (*Amoreuxia palmatifida*) | 104 |
| A male *jito* (*Forchhammeria watsonii*) tree in flower near Huebampo | 105 |
| Vicente Tajia beneath an enormous *pitahaya* | 108 |
| *Criollo* (creole) cows | 111 |
| Corral of *brasil* trunks (*júchajco* in Mayo; *Haematoxylum brasiletto*) | 112 |
| Aurora Moroyoqui weaving a *cobija* (blanket) | 114 |
| Dried flowers of *sanjuanico* (*tásiro* in Mayo; *Jacquinia macrocarpa*) | 121 |
| *Baogoa* (*mezcal*; *Agave* cf. *colorata*) from Sibiricahui | 136 |
| Doña Artemisa Palomares of Los Capomitos, Sinaloa | 140 |
| Mabis, fruit of *mabe* (*tonchi*; *Marsdenia edulis*) | 142 |
| Woven wall of *batayaqui* (*Montanoa rosei*) poles | 147 |
| *Sonajas* (rattles) made of *ayales* (fruits of *Crescentia alata*), Los Capomos | 150 |
| Young *pochote* (*Ceiba acuminata*) near Yocogigua | 154 |
| *Cuajilote* (*Pseudobombax palmeri*) in dry season near Güirocoba | 155 |
| A leafless *palo mulato* (*Bursera grandifolia*) | 159 |
| *Sahuaro* (*saguo* in Mayo; *Carnegiea gigantea*) on Mesa Masiaca | 162 |
| *Nómom* (*Peniocereus striatus*) cactus | 168 |
| Domingo Ibarra with *sina cuenoji* (*Selenicereus vagans*) on *palo chino* near Huasaguari | 169 |
| Francisco Matus and *sinaaqui*, an apparent hybrid between *pitahaya* and *sina*, near Piedra Baya | 170 |
| An exceptionally large *sahuira* (*Stenocereus montanus*) near Baca, Sinaloa | 172 |
| *Jútuguo* (*Ipomoea arborescens*) in leafless transitional tropical deciduous forest near Yocogigua, March | 179 |
| *Güerequi* (*choya huani* in Mayo; *Ibervillea sonorae*) near Teachive in the dry season | 181 |
| Stakes of *vara blanca* (*cuta tósari* in Mayo; *Croton fantzianus*) near Alamos | 185 |
| *Ébano* (*Caesalpinia sclerocarpa*), Arroyo Camahuiroa | 197 |
| Trunks of *brasil* (*Haematoxylum brasiletto*) | 203 |
| A *palo chino* (*Havardia mexicana*) near San Bernardo | 205 |
| *Brea* (*choy* in Mayo; *Parkinsonia praecox*) near Teachive | 212 |

Balbina Nieblas sweeping with an *escoba* of *jíchiquia*
(*Abutilon incanum*), Teachive                                                                225

*Tescalama* (rock fig, *báisaguo* in Mayo; *Ficus petiolaris*) near Chorijoa    230

Fruits of *papache borracho* (*pisi* in Mayo; *Randia obcordata*),
Huasaguari                                                                                    247

*Chiltepines* (*có'cori* in Mayo; *Capsicum annuum*) being dried near
Nahuibampo                                                                                    255

Chair of *guásima* (*agia* in Mayo; *Guazuma ulmifolia*) made
by Marcelina Valenzuela, El Rincón                                                            262

Leaves and fruits of *sanjuanico* (*tásiro* in Mayo; *Jacquinia macrocarpa*)   263

*Orégano* (*Lippia palmeri*) being stirred by Berta Valenzuela,
San Antonio                                                                                   268

# Preface

Tom and I happened into the Río Mayo region independently and explored the land for several years unbeknownst to each other. We became caught in the same intellectual maelstrom, undertook this project, and have emerged deeply changed by the work we can never fully complete. This study occupied us for six years, and our studies continue. During that time we developed strong new ties to the land, to the people who live on the land and know it as part of their being, and to the plants of the region, which have become seeming extensions of our being. Tom found a bride; I found a best friend.

The Río Mayo and the lands south to the Río Fuerte drew us because here the Sonoran Desert, where we have lived and worked for decades, dwindles away and merges into more tropical regimes. In the Río Mayo region the northernmost tropical deciduous forest flows from the foothills into the valleys. *Sahuaro* cacti give way to *etchos;* ironwoods disappear and kapoks appear; creosotebush vanishes and *mautos* pop up. At elevations where we find desertscrub and desert grasslands in the Sonoran Desert, we find tropical deciduous forest and pockets of oaks in the Mayo region. Not only could we head southward from Tucson, Arizona, and the Sonoran Desert and in eight hours or so find ourselves in the exotic forests of the tropics, but we could also venture to the east and climb into the junglelike slopes of the Sierra Madre Occidental. Higher in the mountains we found life forms more typical of Chihuahua to the east. We could hardly lose.

Yet while we were both drawn to the area for its natural history, we also came to know the people of the area. We established contacts, then friendships, and found ourselves always wanting to know more of the people and more of the landscapes.

I first visited Navojoa, the largest city on the Río Mayo, in 1961, putt-putting along Mexico's narrow highways on a motor scooter. Even then I had heard of Mayos. They are southern cousins of the better-known Yaquis, who lived along the Río Yaqui some fifty miles north of Navojoa. Many Yaquis who fled as refugees to the United States sixty years earlier now live in four barrios in Tucson. Their numbers included some Mayos, one of whom I met in Tucson.

From Navojoa I visited nearby Alamos in Mayo country and, shortly thereafter, Las Bocas on the coast and the Río Fuerte in Sinaloa. I bought some Mayo rugs and *taburetes,* hide-covered stools. I came to know a small Mayo community in the slums of Navojoa and spent several days there. I also developed a vague curiosity about the region's tropical vegetation, knowing that it was rich enough to support kapok trees and boa constrictors. Over the next few decades I made repeated visits to Alamos, to Los Tanques to the north, and to the Río Cuchujaqui to the east. It was not until 1990, however, that I began studying the Mayo region—its people, landscapes, and plants—in earnest. Coming back to the region to do more intensive study was like coming home. Tom and I teamed up on this study in 1993 and soon realized that we shared a passion for the Río Mayo and its people.

Since then we have logged more than fifty trips to the Mayo region. Several Mayos have become good friends of ours, their families nearly extensions of our own. We have spoken with as many native Mayo plant consultants as possible, comparing plant names and uses and noticing the variation in cultural practices. This volume is a compilation of what we found and what further library and herbarium research revealed. We were intimately involved in the publication of *Gentry's Río Mayo Plants* (Martin et al. 1998) and benefited from our many associations in compiling that volume. Scholars have for decades predicted the demise of Mayo culture. We, too, found in our fieldwork many distressing indications of erosion of Mayo cultural values, and of loss of identity and language. At times we have despaired that the Mayo culture could not possibly survive for another decade. But we were also heartened by the strength of much that we found.

David Yetman

# Acknowledgments

Many individuals and organizations provided assistance in the preparation of this manuscript. Agnese Nelms Haury, Native & Nature, the University of Arizona–University of Sonora Collaborative Grants, and the Jacobs Research Funds gave major financial support. The Southwest Center of the University of Arizona also provided funding, and the director, Joseph Wilder, offered ongoing encouragement as well. A grant from the Provost's Author Support Fund of the University of Arizona helped cover publication costs. We wish especially to thank Francisco Valenzuela Nolasco of El Rincón and Vicente Tajia Yocupicio of Teachive and their families for the hospitality rendered to us over several years' time and for sharing with us their vast knowledge of the region's plants. Ana Lilia Reina provided companionship in the field and shared her general knowledge of Sonoran plants and culture. We could not have accomplished the plant work without assistance from the various botanists in the Río Mayo Plants Project who shared their knowledge in *Gentry's Río Mayo Plants* (Martin et al. 1998) and assisted in determinations of difficult plants. We also relied heavily on the rich legacy of specimens housed in the University of Arizona Herbarium. Finally, we wish to acknowledge the contribution of our editor, Rose Vekony, who made systematic suggestions for the manuscript and shepherded it through the long process of publication.

# PART ONE

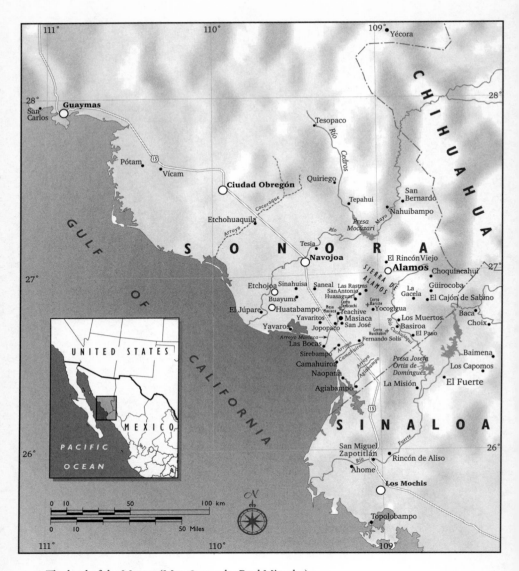

The land of the Mayos. (Map © 2002 by Paul Mirocha)

# 1 The People and the Land

BASEBALL, PASCOLAS, AND PLANTS

In 1980 one of the most popular major league baseball players was a Mexican named Fernando Valenzuela. On the nights he pitched his legendary screwball for the Los Angeles Dodgers, management could count on as many as ten thousand additional admissions. For five years bleacher seats were sold out for every home game in which Valenzuela pitched. He is a Mayo Indian from southern Sonora.

While public relations technicians extolled Fernando's humble peasant origins in promoting his image, little was made of the fact that in his youth Valenzuela had danced the *pascola,* the traditional Mayo dance, and that he spoke *la lengua,* the tongue of the Mayos. Nor did the North American press recall his Mayo origins and talk about the Mayos' pre-Columbian presence and their forcible eviction from their aboriginal lands.

Local legend has it that Cerro Bayájuri, a volcanic inselberg, or mostly buried mountain, jutting from the coastal plain near Etchohuaquila, Valenzuela's native village, is home to powerful spirits. Those Mayos who embark on important missions visit the steep, rocky hill to receive inspiration and benediction for their undertakings. The supplicant of good heart may be blessed. Others may be frightened by demons. Valenzuela, it is said, climbed Bayájuri and asked the spirits for power to be a good baseball player.

Etchohuaquila is a nondescript town whose name in Mayo means "skinny etcho cactus." It is one of dozens of small towns in the agriculturally fertile delta region of the Río Mayo, which empties into the Gulf of California.[1] It has little to offer pilgrims who visit the sacred birthplace of their sports hero. To the east and north is low thornscrub, an apt name for the short leguminous trees, cacti, and shrubs that scratch, prickle, pierce, and tear. It is heavily overgrazed, and overcut by those in search of firewood and fence posts. To the west and south lie hundreds of square miles of irrigated fields that look no different from those in California's San Joaquin Valley. Etchohuaquila itself is a drab, dusty town where the air is perpetually tinged with the acrid smell of agricultural chemicals. One might wander for many miles through the irrigated lands without finding a single native tree.

In the 1930s, Etchohuaquila, like most towns in the delta of the Río Mayo, Sonora's second largest river, was surrounded by rich coastal thornscrub. Scattered farms exploited the deep delta soils, irrigating with water pumped from deep wells or siphoned from a few canals connected with diversion dams that washed out every time the river flooded. Farmers grew cotton and wheat, for Mexican consumption, and garbanzo beans, most of which were sold to consumers in Spain and the Middle East. The natives of Etchohuaquila and most of the more than a hundred other Mayo towns wandered routinely in this thorny landscape (see, for example, Valenzuela Y. 1984). Here they spent countless hours gathering, gleaning, harvesting, cutting, chopping, and picking various roots, trunks, limbs, leaves, flowers, fruits, sap, and bark for an almost unending variety of purposes—for food, medicine, and firewood, of course, but also for tending to and feeding their livestock, and for dyes, mordants, fixatives, glues, adhesives, caulks, catalysts, cements, and excipients; for boards, logs, beams, joists, posts, rods, poles, sticks, timbers, and clubs; for beads, crucifixes, rosaries, talismans, nosegays, and festoons; for spoons, bowls, trays, weaving sticks, tripods, bags, rope; and for myriad other items whose need arose in their day-to-day lives. Natives would even cut trunks of *mezquite*, chop wedge-shaped steps, and lean them against a tree so that chickens could have predator-safe roosts, or against a house so people might more easily climb to the roof.

For many centuries this low, wild semiforest of small trees, bushes, and cacti served the Mayos well. They sweated through the scorching summers—choking on the dust of May and June, and sweltering in the steamy vapor of the July and August rainy season. The Río Mayo flooded twice each year, leaving behind fertile silt in which the Mayos sowed their seeds. Combining their crops of corn, beans, squash, and chiles with the free harvest of the *monte,* the Spanish term for the natural vegetation, they were strangers to hunger.

The lands of the Mayos and those of the Yaquis to the north were too fertile, too arable, and too lucrative to escape the eyes of Spanish, then Mexican and North American, colonists. After achieving independence from Spain early in the nineteenth century, Mexico developed ambitious plans for tapping the region's agricultural potential. For nearly a century Mayos and Yaquis resisted these plans with armed struggle, but by the early 1900s these rebellious natives had been "pacified"; that is, subjugated. Nevertheless, the chaos surrounding

Coastal thornscrub near Jambiolobampo.

the Mexican Revolution made orderly development impractical, if not impossible.² Finally, in the 1940s—with the revolutionary days finally over, power consolidated, Indians crushed, the great depression easing its grip on the nation, and the popular quasi-socialist regime of Lázaro Cárdenas behind—the Mexican government became serious about shaping the Mayo Delta into an irrigated empire. In an alliance with wealthy landowners, the government put up the funds to scrape most of the delta clear of vegetation. The Mocúzari Dam, about fifteen miles upstream from Navojoa, was completed in 1951. Workers also constructed a gridwork of canals, sank a few hundred deep wells, installed modern pumps, and built a network of paved highways to ensure the growers' access to local, national, and, they hoped, international markets.

By the mid-1950s most of the delta's thornscrub had been replaced by carefully tended fields that stretched across miles of unvarying flatness.³ The thornscrub that had provided food, fiber, forage, fence, and fortress was gone, and with it, part of the Mayos' way of life. In the delta region today only a few

6 / The People and the Land

*Pascola* and deer dancers, Masiaca, during a break from the rituals.
(Photo by T. R. Van Devender)

volcanic hills, jutting like icebergs from the flat fields, harbor original vegetation, and they are heavily grazed by goats. The ritual Mayo dancers—pascolas and *venados*—now must journey to upland villages such as Masiaca that are still surrounded by monte in order to purchase masks, drum materials, and *ténaborim* (leg rattles of sewn moth cocoons). The raw materials for these accoutrements have slowly vanished from their backyards.

MAYO LANDS

Fortunately for our botanical interest, Mayos occupy other habitats, not simply the vanishing plant communities of the delta's flatlands and beaches. Villages are scattered along more than 150 km of the Gulf of California coast from northwest of Huatabampo to near Topolobampo, Sinaloa, itself an aboriginal Mayo, or Cáhita, settlement. Mayos long ago became skilled fishermen, and the tradition continues in Agiabampo, Camahuiroa, Las Bocas, and other places. Other Mayo villages are found inland as far as sixty miles from the coast. A few villages upstream on the Mayo, including Camoa and Tesia, are located in even richer foothills thornscrub, and others such as Macoyahui, Mocúzari, and Nahuibampo in tropical deciduous forest. To the southeast of the Río Mayo,

Nelo López, Yavaros. Many Mayo men of the coast carry on the fishing tradition of their ancestors. (Photo by T. R. Van Devender)

in the long *bajadas* and coastal plains that extend to the Río Fuerte, a rich variety of villages, *ejidos,* and *comunidades* populate the beaches, coast, flats, and foothills. As we discuss below, not all of this territory was aboriginally Mayo, but today all its indigenous people are collectively referred to as Mayos.

Annual rainfall in the coastal region averages less than 350 mm (14 inches). In the foothills it may average more than 600 mm (24 inches). Villages inland on the Río Fuerte receive considerably more. In most of the area, irrigation is necessary to raise reliable crops, but *temporales* (rain-fed fields) abound in the uplands, especially in Sinaloa.

Water for irrigating lands in Sonora south and east of the Río Mayo has been allocated and overallocated. There is no surplus. In 1999 angry politicians from water-poor Hermosillo denounced greedy politicians from water-rich

Ciudad Obregón for refusing to permit construction of an aqueduct from the Río Yaqui to Hermosillo. Armed confrontations over water rights have occurred recently between states elsewhere in Mexico, and such a confrontation may one day take place between Sonora and Sinaloa. Construction of Huites Dam on the Río Fuerte in Sinaloa, completed in 1995, inundated two Mayo villages. Although its reservoir promised a new source of irrigation water, only a small portion of the region has benefited. Peasants have taken to the strategy of clearing vegetation from their ejidos in apparent preparation for canals and water, then lambasting the government for failure to deliver them. The tactic sometimes works. It has also vastly decreased and fragmented the remaining thornscrub, in which bare patches gleam like empty beggars' bowls.

Away from the delta region, where wholesale bulldozing of natural vegetation has not yet occurred, the Mayos still live on intimate terms with the monte. This dense vegetation varies from thick, thorny, cactus-dominated coastal thornscrub to tropical deciduous forest of myriad species of trees attaining heights of ten meters or more. During the rainy season, the vegetation explodes into an almost impenetrable low jungle, the trees high enough to limit visibility to a dozen meters. In those humid times it is nigh onto impossible to venture into the monte anywhere in Mayo lands without the aid of a machete (which every Mayo peasant carries and brandishes with dexterity and strength), and it is foolhardy to proceed without a guide.

The Mayos dwell in a region that is smaller than New Jersey but which has great biological diversity. More than 1,800 species of native plants are found among the area's thornscrub, tropical deciduous forest, and limited oak woodland.[4] Perhaps a fifth of these are used by natives in one way or another. Species from the Sonoran Desert to the north and the tropical deciduous forests to the south, from the oak woodlands and pine-oak forests to the east, and from the shore/dune environments and mangrove estuaries on the west, converge in the region. Only a hundred kilometers to the north, the Sonoran Desert ends (or begins, depending on one's perspective). And a couple of hundred kilometers to the northeast lie the boundaries of the tropical forests—whether thornscrub or tropical deciduous forest. The Mayos of Sinaloa and a few villagers in Sonora know the tropical deciduous forest to its upper limit where oaks begin, but most of those in Sonora are most familiar with the thornscrub, and they know it intimately.

CLIMATE

The climate of the Mayo region (which includes the Río Fuerte below Huites Dam) is for the most part hot and arid or semiarid (annual evaporation vastly

View of coastal flats and thornscrub. The low mesa in the distance is Mesa Masiaca. (Photo by T. R. Van Devender)

exceeds precipitation).[5] Rainfall is lowest near the coast: Huatabampo, Sonora, in coastal thornscrub near sea level receives 277 mm (11 inches) per year,[6] while Masiaca, also in coastal thornscrub but 20 km inland and at 70 m elevation, receives 350 mm (14 inches).[7] Topolobampo, Sinaloa, on the Sea of Cortés hosts a coastal thornscrub superficially similar to that found as far north as Guaymas, Sonora, but more dense. Ahome, Sinaloa, in richer coastal thornscrub near the coast, averages 320 mm (13 inches) of rainfall. Further inland, rainfall increases rapidly. At 120 m elevation, Mocúzari Dam, upstream from Navojoa in transitional tropical deciduous forest, receives more than 500 mm (20 inches), and farther upstream, San Bernardo at 220 m in tropical deciduous forest near the Río Mayo receives more than 600 mm (24 inches). Minas Nuevas, in rich tropical deciduous forest in the foothills of the Sierra de Alamos at 520 m elevation, receives 664 mm (26 inches). In Sinaloa rainfall increases dramatically inland toward the Sierra Madre. El Fuerte at only 100 m elevation in tropical deciduous forest receives more than 700 mm (28 inches), and Huites, slightly higher, 796 mm (31 inches). Typically, foothill villages receive several summer rains before the coastal villages receive a drop.

These rainfall values are dramatically higher than those for equivalent elevations in the southwestern United States. For example, Yuma, Arizona, at 37 m elevation in the Lower Colorado River Valley subdivision of the Sonoran Desert, receives only 65 mm (2.6 inches) per year, and stations near Tucson at 745 m in the Arizona Upland subdivision receive 272 mm (10.7 inches; Sellers and Hill 1974). Thus the habitats and vegetation vary greatly depending on elevation and rainfall. In general, natives' botanical knowledge reflects the diversity of their surroundings. Natives of coastal habitats have limited knowledge of foothills thornscrub, and those familiar with the thornscrub are at a loss to identify some trees of the well-developed tropical deciduous forest. Mayos who had lived in the thornscrub at Masiaca were as impressed as we were with *cedros* (*Cedrella odorata,* Meliaceae), *moras* (*Chlorophora tinctoria,* Moraceae), and *sahuiras* (*pitahaya; Stenocereus montanus,* Cactaceae) in Sinaloa.[8] One of our most sophisticated Mayo consultants, a native of Teachive, Sonora, begged to be photographed beneath a huge sahuira near Baca, Sinaloa, so that his family could see it.

Slightly inland from the coast in Mayo country, beyond the reach of tempering sea breezes, temperatures approach 38°C (100°F) daily during the hot months of May through September. Sometimes they rise as high as 45°C (113°F), occasionally higher. The settlements closer to the mountains do not experience the burning temperatures of the coastal plain, but with their more luxuriant vegetation and lack of maritime breezes, their higher humidity makes them nearly as uncomfortable. During the summer months (late April through mid-October) the heat ranges from hot to oppressive to debilitating. Winters, however, are mild and pleasant. Freezing temperatures in the lowlands almost never occur, but one account describes a freeze in 1937 on the lower Río Mayo (Martin et al. 1998:127).[9] Anecdotes tell of freezes along the drainages of the Río Mayo, and even more on the Río Fuerte, but in general the foothills and the coast away from the drainages (down which cold air may flow) are frost-free. Some inland mountain stations and localities in large drainages may experience brief freezes and frosts (Krizman 1972). We investigated a freeze reported for the Alamos area and parts of Sinaloa in December 1997 but found little evidence of extensive damage in thornscrub habitats.[10]

The four natural seasons in the Mayo region are characterized primarily by rainfall distribution (Van Devender et al. 2000). The tropical dry season begins in February or March at the end of the *equipatas,* named for the sound of horse hooves that these gentle rains evoke. It includes both the spring wildflower and arid foresummer seasons of the northeastern Sonoran Desert in southern Arizona. The hot, dry conditions of the dry season are very similar to those in the deserts to the north.

The bulk of the rainfall in the Mayo country traditionally falls in *las aguas,* from July through mid-September, when monsoonal winds usually bring abundant moisture northward from the tropical oceans. Almost daily thunderstorms saturate southern Sonora from air masses that originate in the Gulf of Mexico or in the Pacific Ocean west of Central America.

Fall begins when the monsoonal rains end in mid- to late September. This is a variable season in which drought is often interrupted by intermittent tropical storms or hurricanes that move generally westward along Mexico's Pacific coast and may veer inland (northeastward) into southern Sonora. These sporadic *ciclones* occasionally dump up to a third of a meter of rain in a short period, causing considerable flood-related property losses. Hurricanes Ismael in September 1995, Fausto in September 1996, and Isis in September 1998 followed disappointing monsoonal seasons and rejuvenated the parched region, bringing to native pastoralists a huge boon in the form of forage for their livestock.

Although at 27° N latitude the daytime temperatures from November through February are much cooler than during the rest of the year, winter is characterized as the time of the equipatas. These light rains are the southern extension of frontal storms originating in the North Pacific that bring winter rainfall to the Mediterranean vegetation of coastal California, the Mohave Desert, and the biseasonal northeastern Sonoran Desert. The equipatas normally end by mid-February because March and April storms are generally not strong enough to reach southern Sonora. Although not sufficient to support the diversity of spring annuals or the wildflower displays typical of deserts to the north, the equipatas are still important. In the dry season in years like 1998–99 and 1999–2000, when the equipatas failed, the biota and people alike suffer as wells dry up and cattle die. South of the Río Mayo region, the Pacific storms do not exert much influence. Tropical high pressure nudges them to the north, extending the tropical dry season to eight or nine months.

Plants in high-humidity environments close to the sea generally have reduced moisture stress due to lower evaporation rates. In addition, they receive substantial (but unmeasured) moisture from dew, primarily in the cooler months. Near the coast, woody plants may be larger, and vegetation denser, with epiphytes such as *huírivis cu'u* (*mescalito; Tillandsia exserta* and *T. recurvata,* Bromeliaceae) growing abundantly on shrubs. Heavy humidity and dews may also explain the local abundance on the coastal plain of such cacti as *ónore* (*biznaga; Ferocactus herrerae*), *chicul ónore* (*biznaguita; Mammillaria grahamii, M. mazatlanensis,* and *M. yaquensis*), *jejeri* (*Pereskiopsis porteri*), *jijí'ica* (choya; *Opuntia leptocaulis* var. *brittonii*), and *sina* (*nómom; Peniocereus striatus*). The *murue* (*jaboncillo,* Baja tree ocotillo; *Fouquieria diguetii,* Fouquieriaceae), until recently thought to be a Baja California endemic confined to a narrow disjunctive strip on the mainland near Guaymas (Turner et al. 1995), is locally common in parts of the coastal plain subject to heavy dews.

Many plants are rainfall sensitive, and most trees and shrubs leaf out only during las aguas, or slightly before. Similarly, many plants defoliate or wither away with the cessation of rains, so that most trees drop all or part of their leaves beginning in October. Most (but not all) Burseraceae and *sato'oro* (*torote panalero; Jatropha cordata,* Euphorbiaceae) drop their leaves in October with the onset of the fall drought and remain leafless until the return of las aguas the following year. In contrast, tree legumes such as *mayo* (*mauto; Lysiloma divaricatum*), *macha'aguo* (*tepeguaje; L. watsonii*), *chírajo* (*güinolo* or *chírahui; Acacia cochliacantha*), and *baihuío* (*guayavillo; A. coulteri*), all Fabaceae, and others drop some or most of their leaves in October, retaining a few through the winter. Still others, notably *báis cápora* (*saituna; Ziziphus amole,* Celtidaceae) and individual specimens of *to'bo* (*amapa; Tabebuia impetiginosa,* Bignoniaceae) retain their leaves well into the equipatas and beyond. On the other hand, *to'oro chucuri* (*torote prieto; Bursera laxiflora,* Burseraceae) and *murue* (*jaboncillo,* tree ocotillo; *Fouquieria macdougalii*), species that extend well into the Sonoran Desert to the north, leaf out in response to rain and may be in leaf in any season. Variably and gradually, by the end of the equipatas nearly all the leaves have fallen, and by mid-March the landscape takes on a dreary gray-brown color locally called *mojino*. This pattern of deciduousness suggests that throughout the evolutionary history of these plants, summer rains have been the most dependable (see Axelrod 1979).

The flowering sequence varies among species as much as or more than the deciduation. Throughout the year some tree or other is in flower. Even in the starkest drought of May and June, *cuta nahuila* (*palo piojo; Brongniartia alamosana,* Fabaceae), *jévero* (*chilicote; Erythrina flabelliformis,* Fabaceae), *júyaguo* (*guayacán; Guaiacum coulteri,* Zygophyllaceae), and *nesco* (*Lonchocarpus hermannii,* Fabaceae) produce flowers of brilliant color. The dry season flowering of tropical trees produces a distinct advantage: fruits and their seeds will have fallen to the ground at the arrival of las aguas, increasing the likelihood of germination and rapid seedling growth.

Drylands farming formerly took place each year upriver on both the Río Mayo and the Río Fuerte but is now far more reliable on the wetter Fuerte, where it forms a vital part of the local economy. The temporales, or rain-fed *milpas* (cornfields), rely on adequate moisture in July and August. When las aguas fail, as they have in some areas, and sometimes the entire region, in recent years, crops cannot grow, and a general impoverishment of upland natives (to whom irrigation is not available) occurs. Milpa farming of corn and beans has been abandoned in recent years along large portions of the Río Mayo and elsewhere because, as natives lament, there is insufficient rain to produce a crop.[11] When rains are sufficient, the principal crops are beans, corn, squash, and melons for subsistence, and *ajonjolí* (sesame; *Sesamum orientale*) as a cash crop.

Mayos are primarily people of river, sea, or streamside environments with a long history of agriculture. The village of Nahuibampo on the Río Mayo above Mocúzari Dam relies on livestock grazing and rain-fed milpa farming. The harvests, which are erratic due to unreliable July rain, are used for subsistence only. The river bottom is generally too narrow to afford opportunities for irrigation. Los Capomos, Sinaloa, situated in granitic uplands at roughly the same elevation, relies exclusively on rain-fed farming, usually producing a marketable surplus. Both areas have coarse granitic soils, but the greater rainfall of Los Capomos (ca. 700 mm) has resulted in greater prosperity than in the drier Nahuibampo (ca. 600 mm). The additional 100 mm (4 inches) apparently provides Los Capomos a certain cushion of safety over Nahuibampo's marginal climate for milpas.[12]

Mayos in the region are unanimous in their assertion that las aguas used to be more reliable. As evidence, they point to the slow disappearance of milpas, especially in the northern portions of Mayo country. They also attribute lowered carrying capacity for livestock (and a corresponding loss of biomass) to a general trend toward less rainfall, especially a failure of las aguas since the early 1980s. Other factors in the disappearance of milpas include Mexico's incorporation into international trade and the relative decline of grain prices on world markets. Increased emphasis on livestock grazing by the Mexican government is also an important variable, though to what extent is difficult to ascertain.

GEOLOGY

Geologically, the area occupied by Mayos is simple compared to many other parts of Mexico. The coastal plain, where nearly 90 percent of Mayos reside, is a product of terrestrial tearing. During formation of the Basin and Range Province, which began in Sonora between 10 million and 12 million years ago (Ma), Baja California began rifting from the mainland and drifting to the northwest, forming a depression known as the Proto-Gulf. The Sea of Cortés appeared some 6 million to 5.5 million years ago (Lonsdale 1989). The deltas of the Río Yaqui, Río Mayo, and Río Fuerte are a result of the newly opened gap. The great rivers bore to the sea the tiny bits of rock—clays, silts, and sands derived from weathering of the 2 km thick layer of volcanics that once covered the Sierra Madre to the east. Wind and water deposited them, along with sediments from the Colorado River delta, 700 km to the north (R. Scarborough,

The upper Río Mayo north of San Bernardo. The river is dammed at Mocúzari, some twenty miles south of this location.

pers. comm., 1999), covering the coast and inland with deep sediments. These fine particles buried most of the mountains and hills of the now level plain. The resulting soils are rich in clays. Thousands of feet thick in places, the fine-textured dirt is remarkably fertile and yields readily to the plow. Only water is needed to make the land produce abundantly.

The valleys of the Río Mayo and Río Fuerte flow from deep canyons in the Sierra Madre, some of whose sides were shaped by titanic volcanic explosions and whose courses were carved by wind and water from mid-Tertiary andesites and younger rhyolites.[13] Over a 35-million-year period (40–15 Ma) immense volcanoes and calderas deposited nearly 2,400 m of ash and ignimbrites on the range's granite foundation (R. Scarborough, pers. comm., 1999). Reconstruction of the forces and timing that produced the uplift of the Sierra Madre Occidental is still evolving, but it appears that by the end of the mid-Tertiary Orogeny (35–15 Ma) the basic outline of the cordillera was in place.

Basin-and-range faulting of the Sierra Madre began in the Miocene 17 million years ago, initiating the structural and erosional processes that resulted in the complex geology, topography, and drainage patterns of the modern Sierra Madre Occidental. The uplift of the Sierra Madre produced the canyons and valleys that define the drainages. The rugged topography begins well inland from the coast, some 50 km in the case of the Río Mayo and 120 km in the case of the Fuerte, as the coastal plain ends in low mountains that become increasingly rugged and lofty as one travels eastward. As moist air masses move inland from the Eastern Pacific and tropics and are forced up into the Sierra, rapid cooling draws the moisture from them, rather more in Sinaloa than Sonora. The increased precipitation in the mountains falls primarily as rain, with only occasional snows in the highest regions of the Río Mayo and a limited area of the Río Fuerte drainage in Chihuahua. The Sierra Madre also is a formidable barrier blocking storms from penetrating the highland Mexican Plateau to the east. Stable descending air masses provide much less rain to the desert grasslands and Chihuahuan Desert in Chihuahua, Coahuila, and Durango.

Thus the geological processes that formed the Sierra Madre created the regional climatic regime and the hydrological conditions for development of the agriculturally rich coastal lowlands. Both rivers have been dammed to trap water originating in the highlands of the Sierra Madre: the Mayo at Mocúzari and the Fuerte at Huites (Presa Luis Donaldo Colosio) and El Máhone (Presa Miguel Hidalgo). The Río Cuchujaqui, originating in the Sierra Madre in western Chihuahua and Sonora east of Alamos, is also dammed near its ancient confluence with the Río Fuerte. Water from these reservoirs is released to irrigate lands in the deltas.

Geology casts long shadows over the local economy. Most Mayos work as day laborers in fields irrigated by Madrean waters (Yetman 1998). In many cases they are low-paid workers in fields that belonged to their ancestors. In a few cases they labor in fields they own but have been forced to lease out. Geological and climatic forces created the fabulous fertility and productivity of Mayo lands. They permitted the rise of Mayo culture as well and, to an extent, led to the Mayos' cultural and economic downfall, as we shall see.

## 2  A Brief Ethnography of the Mayos

### CÁHITAS AND MAYOS

Mayos have inhabited the region since well before the Conquest, though for how long before is unknown. According to Almada (1937:19), a prestigious regional historian, Cáhitas (Mayos and Yaquis) penetrated well into the Sierra Madre. Describing their presence in the region, Almada (1952) wrote:

> The Cáhita civilization ruled in what is now the state of Sonora until the arrival of Aztecs as they ranged southward in the twelfth century, remaining there during their forced stay. The Aztecs, the bigger and stronger tribe and forced to remain in the region of Sinaloa, came to dominate the Cáhitas, absorbing them and completely breaking up their civilization, overwhelming it with Nahuatl culture. Many groups or clans of Cáhita origin took flight to escape the yoke of the Aztec imperialism, establishing themselves in other parts of what is now northern Mexico. When the Aztecs resumed their wanderings toward the south, only remnants of the Cáhita civilization endured, the Tehuecos in Sinaloa and the Yaquis and Mayos in Sonora. In this dispersed form the Spaniards found them during their continuing conquest of the Northwest.

Almada's historical pronouncement may be fact or fancy but is intriguing. We will confine ourselves to documented history and hope that Almada was right.

At the time of first contact with Spaniards, Mayos were settled enough to be more or less constantly at war with the Yaquis, their neighbors to the north (with whom they share a common Cáhita language). They also were involved in periodic skirmishes with other Cáhita speakers to the south—Ahomes, Basiroans, Choix, Guasaves, Huites, Sinaloans, Tehuecos, Zuaques, and other extinct groups. They may have squabbled with Guarijíos and Guazapares as well, but Yaquis were their primary opponent. The boundary between Mayos and Yaquis has for many centuries been the nondescript Arroyo Cocoraque, which flows from the Sierra Baroyeca. It crosses Highway 15 near Fundición, Sonora, between Navojoa and Ciudad Obregón, Sonora, and would be noticed only by those specifically looking for it. In contrast, no distinct geographical feature marks the limit of Mayo lands to the south, even into Sinaloa, suggesting that assimilation of previously autonomous groups with clearly demarcated political (as opposed to physical) boundaries took place in that area. Cáhita speakers lived from the Río Mocorito in central Sinaloa to the northern boundary of Yaqui lands around Guaymas (Beals 1943).

Whether all those who today identify themselves as Mayos (*yoremem*) are descended from natives of the Río Mayo region is difficult to determine. The early Jesuit missionaries considered only those people living along the lower Río Mayo (roughly, below present-day Mocúzari Dam) to be Mayos. Their generally cooperative and peaceful nature, which the missionaries admired, distinguished them from the contentious Yaquis (Pérez de Ribas 1645). The Jesuits gave different names to Cáhita speakers living on the upper Mayo and along other drainages based on how these peoples referred to themselves (for example, Basiroas, Macoyahuis, and Zuaques) and endowed each group with a different personality, such as the "fierce Zuaques" and the "arrogant Yaquis." The Mayos of the Masiaca comunidad of Sonora may only later have come to be called Mayos (Yetman 1998).

Jesuits also recognized inhabitants of some southern Sonoran towns as separate nations. These included the Bacabachis, Conicaris, and Tepahuis. The latter two lived on the Río Mayo and today are considered Mayo.[1] The fact that at the time of Spanish contact they considered themselves separate suggests that these people probably would have been incensed at being labeled Mayos. Similarly, there seems to be no historic reason why peoples living on the Río Fuerte in villages such as Baca, Baimena, Los Capomos, Choix, and San Miguel should be considered descendants of Mayos rather than of the Ahomes, Choix, Huites, or Zuaques who inhabited the lower river at the time of contact.[2] They may all be called Mayos today simply because the Jesuits were partial toward the Río Mayo people for their accommodating dispositions. Another possible explanation is that the Spaniards frequently rounded up indigenous groups and located them in "reserves" near missions. This practice may have produced a uniformity of cultures and languages on the Río Fuerte. For example, by the time of Mexican independence, the people of Masiaca were already being referred to as Mayos even though they lived far from the Río Mayo (Yetman 1998).[3]

The linguistic differences among the geographic divisions of Mayos are substantial. Mayos of Masiaca (which lies on an independent drainage south of the Río Mayo) assert that those of the Río Fuerte sound more like Yaquis than

Mayos. Masiacan Mayos consider the language of Mayos living near the Río Mayo to be rather rude and uncivilized. Mayos of the Río Fuerte find those from the Río Mayo to have a strongly accented speech, and they find Yaquis difficult to understand. Linguists will perhaps one day sort out the origins of the linguistic variation in the small area inhabited by Mayos and assist in untangling the origins of the various peoples who now are called, and call themselves, Mayo. Their task may be formidable, however, for the tribes mixed over the centuries as some fled persecution, war, famine, and disease, and others joined groups north and south in rebellions and the Mexican Revolution, homogenizing their previously diverse dialects.

Mayo towns and villages are generally organized into ejidos or comunidades. Ejidos are cooperatively owned lands with titles from the federal government. They are usually divided into common lands and individual parcels, which, under recent changes, members may vote to own outright. Ejido membership is determined by the ejido assembly and is unrestricted. In the early 1970s, when many ejidos were created, *ejidatarios* in northwest Mexico, including those on the southern Sonoran coastal plain, found themselves owning land collectively with people whose accents were strange and customs unfamiliar. However, the older, predominantly Mayo ejidos in the vicinity of the Río Mayo did not have to face this challenge.

A comunidad imposes more restrictions on its members than an ejido does. Membership is fixed, and the assembly cannot alter it. Only the federal government determines a member's legitimacy. To be eligible, members must demonstrate ancient or historic residence of ancestors within the community. Although comunidad lands are similarly divided into individual parcels and common lands, the parcels may not be bought or sold and are used only with permission of the assembly; that is, they are usufruct (meaning property belongs to an individual only as long as the individual uses it), and tenure theoretically can be revoked by the assembly at any time. In practice, however, this is never done.

There are two types of comunidades in Mexico: *comunidades agrarias,* agrarian communities whose members have a long history of agrarian tenancy, and *comunidades indígenas,* communities of natives with pre-Columbian roots who also have a long history of tenancy.[4] Comunidades in the Mayo region are comunidades indígenas made up primarily of Mayos, but usually with substantial numbers of non-Mayos as well. This membership of non-Mayos is controversial and the source of ongoing conflicts, especially in Baimena and Masiaca, where non-Mayos tend to dominate in the comunidades' politics and receive a disproportionate share of benefits.

In the Río Mayo area, Buaysiacobe, El Júpare, and Nahuibampo are examples of ejidos, while Camahuiroa, Bachoco, and Masiaca are comunidades. In Sinaloa, Los Capomos is an ejido, while Baimena and Choix are comunidades. Because of the restrictions on membership, comunidades tend to be older and more stable than ejidos. In 1997, members of the Los Capomos ejido voted to privatize individual parcels. Each ejidatario was granted thirteen to sixteen hectares of prime farmland and received title to the land. The comuneros of Baimena have no legal means for such an election.

Both systems of land tenure assume that members will earn their livelihood from the soil either by grazing livestock or planting crops. Nevertheless, in the Masiaca comunidad a successful cooperative at Las Bocas makes fishermen its most affluent members, and at Buaysiacobe on the Río Mayo, ejidatarios illegally leased their parcels to agribusiness and lived off the proceeds until 1996. Changes in the Mexican constitution that permitted the outright sale or lease of lands allowed the ejidatarios to perpetuate what had previously been forbidden.

The people who call themselves Mayos are united only by their common language and the fiestas they celebrate.[5] No pan-Mayo organization exists, and no village or ethnic government unites Mayos, unlike the Yaquis to the north, who retain a strong organization of village and tribal governors. No means exists by which anything like the Mayo point of view can be expressed. Most adults in Mayo villages still converse in la lengua, but nearly all of them speak Spanish in varying degrees of fluency. According to the 1990 census, 234 monolingual Mayo speakers were living in Sonora, while over 27,300 residents were bilingual Mayo speakers, making Mayos the largest indigenous group in Sonora (although constituting little more than 1 percent of the total population) (Instituto Nacional de Estadísticas, Geografía, e Informática). Roughly 11,000 residents spoke Yaqui, of which nearly 600 spoke Yaqui only.[6] In Sinaloa fewer than 100 residents were listed as monolingual Mayo speakers, while 9,689 were listed as speaking Mayo and Spanish.[7] The high incidence of bilingualism is a strong indicator of assimilation into the dominant mestizo culture.[8]

Only a small percentage of the region's children and teenagers are fluent in the Mayo language, and we encountered few children during our research who openly acknowledged fluency in la lengua. In a few towns (Etchojoa, Sonora, and Baimena, Sinaloa) Mayo is taught in the schools through the efforts of Mexico's Instituto Nacional Indigenista to encourage retention of native languages. For the most part, though, children find speaking Spanish to be socially acceptable, while conversing in Mayo is considered a sign of backward-

Norma Humo and her son, José Luis Baisegua, Sirebampo.

ness and lack of culture, and their parents often agree. Our experience also suggests that young people are less sophisticated in identifying and naming plants and are far less familiar with their Mayo names than older people.

The other primary indicator of "Mayo-ness" is participation in fiestas, and most villages honor their patron saint with a fiesta that lasts two or three days. Large villages and towns also have elaborate ceremonies during Semana Santa (Easter week). The patron saint's days are marked by the dancing of pascolas (old men of the fiesta) and venados (deer dancers). The fiestas also include large amounts of traditional foods (especially *guacabaqui*, a beef-vegetable stew, and tortillas) and usually public dances.[9] Most important, the fiestas are a time of renewing personal and family ties among and also inside villages. Even though fiestas sometimes coincide with religious or national holidays, their deeper meaning is usually independent of the wider celebration.

Semana Santa festivities feature elaborate rituals involving *judíos* (Jews) or *fariseos* (Pharisees) dressed in fantastic costumes, and *contis*, processions in which a ritualized version of the Passion of Christ is carried out (Crumrine

Holy Saturday in Masiaca. Processions and pageantry mark this holiest of Mayo days.

*Fariseos* (Pharisees), Benito Juárez. During Lent the fariseos are mendicants, traveling on foot from village to village.

*Fariseo*, Etchojoa. Each member of the brotherhood makes his own mask. At the culmination of the ceremonies, he will burn it and his lance in a bonfire. (Photo by D. Burckhalter)

1977).[10] The culmination of Semana Santa is the *gloria,* a procession and drama enacted throughout Holy Saturday morning in which the fariseos attack the church, and the people defend it. After the evil forces are defeated, the penitent fariseos burn their evil masks and lances in a huge bonfire along with an effigy of Judas.

The fiestas are often connected with the village church, but except for obvious symbolism and overt Christian elements, they are purely Mayo, excluding even the clergy at times.[11] Mayos, except for converts to Protestantism, consider attendance and participation in fiestas to be the most important and rewarding activities in their lives. Villagers, especially men, visit fiestas in other villages as much as possible.[12]

The fiestas are organized by a group called *fiesteros,* whose members have

Men watching *pascolas*, Masiaca plaza, Holy Week. The dances continue off and on for forty hours. (Photo by T. R. Van Devender)

taken a vow to carry on the important traditions. To be a fiestero is an honor but also a burden, for it is time-consuming and expensive to sponsor fiestas. Fiesteros are viewed as exemplary citizens by other Mayos and thus are likely to be forgiven for otherwise grievous faults. Participating in the fiesta honoring the patron saint or becoming a fiestero usually marks one as a traditional Mayo. The fiesteros meet monthly to plan their activities (which include maintaining the village cemetery and keeping it free of weeds, trash, and clutter). In most villages the Mayo language is spoken almost exclusively among older folk and at these meetings.

An important part of fiestas, the deer dancer represents a conservative and ancient tradition, probably a fusion of pre-Columbian and Christian elements. In the Mayo and Yaqui religious history, the deer is identified with Christ, and the deer dancer wards off evil. During fiestas the deer dancer performs certain dances simultaneously with the pascolas, who are usually associated with dark forces. The deer dancer keeps his back to the pascolas, and the accompanying musicians attempt to drown out the music played by the pascola musicians. The deer dancer must be familiar with a litany of songs in the Mayo language and therefore must be fluent in la lengua.

Included in the songs performed by the deer dancer are those casting mestizo culture in a negative light. One such song, which ridicules a Mexican cowgirl, is known to nearly every traditional Mayo. It was sung for us in 1998 by Don Reyes Baisegua of Sirebampo.

Don Reyes Baisegua, Sirebampo. Reyes is known for his knowledge of Mayo traditions.

| JUANITA VAQUERA | JUANITA THE COWGIRL |
|---|---|
| Juanita Vaquera ranchóme si'ica | Juanita the Cowgirl ran off to a ranch. |
| Baquíata nuuseca | She was supposed to take along a heifer. |
| Tolótasu nuupaka | How could she take it to the ranch? |
| Jítasu junta jíata casica | But she brought a bull instead. |
| Tóloco bacota | She used a vine snake for a lariat. |
| Jítasu espuela casica | Where will she find a spur? |
| Mó'ochocol cóbata espuela casica | She'll use a horned lizard head for a spur. |
| Mosen behuata sila casica | She'll use a turtle shell for a saddle. |

The pascolas, on the other hand, need not speak the language. One pascola lamented that there are plenty of pascola dancers around, but few who speak la lengua and appreciate the seriousness of the dancing tradition. Most well-known pascolas are traditional Mayos, however. Their dances require elaborate footwork, and they crack earthy jokes in Mayo and poke fun at non-Mayo

Doña Amelia Verdugo, Los Muertos. Amelia is recognized in her village as one who knows traditional herbal cures. (Photo by T. R. Van Devender)

targets, all to the delight of the audience. During the rituals that open the fiesta, they apologize to Mary and Jesus for their upcoming ribald behavior.

DRESS

Mayo clothing is indistinguishable from that of mestizos, probably because of the persecution of Mayos during the *porfiriato* (the thirty-year reign of Porfirio Díaz, roughly 1880–1910) and after the revolution. Women formerly wore rebozos and *ropa de manta* (calico) blouses and long skirts, while men wore *calzones*, peasant-style pants and loose shirts of white muslin of local manufacture. During the virtually continuous unrest between 1880 and 1910, anyone resembling a Mayo was subject to harassment and arrest. Since that time, younger Mayos have gradually adopted the style common to mestizo peasants throughout northwest Mexico as a sort of protective coloration to keep them unnoticeable in a crowd.

Women, especially older women, often wear long cotton dresses or simple blouses and skirts. Men wear nondescript pants and long-sleeved shirts, preferably close-fitting, snap-buttoned cowboy shirts. Men also wear the mass-produced plasticized straw hats preferred by mestizos. The preferred footwear is leather *huaraches* (sandals) made locally. Most Mayos also possess shoes for more formal occasions. Young men who can afford them wear cowboy boots purchased in the cities. Young people purchase their clothing in cities as often as possible and are indistinguishable in their dress from mestizos. In many villages women cover their heads with a towel or scarf when they leave the house, probably a remnant of the traditional rebozo, still worn by Yaqui women. This type of head covering is the only item of Mayo clothing that hints of an earlier cultural style of dress. Yaqui women, in contrast, wear long dresses and brightly colored shawls in which they carry many items, from babies to groceries.

## THE MAYO LANGUAGE

Mayos and Yaquis are able to converse with each other in their native languages without an interpreter. Both languages are labeled Cáhita, which means "there is nothing." The Mayo language has been extensively studied and described (Collard and Collard 1962; Germán E. et al. 1987; Freeze 1989; Zavala C. 1989). Textbooks are available for children and a modest number of Mayo writings have been published, mostly through the Secretaría de Educación Pública.

Mayo shares with Spanish the sounds *a, b, c, e, gu, hu, i, j, l, m, n, o, p, qu, r, s, t, u,* and *y*. The vowel sounds, as can be seen, are close enough to Spanish to permit any Spanish speaker to pronounce most Cáhita words adequately. The consonants are sounded more like their Spanish counterparts than those in English. Whether this is a result of four centuries of Spanish influence or merely a happy coincidence is difficult to determine. In addition to the common sounds, however, the glottal stop (') functions as a separate letter; including it or failing to do so may change the meaning of a word. The language lacks the letters *d, f, k, ll, rr, v, w, x,* and *z*. The Spanish *gue-, güi-,* and *güe-* sounds do not exist; they are replaced with the gentler *hue-* or *hui-*.

The Mayo language is highly inflected. Its verb structure is complex in that verbs can be completely altered by the addition of suffixes, which abound; root forms of verbs are often altered with change of person. To show endearment or affection, an *r* may be changed to an *l*. This practice was most confusing until Vicente Tajia explained it for us one day.

Plaza, Nahuibampo. Mayo villages traditionally mark pedestrian and traffic boundaries with whitewashed rocks, which are also visible in the dark and help prevent stumbling and falling. (Photo by T. R. Van Devender)

## MAYO NAMES

Although Mayo names include many common Mexican surnames, several are distinctively Cáhita, either peculiarly Mayo or shared with Yaquis. Among these distinctive names of Cáhita origin are *Anguamea, Bacasegua, Baisegua, Buitimea, Josoina, Tajia, Yocupicio,* and above all *Moroyoqui,* the quintessential, purely Mayo name. Legend has it that Masiaca was founded in the year 1400 by one huge family named Moroyoqui. They interbred, so the story goes, thus giving rise to the town, which was originally called Masiaca *primo hermano* (Masiaca first cousin). The surname *Valenzuela,* while not uniquely Mayo, is extremely common (see appendix B, Yoreme Consultants). Other common surnames among Mayos are *Estrella, Gámez, López, Piña,* and *Zazueta.*

How and when Mayos began adopting Spanish names is difficult to ascertain. Archival documents from Masiaca in 1834 list a Valenzuela and a López as *naturales* (natives), indicating that Spanish names were already being widely used by Mayos. Today all indigenous peoples of Sonora retain Hispanic given names and apply their surnames following Spanish custom, with children bearing their father's surname followed by their mother's. As is the

Mayo transportation, Masiaca. Few Mayos own automobiles.

Burros roam freely in the vast Masiaca *comunidad*. Many of them are used to pull carts.

case with mestizos, a woman does not take her husband's surname except in formal situations, when she may attach it to her own surname following the preposition *de*.

In spite of the decline of "Mayo-ness" in the region, older villages from Huatabampo south to the Río Fuerte still exude a timelessness and an atmosphere of the traditional Mayo culture. The older houses of wattle and daub with cane or organpipe ribs woven into walls, roadways and pathways lined with whitewashed rocks, mule- or burro-drawn carts, and the smoke of cooking fires rising from the homes still combine to present a Mayo ambience. As humble as these villages are, places such as Jambiolobampo, Bachoco, Sirebampo, Nahuibampo, and Yocogigua reek of tradition and fading history. To understand why they are there and how they operate, it is necessary to review Mayo history.

# 3  Historical and Contemporary Mayos

A HISTORY OF THE MAYOS FROM A MAYO PERSPECTIVE

While the focus of this work is primarily ethnobotanical—the plants used by Mayos—we present the following history in an attempt to clarify the historic and present land tenure in the region in a way that is sympathetic to the Mayos. The fact that most Mayos live today in scattered small villages over a large region and are familiar with a variety of habitats can be attributed to historical forces that impelled or drew them in one direction or another. They defended their traditional lands fiercely, and their history is largely one of a defense against encroachment.

The Mayos were preceded in the region by what is now called the Huatabampo culture, which appeared in the lower Río Mayo and Río Fuerte valleys between 140 B.C.E. and 900 C.E. (Villalpando 1985). These people were hunter-gatherers but also agriculturalists who raised corn, beans, and squash. They produced high-quality pottery, worked seashells into fine jewelry, and carved stone into polished implements and figurines. By 700 C.E. the population was sufficiently organized and urbanized to conduct systematic trade within a network that extended to Casas Grandes, far to the northwest in what is now Chihuahua, and with Sinaloan cultures to the south.

In about 1000 C.E. the Huatabampo culture was disrupted, perhaps in part by warfare, but also apparently due to floods and hurricanes that forced a major realignment of the lower Río Mayo. This natural catastrophe wrought havoc on the established communities, and the Mayos may have arrived in the region soon after. Although one popular mythology has the Cáhitas descending from the Toltecs, there is no firm basis for this.

If Almada's historical vignette (quoted at the beginning of chapter 2) is correct, the Aztecs quickly forgot their brief stop in Mayo country. During their forty years of hegemony in the Valley of Mexico, they did not accomplish mil-

House of woven *pitahaya* arms, Bachomojaqui. This traditional structure has changed little over the centuries. Cooling winds penetrate between the ribs.

itary suppression of the distant Cáhitas of the Chichimeca and engaged in only limited trading with them. Whether the Aztecs would even have been interested in a people of such limited material wealth as the Cáhitas is questionable (see Padden 1967).[1]

What the Mayos were like prior to contact with Europeans must be reconstructed from brief Jesuit accounts. At the time of contact, the Mayo economy was based on corn, beans, squash, cotton, game, and a few other domesticated crops that they raised in the fertile delta floodplain (Obregón 1988). They carried out limited trading with inland peoples. They had no impressive public buildings, and there is no evidence of complex social organization. Regular flooding of the Río Mayo meant the Mayos had no need for an elaborate system of irrigation ditches and a highly organized social force for ditch maintenance, and the archaeological record does not reveal production localities. Their houses were constructed much as they are today, with woven cane walls covered with daub, and roofs of light wooden beams covered with crisscrossed branches. Houses were covered with earth in the lowlands, while those near the foothills had palm roofs. Outdoor ramadas sufficed for cooking.

Evidence of the Mayos' social organization is scant, especially in comparison to the archaeological records left by more architecturally oriented groups in the New World such as the Anasazi, the people at Casas Grandes, the Mayas,

House of adobe walls and dirt roof, Teachive. Inland homes require more protection from heat and cold than those on the coast. This home is more than a hundred years old.

and the Toltecs.[2] Virtually nothing was written about their infrastructure. The early Jesuits concentrated far more on converting the "heathen" and conducting military expeditions than they did on describing the cultures they encountered in northwest New Spain.[3] The absence of descriptions of their pre-Columbian culture is frustrating to historians as well as to the contemporary Mayo.

While it is clear from early descriptions that local *caciques* (chiefs) and *hechiceros* (sorcerers or, perhaps more accurately, medical practitioners) exerted considerable influence, we know little about how they assumed power and influence. Pérez de Ribas (1645) notes only that caciques appeared to gain power more from their prowess in battle and their ability to engender large families than from any hereditary ties.[4] Whether Mayos and other groups were sufficiently organized to be considered class societies or should be considered tribes cannot be determined. The fact that by the early seventeenth century Spaniards were able to recruit thousands of Mayos to battle Yaquis indicates that some type of organization beyond the level of autonomous *ranchería* must have existed.[5]

The post-Conquest historical record yields little insight into Mayo material culture. The Spanish invaders, to their dismay, found no gold or silver on Mayo farmlands. Deep coastal sediments cover any mineralized bedrock deposits over much of the Río Mayo region. The volcanic inselbergs protruding from the sediments are generally lacking in mineralization. The relative lack of riches may have been the Mayos' salvation, for the rapacious conquistadors found nothing worth pillaging among them, even despite reports of gold upstream such as had tempted them in Sinaloa's coastal region. Silver was not discovered until nearly the end of the seventeenth century.[6] As for making slaves of the Mayos, their lands were sufficiently distant from the capital in Mexico City (1,600 kilometers) that taking large numbers of Mayos back as slaves to plantations or haciendas would have proved inconvenient, especially with so many other Indians living closer. The early slave-raiding parties apparently captured few, if any, Mayos. Nevertheless, when silver was discovered seventy years after the Jesuits arrived, Indian labor became indispensable for extracting wealth from the ground.[7] Furthermore, Mayo farms were capable of producing surpluses, which led to the Spanish theft of Mayo lands as agricultural trade developed, and would lead to the Mayos' downfall in the nineteenth century as the Industrial Revolution vastly expanded world agricultural markets.

The Mayos' early contact with Spaniards is well documented. In 1533 Diego de Guzmán, nephew of the notorious slave gatherer Nuño de Guzmán, crossed the Río Mayo while marching northward (Troncoso 1905). He turned back at the Río Yaqui after Yaquis refused to allow him to cross their lands and engaged him in battle. The Yaquis proved to be such valiant warriors that Diego lost his enthusiasm for proceeding farther north. He made note of the Yaquis' prowess as warriors but made no mention of resistance by the Mayos, even though, given his slave-gathering activities, he must have enslaved some. The Mayos' first contact with the Spanish, then, can hardly have favorably impressed them. If Mayos believed the obnoxious Spaniards would defeat their enemies the Yaquis, they were disappointed.

The goal of the early expeditions in northwest Mexico was ostensibly to bring Christianity to the savage natives. In reality, they were expeditions of conquest that used religion as an excuse for plundering. After the first murderous expeditions, many groups chose to flee rather than face cavalry, muskets, torture, and disease. Only the Yaquis seemed to possess the organization and sophistication to stop the inexorable tide.

One early account of Sonora was written by Cabeza de Vaca, the mystical philosopher-explorer who shipwrecked in Florida in 1526 and was then marooned in Texas. Later he wandered westward for nearly eight years, living with natives, learning their languages, and treating their illnesses. In 1536 he encountered Spaniards in southern Sonora or northern Sinaloa. They were as surprised to see him as he was to see them. Later he described what he found in what is now Sonora:

> We traveled over a great part of the country, and found it all deserted, as the people had fled to the mountains, leaving houses and fields out of fear of the Christians. This filled our hearts with sorrow, seeing the land so fertile and beautiful, so full of water and streams, but abandoned and the places burned down, and the people, so thin and wan, fleeing and hiding; and as they did not raise any crops their destitution had become so great that they ate bark and roots . . . They brought us blankets, which they had been concealing from the Christians, and gave them to us, and told us how the Christians had penetrated into the country before, and had destroyed and burnt the villages, taking with them half of the men and all the women and children, and those who could escaped by flight. (Cabeza de Vaca 1972:133–34)

Cabeza de Vaca wrote no names to associate with the peoples he described. His route, though, took him through Mayo lands, so he could well have been describing (among others) the Mayos.

Neither Cabeza de Vaca nor any contemporaries described resistance by Mayos. At that time numerous Cáhita speakers lived along the rivers and were identified as such, including the Sinaloa, the Zuaque (Fuerte), the Mayo, and the Yaqui (Sauer 1935). Sinaloas and Zuaques offered tenacious resistance to the invading Spaniards (Pérez de Ribas 1645), but apparently the Mayos did not.

The next historical mention of the Mayos was in 1539 in connection with Coronado's unsuccessful expedition in search of Cíbola and its gold and converts (Acosta 1949). Coronado chose a route ascending a northern tributary of the Mayo (the Río Cedros, whose lower part is now an arm of Mocúzari Reservoir), past present-day Tepahui and Quiriego and into the higher mountains. His expedition skirted the Yaqui lands, quite possibly as a result of Guzmán's sobering experience of Yaqui ferocity. Coronado did not want his search for riches distracted by skirmishes with natives. The fact that the expedition moved directly through Mayo lands (they almost certainly passed by Conicárit, an important town) indicates the Spaniards did not expect hostility from the Mayos.

Another expedition into what is now northern and eastern Sonora was undertaken in 1565 by Francisco de Ibarra, a governor of the province of Nueva Vizcaya. Ibarra succeeded in advancing only to a point near present-day Casas Grandes in Chihuahua, via the upper Río Sonora and Río Bavispe (Pérez de Ribas 1645; Sauer 1932; Acosta 1949:28–30). On his return, he spent enough time among the Yaquis for his troops to help them defeat the Mayos in a skirmish, and pillage and burn one of their villages. His brief description makes it clear that even though their languages were nearly identical, the Mayos and Yaquis were enemies who engaged in frequent warfare. The fact that the Yaquis recruited Ibarra's assistance suggests that there was a semblance of military equality between the two peoples. Neither was strong enough to inflict a permanent defeat on the other, but perhaps neither wished to do so. It is possible that both the Mayos and Yaquis simply wanted to maintain forces strong enough to prevent others from seizing their lands.

The Spaniards and Mayos continued to have intermittent contact, mostly to the Mayos' detriment. One especially unfortunate episode involved the expedition of 1585 led by Hernando de Bazán, an irascible governor of Nueva Viszcaya. He used the excuse of the death and mutilation of two of his soldiers at the hands of some indigenous people, perhaps Sinaloans, to go on a rampage against all native peoples in the region. Bazán spurned gifts of food and supplies from the Mayos, which they offered innocently and generously, and then he invaded their lands, seized a few men, put them in chains, and dragged them back with him as slaves. For this his superiors removed him from his office, not only because the Mayos were noted for their peaceful, generous nature (Pérez de Ribas 1645; Troncoso 1905) but also because Bazán's cruelty was excessive, even for those ruthless times.

After seventy-five years of military and quasi-military expeditions had failed to pacify the indigenous groups of northwest Mexico, the Crown resorted to a new strategy. Beginning in the seventeenth century, missions and missionaries were used to achieve what brute force had failed to achieve.[8] A corollary of the strategy was to recruit converts and conquered people to assist in the conquest of their ancient enemies, for example, recruiting Tehuecos of the lower Fuerte to defeat their old foes, the fierce Zuaques of the middle reaches.

This ecclesiastic-military strategy bore fruit in 1605 when Diego Martínez de Hurdaide was appointed captain of the Spanish forces of the northwest. Hurdaide demonstrated his military prowess at an early age. "The theater of his real exploits was the Province of Sinaloa, where he 'pacified and reduced nearly twenty different nations, some of them isolated and fierce'" (Acosta 1949:39). At the time the Río Fuerte was rather densely populated with Cáhita-speaking peoples. According to Acosta,

> thus was the conquest of the pueblos situated on the Río Zuaque or Sinaloa [Fuerte] whose principal nations were: Sinaloas [Huites], Tehuecos, Zuaques, and Ahomes. The first [the Huites, not the Sinaloas] lived upriver where it emerges from the sierras. That tribe consisted in more than a thousand families or more well armed with bows and arrows. Six leagues below dwelt the Tehueco nation, fearless and warlike, who could easily mount a force of a thousand five hundred warriors. Ten leagues downstream from them were the Zuaques who occupied several towns and could mount a force of more than two thousand warlike and brave fighters, and finally, four leagues below, occupying the course of the river until it reached the sea, dwelt the gentle Ahomes, who comprised about a thousand families. (41)

Hurdaide used a bold strategy to defeat the powerful Zuaques, the most formidable obstacle to the Spanish and Jesuit conquest of the region: he made a sortie into the center of Mochicahui, their most important settlement, kidnapped and held hostage the leaders of the Zuaques, attacked their unprepared army, and took the women and children as hostages. With the Zuaque leaders, women, and children in his control, it was not long before he received the submission of the Zuaques and Tehuecos as well. He had their leaders publicly whipped. Their warriors cut off their long hair as a sign of submission. The other nations followed suit, and Hurdaide proceeded to secure the Río Fuerte for the Crown and the Jesuits. It helped Hurdaide that the Ahomes and Zuaques were frequently at war, and that the Tehuecos and Zuaques viewed each other with suspicion, despite being able to understand each other's languages. The descendants of these Río Fuerte peoples are the Mayos who live in towns and villages along the Fuerte today.

Hurdaide was tough but seems to have been a good administrator and a relatively benign conqueror.[9] In 1608, after three years of aggressively subduing the natives, putting down mutinies, and installing missions, Spanish forces under his direction entered Mayo lands. News of his military successes must have preceded him, for he received the Mayos' submission without bloodshed, an achievement that brought great satisfaction to Hurdaide and his Jesuit contemporaries (Pérez de Ribas 1645). The Spaniards were sufficiently impressed with the Mayos' military and provisioning skills that they recruited them by the thousands for campaigns against other indigenous groups. Mayos were "more than thirty thousand strong, and could launch ten thousand warriors into battle" (Acosta 1983:46). That the Mayos were willing to provide their young men in such numbers suggests that they did not view themselves as a conquered people nor the Spaniards as conquerors, but rather as allies. It also presupposes a remarkable degree of cohesion among a group said to have achieved only tribal organization (Kirchhoff 1953).

For the Jesuits and Spaniards, the Mayos' willingness to accept the Church's authority without resistance established the obvious truth of the Catholic faith. In reality, the Mayos' acceptance of Catholicism was motivated by hope of deliverance from disease and their respect for the Spaniards' superior arms. They probably also hoped to enlist the new technology in battling their ancient enemies, the Yaquis, and, to a lesser extent, the Tehuecos and Zuaques to the south (Spicer 1962).

In 1609 Hurdaide recruited Mayos for a major offensive against the Yaquis, who had incurred his wrath by granting asylum to two Tepehuanes accused by Hurdaide of murdering two Spanish soldiers. The Yaquis steadfastly refused to turn the refugees over to Hurdaide. The Mayos joined the battle with en-

The Catholic church in Masiaca. Although much of the structure is relatively new, parts are nearly two hundred years old. Occasional disputes arise between traditional Mayo societies and the clergy over control of the building.

thusiasm (Pérez de Ribas 1645),[10] but the first attack on the Yaquis was unsuccessful. A second, better-armed, sally also failed, and the Yaquis resoundingly defeated the combined forces of Spaniards, Mayos, and other indigenous forces. In spite of the Europeans' cavalry, superior arms, and many hours of training, the Yaquis nearly exterminated the Spanish forces.[11] Hurdaide barely escaped with his life.

The defeat not only was a military embarrassment to the Crown (it discredited once and for all the myth of Spanish invulnerability) but also jeopardized the Spaniards' ability to intimidate other indigenous people, which was critical to the enduring presence of the Spaniards. To Hurdaide's surprise (and immeasurable relief), however, the Yaquis approached him the following year (1610), offering submission and requesting peace. One of the conditions Hurdaide imposed was that the Yaquis cease their warring with Mayos, a condition most welcome to the latter (Acosta 1949). Although the Mayos had lost many soldiers in the Yaqui wars, their alliance with Hurdaide had paid off.

Hurdaide used his defeat by the Yaquis, and his ultimate victory, as a tool to gain the Crown's approval to construct a fort as a northern outpost of Spanish power. In 1610 he oversaw the building of a permanent garrison at Montesclaros (El Fuerte) on the Río Fuerte (where it still stands) and used it as a base to subdue those groups who impeded the extension of the Spanish Empire to the north and west. In all his efforts, Jesuits were virtually at his side. When tribes were conquered (either pacified or beaten into submission), he left the clerics to maintain the appropriate rule of law and order.

The Mayos appear to have accepted the Spaniards and the Jesuits readily, even enthusiastically.[12] They apparently sent a welcoming host to greet the Jesuits several leagues from the Río Mayo. Pérez de Ribas was impressed indeed with the piety and devotion of the Mayo converts, noting in 1614, "I have never seen in any converts so clear a sign of the grace and presence of the Holy Sprit as among these Mayos. Even those still set in old customs, on being baptized, appear overcome with such extraordinary joy that the lame and aged seem to recover their feet and their youthful energy, and the mute recover their speech, so that they can run to the church and to their priest and give thanks for his merciful gift to them" (Pérez de Ribas 1645, translation ours).[13]

Pérez de Ribas assumed naively, as did most of his contemporaries, that the Mayos received the Jesuits warmly due to the self-evident truth of the (European) Catholic way of life. So confident was he of the superiority of European culture and its manifestations that he and his colleagues interpreted the Mayos' responses as the logical reaction of people who witness their superiors and hope to copy them. Except for Cabeza de Vaca, the Spaniards interpreted all the events through the narrow lens of the Eurocentric view. As we shall see, the Jesuits might have been a good deal less euphoric had they attempted to view events from a Mayo perspective.

It may have been accounts such as that by Pérez de Ribas that established the reputation of the Mayos as being gentle and peace loving. Villa (1951:31), writing in the 1930s, compared the Yaquis and Mayos: "as much as the Yaquis are characterized by their warlike and irascible temperament, the Mayos are of a peaceful nature, having shown themselves for some time to be submissive and obedient to the constituted powers." Vázquez (1955:128) describes them in a similar vein: "Mayos are of a gentle character and have shown themselves since long ago to be submissive to the Government. They have dedicated themselves to agriculture, to raising cattle and a few small industries, but never reached the level of resistance and political agility of the Yaquis." But Vázquez was mistaken, as we shall see.

Due to the scarcity of friars, the Mayos had to wait five years following their agreement with Hurdaide before priests were available to convert the Mayos and minister to their putative spiritual needs. When the clerics arrived, they plowed with boundless energy into transforming the Indians into Europeanized Catholics. First, they "reduced" the Mayo population, herding all of them to eight pueblos where they could more easily control and observe them.[14] The first Jesuits among the Mayos, Padres Pedro Méndez and Diego de la Cruz, reported baptizing "around thirty thousand Christian souls" by the year 1620 (Pérez de Ribas 1645), an average of about fifteen per day, year in and year out.[15] This figure gives us a fair estimate of Mayo population at the time. Padre Méndez claimed to have founded seven Mayo pueblos and baptized 3,100 children and 500 adults in his first fifteen days (Alegre 1888). Many of the baptized were ill and died soon thereafter, indicating that an epidemic was sweeping the Mayos at the time (Pérez de Ribas 1645). Neither Pérez de Ribas nor Alegre (writing two centuries later) mention what explanations, if any, were offered to the Mayos about the nature of the ceremony to which they were subjecting themselves and their children. Perhaps the incentives made explanations unnecessary.

The European view of New World natives as savage, nonrational barbarians is reflected in the writings of the Jesuits, who repeatedly characterized the Mayos, young and old alike, as childlike and in need of stern instruction. The Jesuit Ignaz Pfefferkorn's 1794 characterization of Sonoran Indians is most revealing:

> Imagine a person who possesses all the customary qualities which make one disgusting, base, and contemptible; a person who proceeds in all of his actions blindly, without consideration or deliberation; a person who is untouched by kindness, unmoved by sympathy, unshamed by disgrace, not troubled by care; a person who loves neither faith nor truth and who has no firm will on any occasion; a person not charmed by honor, not gladdened by fortune, or sorrowed by misfortune; finally, a person who looks only at the present and the sensual, who has only animal instincts, who lives indifferently, and who dies indifferently. Such a person is the true picture of a Sonoran. (Pfefferkorn 1949:166)

Pfefferkorn goes on to lament the Sonorans' "natural stupidity, their complete neglect of themselves, the baseness of their spirits." A Jesuit contemporary of Pfefferkorn, Juan Nentvig, makes a similar pronouncement: "The disposition of the [Sonoran] Indian rests on four foundations, each one worse than the other, and they are: ignorance, ingratitude, inconstancy, and laziness. Such in truth is the pivot on which the life of the Indian turns and moves" (1951:54).

If we consider the continuing epidemics of diseases already rampant in the region, the Mayos' enthusiasm for the Jesuits' arrival appears in a different light. At least two epidemics, one in 1602 and one in 1606–7, decimated the Cáhitas, probably including Mayos (Reff 1991). An epidemic in 1593 had affected peoples to the south and may have affected the Mayos as well. Reports of diseases among the indigenous groups of northwest Mexico spread almost as rapidly as the diseases themselves, so the Mayos were undoubtedly aware of them and, in all probability, terrified by the news. These episodes were among dozens of epidemics that reduced the indigenous population of New Spain by as much as 90 percent in the first one hundred years of Spanish occupation. Natives, noting that the Jesuits seemed far less prone to infection than they were, concluded logically that the priests' immunity had something to do with their faith—that is, the stories the Spaniards were promoting and the rules they were laying down—and they naturally signed up in droves, hoping to be granted the same relative freedom from disease as the blackrobes. They had two choices: sign up and receive the priests' immunity, or take military action to drive the source of the pestilence forever from their lands. Reff (1991) cites contemporary sources who told of natives traveling great distances to seek blessing from the Jesuits in the hope that they could ward off or cure the new plagues. The Jesuits were concerned lest the natives misconstrue the spiritual intent of baptism as a sophisticated medical procedure.

It is important, then, to view the warmth with which the Mayos apparently welcomed the arrival of the Jesuits (if not of the Spanish in general) in the context of disease and international competition. The Mayos' enthusiasm for the new faith and the benefits of rapid-fire baptism must have soon begun to diminish. Despite their acceptance of the European religion, the Mayos (and all indigenous peoples) continued to be ravished for the next thirty years by diseases introduced by the same forces that had brought them the new religion. Although Jesuits have been nearly universally praised for the "civilizing" benefits of the mission system and the introduction of European technology and commodities, these "benefits" are seldom spelled out. When examined closely, they usually fail to represent much more than a covert depreciation of native American cultures.[16] One thing is clear: the mission system brought people into crowded conditions, greatly assisting the spread of such diseases as dysentery, influenza, leprosy, malaria, measles, plague, smallpox, typhoid, and typhus.[17]

The Mayos' numbers had probably already been greatly diminished by disease in the first seventy-five years of contact with early Spanish explorers; at least half of those remaining died horrible deaths from smallpox and other epidemics following the Jesuits' arrival (Pérez de Ribas [Robertson 1968, book 2, chap. 48]).[18] The warm embraces with which the Mayos received the Europeans were the Indians' downfall, for the close contact with Spaniards and the native Americans' lack of resistance to alien diseases left them frightfully vulnerable. The priests and soldiers had had many generations to build up immunities and genetic resistance to the diseases to help them ward off calamity. The Indians, unaccustomed to European diseases, had none. As Alfred Crosby (1986) points out, bacteria and viruses must be included among the European fauna and flora introduced into the New World by conquerors and colonists.

Contemporary sources report that for nearly seventy years after the Jesuits' arrival the Mayos lived in relative tranquility apart from the diseases that decimated their numbers.[19] Despite the regimentation imposed by the Jesuits as a price for their ministrations, both spiritual and administrative, no major rebellions stand out in the historical record. Whether the Mayos felt this was a peaceful time or not, the Jesuits seem to have considered it such. The Mayos must have benefited from the end to regional hostilities, but they can hardly have welcomed the judgments of the Inquisition and the imposition of harsh religious requirements. These included whippings and other corporal punishments for petty infractions such as failure to attend mass, recite incantations, refrain from eating meat on Fridays, or supply the required labor to the mission. They also had to tolerate the denunciation, punishment, and labeling as a sorcerer of anyone who questioned the correctness and rationality of the Catholic faith.

When various Mayos objected to the Jesuits' message and practices, Pérez de Ribas (1645) dismissed their objections as the work of resentful sorcerers whose power had been diminished by the truth of the Catholic faith. The priest's pious rejoicing on the defeat of such "sorcerers" appears today patronizing and arrogant. Using the implied might of European ideology, Pérez de Ribas summarily dismissed the Mayos' legitimate objections to the imposition of a foreign way of life.[20] The Inquisition had as little room for questioning natives as it did for questioning Europeans such as Giordano Bruno and Galileo.

Life under the Jesuits can hardly have been idyllic. Arbelaez (1991:369) describes the Jesuits' demands: "A strict sedentary life and Catholic chastity, indoctrination, and social norms were introduced. Autochthonous religious practices were prohibited and hunting and gathering banned unless directed and supervised by the priest. Monogamy, strict control of sexual relations, and discipline were harshly implemented." As one of the missionaries wrote in a letter in 1752: "Every missionary must accomplish three main points in the reduction of the Indians: (1) they must not have more wives than the ones given by God, the Church and the Priest; (2) there must not be any drunkenness or the Indians allowed to have wine, and (3) there must not be any thieves or burglars" (ibid.).

Before the Jesuits arrived, the Mayos had combined agriculture with the free life of hunting and gathering, and these proscriptions on their previously free-spirited existence must have seemed harsh and bewildering. Just why the priests, and not the Indians, should have been allowed to indulge in wine, must have been particularly perplexing. Furthermore, hunters the world over dislike being told that their activities will be curtailed or subject to magisterial supervision. Even today when the Mayo men of rural villages are in the monte, they keep a constant eye out for sign of *venaditos* (deer) to supplement their simple diet, even though such hunting is patently illegal. No respectable Mayo has ever obtained a hunting license.

If European institutions promised greater enlightenment and the civilizing of savages, government by the clergy was a dubious way to demonstrate it to the indigenous people. At least their caciques had maintained power by a general consensus. The Jesuits, in contrast, imposed their will—government from the top down. They were absolute rulers—lawmaker, judge, and jury on all matters—and the stern eyes of the clergy monitored the Mayos' behavior at all times. In addition to the strict religious requirements, the Mayos were required to work at least two days in the church's fields each week without pay, another practice that must have seemed unfair. In spite of these negative aspects, the several decades of relative peace with the Yaquis following the Jesuits' arrival was a favorable development. Many Mayos may have accepted regimentation as the price they had to pay for security from attacks and depredations by other tribes.

In the 1680s prospectors discovered silver at Promontorios near present-day Alamos. Within a few years more than forty mines were operating in the region (Bolton 1936:240). Less than twenty years later the discovery of another lode at Baroyeca to the west of the Río Mayo brought a further onslaught of wealth seekers. The lure of the glittering metal changed forever the social relations in the region and introduced a commodity-driven economy. The silver had to be dug from the ground and, unless it was in metallic form, smelted to obtain the pure metal. To accumulate this treasure, the *colonos* (colonists) needed labor, and lots of it, so they naturally turned to the Mayos. Although some of them willingly went to work in the mines, others were virtually kidnapped and forced to work as part of a *repartamiento* system whereby villages were required to provide laborers (West 1993:62–64). The Jesuits protested, and eventually laws were passed limiting, but not eliminating, the number of months Indians could be required to work.[21] For the most part these laws were ignored and the Mayos were semivoluntary slaves to the mines.[22] The work was dangerous and punishingly hard, the hours interminable, the living conditions deplorable, the pay scandalously low. Still, at least some Mayos were not averse to escaping the stern eyes of the Jesuits.

As more mines opened and the demand for Indian labor increased, mine owners competing for scarce labor offered better pay and working conditions (Brading 1971). Other mines outside the region, some producing gold as well, opened to the east in the Sierra Madre in Chihuahua, including the mines at Santa Bárbara and Batopilas. Some Mayos left their villages voluntarily to work at those distant excavations, perhaps lured by tales of good pay, loose life, and freedom from the religious dictates of the priests.[23] West (1993:63) suggests that many of the expatriates returned to their fields to assist with planting and harvesting, only to return thereafter to the mines, a source of dissatisfaction to both the clergy and the mine owners. Figueroa (1994:68) notes that "escapees" from the missions were frequently returned to their homes by force and, it appears, punished severely for leaving without authorization.

The Jesuits had good reason to fear the rough life of the miners. In addition to offering the Indians freedom and relative affluence, work in the mines provided them with cosmopolitan camaraderie, access to ideas of rebellion, and, quite probably, opportunities to learn the use of firearms, explosives, and military tactics. Lessons learned from the experience of freedom from the clergy or *hacendados* (estate owners), familiarity with wage labor, and ideas of resistance and rebellion gleaned from association with men of diverse backgrounds and beliefs ultimately made their way back to the towns in the Mayo and Yaqui Deltas.

The ore veins were gradually exhausted, but with the influx of miners and associated colonists, the flat and fertile Mayo lands toward the coast increased in value. The Jesuits saw with the first prospectors that a mining economy demanded food. It did not take the clerics long to recognize the potential of Mayo (and Yaqui) lands to produce an agricultural surplus. The expansion of Jesuit missions into Baja California was possible only because they could be supplied with crop surpluses from Cáhita lands with Cáhita labor (Crosby 1994).[24] The policy of reduction (relocating natives near missions) can thus be viewed as a gathering and organizing of the agricultural workforce required for producing a surplus.[25]

The Jesuits were not the only ones to see that good profits could be earned selling agricultural products to the mines. Entrepreneurs realized that the delta soils under irrigation were capable of producing several crops a year, all in great abundance, while the vast foothills could become home to thousands of head of cattle. Nevertheless, while agricultural production may have risen, it remained regional and never became part of national or international markets. Transportation to the populous center of the country was monumentally difficult, so basic food crops seldom made their way into the far interior. (Even in the early twentieth century the 1600-kilometer trip to Mexico City from Alamos required at least a week, usually more.) Certainly there was no hope for producing crops such as cotton, tobacco, or sugar cane for export to Europe.

Crop production for local or regional consumption made excellent sense, however. A reliable crop surplus would benefit the Jesuit order and their Indian charges as well. The increase in production subsequently made food available to the region's mines, creating a bustling trade that also created some friction because the Jesuits were able to undercut the prices charged by merchants who brought food supplies from Mexico City along laborious trade routes.

As the mines expanded at the beginning of the eighteenth century, more and more colonists flocked to northwest Mexico, a trend that continues to the present day.[26] The newcomers were for the most part a hardened bunch seeking to make a fortune. Natives were considered obstacles to their expansionist dreams, and the colonists staked claims for mines and lands with no regard to current occupancy. They built homes wherever they wished, allowed their livestock to roam freely, and refused to recognize indigenous people's rights to any land. To the colonists, the Indians were subhuman savages, even when the Indians insisted otherwise.[27] A Jesuit missionary demanded that secular authorities punish a Yaqui for insisting on being addressed as *señor,* the term the Spaniards required the Yaquis to use for them (Spicer 1980:41).

The increasing numbers of colonists added to the prosperity of Jesuit missions, which benefited by sale of commodities to the miners. Still, the increased numbers of non-Indians posed a continuing threat to the integrity of indigenous cultures. When land disputes arose, the colonists, not the Mayos, usually had the ear of the Spanish Crown. The Jesuits acted as advocates for the Indians, but with varying degrees of zeal, and with religious orthodoxy as the price extracted in return.[28] Even their motives for wanting to prevent colonists from taking Indian lands are suspect, for they feared the colonists would contaminate the natives with worldly notions.[29] The Indians were children to be protected from an evil world.

Cáhitas—Mayos and Yaquis alike—viewed the colonists as invaders who seized their lands, brought new strains of disease, violated women, dragged men off into forced labor, and let their cattle destroy the Mayos' crops. The Jesuits pleaded with the Crown to exercise some control over the settlers, but to no avail. Mayo rights were invariably ignored. Finally in 1740, to the Jesuits' horror, most Mayos joined in a rebellion fomented by a shadowy Yaqui named

Muni (Bean), who may have sought leadership over all Cáhitas (Spicer 1980: 39–50). Little is known of him except that he was a key player in the rebellion and was adept at playing the Crown against the Jesuits. The rebellion extended south to the Río Fuerte and included Zuaques, Tehuecos, and Ahomes. The Mayos' and other Cáhitas' enthusiastic response to Muni's influence indicates that the historic animosities between them and Yaquis had been left behind, and that a Cáhita consensus had begun to emerge. They had come to realize that more than a century of Jesuit teaching and authoritarianism had brought little more than the destruction of their culture.

The uprising of 1740 was not limited to Cáhitas. Other indigenous groups in northwestern Mexico also rebelled. In the north and east, Apaches carried out raids with relative impunity. By the early 1750s Pimas, led by Luis of Sáric, and Seris were also in open rebellion in the north and central parts of the province. The indigenous people of northwest New Spain had evaluated Spanish-imposed institutions for more than one hundred years and realized that instead of the better life promised by the Jesuits, they were faced with disease, exploitation of labor, theft of lands, brutality, and disrespect for native institutions. Spaniards—Europeans—were the problem.

Muni appears to have been a charismatic master of intrigue, playing the Jesuits against the civil authorities. While he was visiting Mexico City (purportedly protesting Jesuit treatment of the Yaquis), a Yaqui follower named Juan Calixto, reportedly following Muni's instructions, led an insurrection among the Mayos, whose festering resentment against both Jesuit rule and Spanish exploitation exploded into an uprising that took the Spaniards by surprise. The so-called Muni rebellion moved far beyond his control and involved more military activity in the Mayo and Fuerte regions than in the Yaqui. Even so, on his return Muni was executed, though to all appearances he was innocent of the charges brought against him.

During the 1740 rebellion all non-Cáhitas were killed or expelled and their institutions banished from the Mayo region. Spanish settlements were sacked and burned, including the mining town of Baroyeca, and all non-Indian property was destroyed (Acosta 1949). The Mayo lands were again free of foreigners, while only a few Jesuits remained in Yaqui territory. The rebellion shattered the myth of Mayo passivity and peacefulness. Although the Yaquis had failed to win a single major battle during the uprising, the Mayos had won a decisive encounter near Etchojoa. Once the shaken Spaniards regrouped, however, the rebellion was suppressed, even though the Crown could not claim that the Mayos or Yaquis had been defeated.

In the aftermath of the revolt, reprisals were fierce and a repressive period ensued as the Crown tightened its grip on the Indians. The population of Mayos (and Cáhitas in general) declined precipitously. By 1767, when the Crown expelled the Jesuits from New Spain,[30] fewer than six thousand Mayos lived along the Río Mayo, one-fifth of their estimated numbers a century and a half earlier (Spicer 1962:53). Spicer suggests that a diaspora of Mayos and Yaquis (who also experienced drastic population decline) may have accounted for part of the population decrease. Cáhitas moved to the various mining towns in the Sierra Madre and to distant haciendas, where Spaniards, desperate for laborers, hired them as cowhands. Yaquis and Mayos since that time have formed the backbone of Sonoran mining and ranching. Some smaller villages south of the Río Mayo may have increased in population as Mayos seeking refuge from Spanish revenge over the uprising fled their homes and took up residence with people who, though of different identity, spoke their language and observed the same general customs. Mayos also may have relocated to villages and towns along the Río Fuerte, thus beginning a process of amalgamation that today blurs the distinction between Mayos and natives of the Fuerte.

After the Jesuits' abrupt departure in 1767—when the blackrobes were led away under arrest and in chains—Spanish officials and colonists scrambled to appropriate their possessions and property (Calderón Valdés 1985, 2:261–69). As part of a plan to break up communal landholdings, colonists were offered up to 3,500 hectares of land by the Crown, while Cáhitas were limited to 5.5 hectares per family (Figueroa 1994:80). The Mayos watched with dismay as government and private interests began surveying the Mayo Delta. When the survey was completed in 1790, much of the land was allocated to non-Mayos. Only the persistent fear that the Mayos and Yaquis would be incited once again into revolt prevented all their lands from being appropriated. Even so, piece by piece, settlers and squatters whittled away at the remaining ancestral holdings.

The final decades of the Spanish colonial era in Mexico were marked by ever-increasing attacks by Apaches who swooped into Sonora from the northeast. Shifting their focus from the Mayos, the Spaniards threw their money and troops into defending against these whirlwind invaders whose raids extended as far south as the Río Mayo. Then, as the Spanish rule of Mexico slowly deteriorated, Mayos and Yaquis watched from the sidelines, hardly comprehending the internecine struggles and, thereafter, the war of independence. They could scarcely have objected since they, at least, were not being plundered systematically at the time. Yaquis appear to have maintained and even strengthened their traditional governments and reinforced their hold on their territory in the last third of the eighteenth century (Hu-Dehart 1984). Mayos, however, already dispersed and lacking the Yaquis' territorial identity, appear to have suffered considerable loss of land and identity (O'Connor 1989).

Mexico won its independence from Spain in 1821. When the new government assumed power, it became the policy (as it is today) to assimilate Indians into Mexican society—to make them into good Mexicans. Even President Benito Juárez, a Zapotec, sought an end to traditional Indian communal farming, believing that the more efficient system of private ownership would increase production and overcome historic blockages to forming a modern industrial capitalist economy. Thus the Mexican government assumed that communally held lands would be divided among individual Yaquis and Mayos who would then become peasant farmers and ranchers, and thus be taxed. Since the Indians were to be treated like everyone else, they should be required to pay the same taxes.

La Ley Almada, decreed in Sonora in 1828, ordered that lands illegally seized be returned to indigenous people, divided, and individually allocated for purposes of taxation. The Mayos took advantage of the former provision to regain authority over their traditional communal lands, but steadfastly resisted the latter provision. They would have nothing to do with enforced land capitalism. They viewed themselves as never having been conquered by Mexicans. The Spanish clerics had resided in their lands by mutual agreement only. The natives had never paid land taxes to Spain for the land that was, after all, theirs, and they saw no reason why they should begin now. They also found the notion of individual land ownership contrary to nature. They concurred with the Yaqui saying "Dios dió a todos la tierra y no un pedazo a cada uno" (God gave the land to all, not a piece to each). Their intransigence thus began a period of resistance to Mexican control of Cáhita lands that was to continue off and on for a hundred years. To some extent it continues in Yaqui lands today.

Settlers continued to pour into Mexico's northwest. By the year 1800, two-thirds of the Sonoran population was mestizo. Forty years earlier it had been less than one-third (Gerhard 1982:285). The arrival of more settlers and the expansion of those now settled ensured that Indians were a minority and would remain so. As the non-indigenous population increased, so did threats to traditional lands. Sonoran Indians were besieged by greedy outsiders, and the Cáhitas, whose lands were most suitable for irrigation and of almost legendary fertility, were the most vulnerable. Without the power of the Spanish Crown to protect communal lands from invaders, uneven though that protection had been, mestizos threatened to overrun the Mayo and Yaqui lands, and the Cáhitas had no one but themselves to turn to. An armed response became inevitable.

In 1824 government authorities announced their intention to survey and measure the Mayo and Yaqui lands to establish a basis for taxation, and to draft Cáhitas into the military to fight Apaches and other insurrectionists.[31] Taxation, the Cáhitas understood, was but a preliminary step toward seizure of their lands, and military conscription was a means of weakening their defenses against invading colonos. Those threats, after decades of seeing their lands overrun by outsiders, left them few options.

The leader of the first sustained rebellion against Mexican intrusion was a Yaqui named Juan Ignacio Jusacamea, whose battle name was Juan Banderas (Zuñiga 1835). In 1825 he recruited Mayos (or, perhaps, Mayos incorporated him into their plans) and other native groups in Sonora and Sinaloa to his cause of driving mestizos from indigenous lands. He nearly succeeded, uniting for the first time Mayos, Ocoronis, Tehuecos, and Zuaques from the Río Fuerte with Yaquis, Seris, and Opatas. Hostilities flared up again in 1832, when non-Indian families from Alamos seized land near Camoa on the Río Mayo (O'Connor 1989:19). Such was the disorganization of the new Mexican government that the Banderas rebellion lasted nearly eight years. The allied indigenous forces were finally defeated in 1833, when Banderas was captured and executed, despite his having been recognized by the Mexican government as the region's governor-general.

Banderas's death temporarily put an end to armed insurrection, but Mayos and Yaquis continued to seethe with resentment at the persistent intrusions into their lands and their treatment by hacendados and *latifundistas* (powerful owners of large tracts of land).[32] The Cáhitas joined whatever insurrections and rebellions they could find.[33] As Mexico struggled to form a stable government, put down rebellions, repel filibusterers (armed paramilitary invaders) from the north, and bring factions of warring strongmen together, it was racked by internecine strife. Federalists (conservatives, those favoring a weak union of states) battled with centralists (liberals, those who advocated a strong central government). Sensing an opportunity to make an advantageous alliance, Mayos aligned themselves with the forces of Manuel Gándara, an opportunistic federalist hacendado whose family controlled Sonoran politics from about 1830 until 1857. His opponent was Ignacio Pesqueira, an equally opportunistic centralist whose family succeeded the Gándara dynasty and ruled until 1879.[34]

The Mayos sided with Gándara only because he appeared to support their land claims. As an hacendado, Gándara tolerated indigenous communities and opposed the general capitalist transformation of Mexico, preferring instead the semifeudal institution of the hacienda as the underpinning of Mexican society. Pesqueira, in contrast, was openly hostile to indigenous communities, preferring to break them up and divide the formerly communal lands into privately owned plots. In the lengthy power struggle, Gándara lost. Pesqueira promised the Sonoran bourgeoisie that he would "open up" the lands of the Río Mayo to private ownership and create a port near El Júpare at the mouth of the Río Mayo (O'Connor 1989:20). In 1858 Pesqueira ruthlessly put down

Mayo protests and rebellions as Sonora's emerging oligarchy (primarily from Alamos) began to divide the fertile lands of the lower Mayo Valley among themselves. Pesqueira "secured" the Mayos' lands on behalf of Sonora's wealthy. In 1861 and 1862 Mayos rebelled, but their resistance was smashed, this time by a private army organized by the Alamos oligarchy to maintain their appropriation of Mayo lands (ibid.).

In 1865, a French expeditionary force landed at Guaymas, seized control of the port, and from there invaded—and conquered—Sonora. For more than a year the French controlled the state.[35] The French courted the Mayos, promising that in exchange for their support, they would return the Mayos' lands to them. Mayos were receptive to the message and allied themselves with the foreigners, who defeated the same private Alamos mercenary army that had stifled Mayo protests. The Mayos were also influenced by José María "El Chato" Almada, who was, ironically, a prominent Alamos scion and supported French rule.

The French may have intended to fulfill their promises, for the French monarchic institutions recognized communal lands occupied by indigenous peoples just as they recognized hereditary landed estates and respected the rights of the Church to own great tracts of land. Elsewhere during his reign, the emperor Maximilian had intervened on the side of Indian communities who claimed forcefully that their communal lands had been taken over by hacendados, thus strengthening indigenous claims to aboriginal lands and opening up to usufruct lands previously claimed by the landed elite (see Krauze 1997:183). Privatization and the creation of a vast host of yeomen—in short, the commodification of land, a key point in Benito Juárez's liberalization of Mexico—was incompatible with the hereditary privilege of the monarchy.

Although the Mayos rallied behind the French cause and were instrumental in helping the French invaders take Alamos, ultimately the French lost.[36] The defeated Mayos were once again subjugated and incurred the wrath of the patriotic Sonorans who viewed them as traitors.

Pesqueira, who led the Sonoran resistance, was a strongman and politician whose enmity the Cáhitas had first aroused because of their loyalty to Gándara, Pesqueira's old enemy (Almada 1952:507–14). Pesqueira's attempt at patriotic leadership was clumsy and lacking in finesse, but after the French were driven from Sonora, he vowed revenge on all those who had aided and abetted the enemy—except for the wealthy, whom he found easier to forgive once their largesse became obvious. When the Mayos continued with armed struggle, mostly on a small scale, harassing mestizos who occupied Mayo lands, Pesqueira swore that he would end the Cáhitas' belligerence once and for all.[37] By the end of 1868 his military crackdown succeeded in ending, at least for the time being, the organized Mayo resistance. His victory, though, did not come easily, and not without a campaign of bloody violence. He inflicted a resounding defeat on the Mayos at Santa Cruz on the Río Mayo. That done, he lost interest in his campaign of submission and turned the chore over to his assistant, Gen. Jesús María Morales, who maintained a reign of terror over the Cáhitas, especially the Yaquis (Spicer 1962:66).[38] Morales routinely massacred villagers as a means of warning Mayos to behave properly.

Pesqueira's suppression of the Mayos and Yaquis allowed outside capital to flow into the Mayo and Yaqui Valleys, and the Mexican government established a garrison near Navojoa to monitor Mayo behavior. By the 1870s Mexican nationals and foreigners alike saw the flat land, fertile soil, warm climate, and abundant water for irrigation, and dreamed of agricultural riches. By the end of the decade, an alliance had formed between two men, Ramón Corral and Luis Torres, who would hold the real power in Sonora for the next thirty years. This was also the beginning of the more than thirty-year reign of Porfirio Díaz as Mexico's president and strongman. Under his dictatorship, guided by the principles of liberal reform (established, ironically, under Benito Juárez), foreign investors were encouraged to establish modern systems of agriculture in Mexico. Mayo and Yaqui lands, the most fertile in the country, were the ripest for the plucking. As the government assisted private interests in constructing canals and irrigation works to bring water from the Río Mayo and Río Yaqui and exploit the fertility of the land, the Cáhitas watched, conferred, objected, and then rebelled.

Their guide through this conflict was the greatest Yaqui leader of all, José María Leyva, better known as Cajeme (Ca'a ja'eme = he who drinks no water). Cajeme united Yaquis and Mayos in a rebellion that nearly succeeded in establishing a permanent joint indigenous nation. The roots of the uprising were much the same as before: the dispossession of the natives' lands and the appropriation of those lands by outsiders, who in this case were encouraged by Porfirio Díaz and Ramón Corral, who served as vice president under Porfirio.

By 1875 Cajeme had organized the Mayos and Yaquis into a virtual independent nation (Troncoso 1905:63). He opened a port, charged duties on ships and tolls on travelers passing through the region, established a central treasury, organized an army, and arranged for the acquisition of firearms. His influence extended south of the Río Mayo to the Río Fuerte, where his appointed lieutenant (*lugarteniente*) oversaw the governing of the Mayos and general preparedness for war, and north nearly to Guaymas. Mexicans watched with dismay as the pan-Cáhita forces won several key battles. Cajeme repeatedly offered to cease the hostilities if mestizos would leave Cáhita lands. His offers were summarily rejected.

As was the case with Juan Banderas, Cajeme's successful campaigns caused pandemonium in Sonora and throughout the republic. By 1880 it was apparent to Mexicans that there would be a general insurrection of Cáhitas, not just a small rebellion of a few Yaquis. In 1882 a combined force of a thousand Mayos and three thousand Yaquis under Cajeme's leadership defeated the government troops at Capetamaya near Navojoa (Troncoso 1905:75). Mexican residents fled the villages or armed themselves heavily and called on the government to protect them. All of mestizo Sonora panicked, fearing a Cáhita victory.

The Mayos and Yaquis, apparently convinced that they had driven mestizos from their lands, lost interest in fighting once the outsiders had fled. The government forces, however, regrouped and responded to the insurrection by capturing and executing anyone who might pass for a Mayo leader or warrior. In 1884 the Mayos were decisively defeated near Navojoa and ceased to be a factor in Cajeme's rebellion. The Yaquis provided little assistance to the Mayos during the critical months before the decisive final battle, and thereafter the Mayos' resistance to the government waned. Many of them were demoralized and weary of fighting, and there were deep divisions between rebels and those wanting peace on the Mexicans' terms. With the defeat and execution of Cajeme in 1887 the widespread Mayo rebellions, after one brief flare-up, were at an end. Cajeme had lost. Lost as well were Mayo dreams of maintaining their lands free of non-Mayos. A city and municipio were founded and named in Cajeme's honor, but the city's name was later changed to Ciudad Obregón.

The Mayos were not yet finished opposing Mexican appropriation of their lands and destruction of their culture. In September 1888, Mexican military officials reported large numbers of Mayos abandoning their work as peons on landed estates and converging on villages, where they listened to a wave of prophets, men and women, who preached the end of the world by flood. The most notable of these took place in a village near Masiaca, where a sixteen-year-old Mayo named Damián Quijano drew large crowds as he spoke of impending doom for those who had usurped Mayo lands. Damián was arrested, along with sixty of his followers and protectors, as a threat to the peace and tranquillity of the land (Troncoso 1905:19).

Mexican military authorities were touchy about these large gatherings, imagining in every group of more than three or four people another incipient rebellion. Soldiers found suspicious-looking saints and images in nearly every Río Mayo village and sensed an imminent insurrection. They were especially concerned about Damián, who was said to be a nephew of one of Cajeme's close associates.

The threat from Damián was insignificant, however, compared to the threat posed by Teresa Urrea, known to her followers as La Santa de Cabora.

The daughter of mestizo parents, she became a charismatic prophetess who reportedly had influenced Damián, and she became enormously popular in her own right. She purportedly performed miraculous cures at her father's ranch at Cabora on the upper Arroyo Cocoraque in Mayo country, and she uttered vague and mysterious prophecies calling on the Mayos to beware of floods that would inundate all gentiles.[39] Prophetic followers of Teresa sprang up throughout Mayo country, but Mexican authorities took little action, reluctant to spark a new rebellion by interfering with charismatic leaders who appeared harmless.

Teresa's influence soon extended throughout the Mayo region. Her charismatic words acted as a slow-moving catalyst on many Mayos. In 1892, to the complete surprise of Mexican authorities (and probably to Teresa), a large band of Mayos rallying behind the cry "Viva La Santa de Cabora" descended upon Navojoa, sacked it, and executed the municipal president, seizing small amounts of money and arms in the process. Elsewhere small bands of Mayos attacked mestizo settlements, then rushed to Cabora to be blessed by Teresa. The rebels were armed primarily with bows and arrows, and only a few obsolete firearms. Mexican military authorities quickly put down the rebellion, executing all the leaders and participants they could find and deporting some forty others to forced labor in the mines of Santa Rosalía in Baja California (Almada 1990:22). Teresa Urrea was arrested, along with her father, and taken to Cócorit, a former Yaqui village, where they were incarcerated. Eventually both were exiled to the United States, even though her father was a prominent and wealthy member of Alamos society.[40] The Mayos were still a threat to *pax mexicana*.

The Mexican army's ferocious actions made the prophet-inspired rebellion short-lived.[41] Mexican soldiers hunted down any Mayos who appeared capable of complicity in the supposed uprising and either executed them on the spot or dragged them off to be shipped south to slave plantations. Bands of soldiers combed the Mayo Valley for signs of rebellion, and the slightest resistance provided grounds for assassination. Military authorities exhorted the Mayos to return to their fields, by which they meant, get back to the haciendas and their masters.

One part of the government strategy to "pacify" the Mayos was to eliminate their unity. They did this by disbanding the system of local *gobernadores* (governors), which had provided cultural cohesion to the Mayos for centuries. By the mid-1890s nearly all the Mayos' traditional governing structures had been disbanded by order of military authorities. Because the Mayos had never been as tightly organized in related villages as the Yaquis, the obliteration of their traditional governors dealt a severe blow to whatever hopes Mayos might have had for establishing a pan-Mayo organization. Many villages ceased to function as families became virtual slaves on haciendas (Crumrine 1983). While prophets continued to appear sporadically, responses to them were more and more fragmentary. More and more villages were swept aside and relocated as modern agriculture spread across the delta.[42]

A diaspora of Mayos set in, and the close community ties and contact among villages necessary for maintaining ancient cultural bonds gradually eroded. The system of village governors faded into obscurity, and the villages were assimilated into the mestizo political system, where the Mayos could be more easily watched. The strategy of dispersing Mayos and neutralizing their traditional government was successful. They were no longer a threat to those who sought to appropriate Mayo lands for their own use. Between 1892 and 1902 a maze of new canals were constructed to bring irrigation water to mestizo lands along the Río Mayo. The Mayos could only watch in despair.

One of the mestizo beneficiaries of the newly "opened" land was the great Sonoran hero of the Mexican Revolution Alvaro Obregón, a small-time hacendado from Huatabampo. During the revolution many Mayos joined Obregón's constitutionalist forces. He was himself a farmer, a harvester of garbanzo beans in the Mayo Delta on lands his family, the Salidos from Alamos, had appropriated from the Mayos. Mayos joined him because he promised that once the constitution was restored, he would support them in their land claims. They formed a solid though small bloc in his forces (Hall 1981), and Obregón freely acknowledged the military presence of his Mayo friends: "This happened on the last days of March, 1912, and by the 14th of April I had gathered three hundred men, mostly natives of the region, of indigenous stock, most of them land owners, including me, who had been cultivating garbanzo beans on my little hacienda that I owned on the left bank of the Río Mayo, which bore the name Quinta Chilla" (Obregón 1959:8).

On April 14, 1912, they strode off to battle in full dress, including their bows and arrows. The host boarded a train in Navojoa and arrived the next day in Hermosillo, where Sonoran officials bestowed on Obregón and his Mayo followers the title of the Fourth Sonoran Irregular Battalion. A month later Obregón had consolidated forces, including "at Navojoa . . . a large number of Indians armed with arrows" (Obregón 1959:34).

During Obregón's early marches, the last armed rebellion of Mayos occurred, this one led by a Mayo who called himself Totoliguoqui (hen's foot). His forces attacked several mestizo towns in the Mayo Delta and attempted to drive out mestizos from the region. Totoliguoqui was captured and executed in 1914 (O'Connor 1989:26).

In December 1912, after only seven months of service, Obregón resigned from the army, but he returned after only two months to head Sonora's War Department. The Mayos, by all accounts, were brave and mighty fighters.[43] They were joined by many Yaquis and followed Obregón from skirmishes in Sonora to battles throughout central Mexico, which he always seemed to win. In an ironic turn of events, his Mayo troops probably participated, on June 5, 1915, in the battle of Puebla, where they engaged, if only on a small scale, with the forces of Emiliano Zapata, who were fighting for precisely the same cause, the right of people to retain their aboriginal lands. Zapata lost the battle and thereafter his military initiative, and his agrarian movement declined (Brunk 1995).

The Mayos' decision to align themselves with Obregón was an unhappy one for them. Obregón originally supported the presidency of Venustiano Carranza, a patrician hacendado ideologically opposed to land reform (and an implacable opponent of Zapata's Plan de Ayala, which would have granted the Mayos all their wishes).[44] Furthermore, Obregón himself, though hardly a latifundista, was of hacendado blood and owned a medium-sized farm. He could hardly have been expected to prove himself a champion of indigenous people's land claims, especially following the revolution, when he acquired huge blocks of land. In the words of Paul Friedrich (1970:105), Obregón, "himself a big landowner, exploited a nominal liberalism to mask the slowdown or stoppage of agrarian and other reform measures." Indeed, he was committed to the capitalist transformation of Mexico, including the "opening up" of Mayo and Yaqui lands to "development."

The Mayos had been tricked by Obregón's empty promises. Their great personal sacrifice gained them nothing. After Carranza was assassinated in 1920, Obregón ascended to national power, and the Mayos returned to Sonora feeling victorious, believing in a new future for indigenous people. They hoped to be viewed as heroes and rewarded with the return of their lands. What they found was even greater mestizo control of their lands, depopulated towns, families missing, and homes leveled. Those who did have land had no water for irrigation.

Obregón turned out to be an even worse traitor than the others. He went on to become president of Mexico and one of Sonora's favorite sons, at least of wealthy Sonorans who benefited from his coziness with the same hacendados against whom he had battled, and his willingness to have the Mexican constitution interpreted in a way most favorable to mestizos' acquiring Mayo lands (Hall 1981).[45] The dismayed Mayos looked on in silence, wondering how they would ever put tortillas on their tables.[46]

Obregón was succeeded as president by a fellow Sonoran and close associate, the rabidly anticlerical Plutarco Elías Calles. He despised the Mayos. In 1926 Calles demonstrated to them just how long a memory he had and how vicious his troops and appointees could be.[47] During Calles's campaign against the Catholic Church, most of the Mayos' churches were burned, and their saints and images destroyed by mestizos whose own sanctuaries were often spared. With the vision of burning churches, desecrated shrines, and defiled saints fresh in their minds, some Mayos renounced their faith and their traditions (Erasmus 1967). If their saints had failed so miserably to protect them, what good were they, and why waste time with them?

The inflated promises and high-sounding slogans of the Mexican Revolution left the Mayos with nothing. They were no better off, and perhaps worse off, than they had been under the thirty-year porfiriato.[48] The military alliance they had forged with the hacendados had failed. Warlords such as Obregón had managed to increase their own landholdings while their wealthy friends helped themselves to the rich lands of the Mayo Delta. The prophets' words had been proved wrong. Their cherished holy objects had been vandalized. And they had little or no land. Although some communal lands were later returned to Mayos, and some ejidos created, most of the best lands remained in the hands of mestizos, as they do today. Perhaps worst of all, the ancient organization of Mayo communities had been destroyed. Nothing was left to replace the village gobernadores, whose wisdom and counsel had been respected and who had represented the villages in regional matters as well as mediating village controversies.

During the porfiriato, especially in the years 1902–9, the Sonoran government of Rafael Izábal (and the real power behind the governor, Ramón Corral) finally dealt decisively with the "Yaqui problem" (in reality a Cáhita problem). Government forces captured as many Yaquis as possible and shipped them to sugar cane plantations in Oaxaca and the *henequén* (sisal) estates in the Yucatán, where they worked as slaves. The deportations continued well into the revolution. Many thousands perished under the brutal working conditions, while countless others were executed. Many escaped to the United States, where they formed colonies in exile, which persist today. Because the government was not interested in making the distinction between Mayos and Yaquis, they were forcibly thrown together, and any descendants found their cultural heritage one of mixed history.

One irony of the mass enslavement and incarceration of Cáhitas was that Sonora was consequently deprived of much of its workforce. Mayos and Yaquis were renowned as industrious, skilled, and competent workers, and as such were in great demand by mines, ranches, farms, and industry. The pogroms against them slowed Sonora's economic development. But, according to such observers as M. M. Diéguez, a former union leader and revolutionary army officer, the only good Yaqui was a dead Yaqui (Calderón Valdés 1985, 4:372).

In a further irony, the final armed struggle by Yaquis occurred in 1927, when

Yaquis briefly held former president Alvaro Obregón captive. The general assigned to smash Yaqui resistance and rebellion once and for all was Román Yocupicio, a Mayo from Masiaca. He was largely successful. He executed hundreds of Yaqui dissenters and went on to become governor of Sonora, officiating at the final dissection and division of aboriginal Mayo lands into the hands of non-Mayos (Almada Bay 1993; Bantjes 1998).

Although Mayos today do not speak of uprisings, their sympathy with eschatological prophets continues. Mayos are far more open to conversion to evangelical and messianic Protestantism than are Yaquis. Most Mayo villages have an evangelical church where *hermanos de la fe* (brothers of the faith) gather to hear predictions of the end of the world.[49] Gone from these fiery preachers, however, is the message that Mayos alone will be spared or saved. Instead, they preach about abandoning the traditional Mayo way and becoming puritanical capitalists (Mary O'Connor, pers. comm., 1995). Authorities no longer view Mayo evangelism with concern.

The final irony of the Mayos' struggles was Yocupicio's division of their lands among mestizos in the mid-1930s. Yocupicio invoked revolutionary rhetoric, reminding the Mayos that they had a patriotic duty to share their lands with other Mexicans in order to promote the national good. He allocated some lands to Mayos as ejidos, but the best lands and the bulk of them went to mestizos, some as ejidos, but mostly as choice parcels to private owners. Funding for constructing irrigation works went primarily to mestizos as well, usually to Yocupicio's wealthy friends. Yocupicio was obsessed with breaking up communal holdings in the Mayo basin and establishing capitalist forms of agriculture there (Almada Bay 1993). Today his reception among Mayos is mixed, at best.[50]

Since the end of the revolution the Mayos' social situation has stabilized somewhat. Only sporadic episodes of Mayo nationalism have occurred. The population has become even more dispersed, but comunidades and ejidos have been formalized, and today most of the region's sixty thousand Mayos are directly or indirectly associated with one or the other. While Mayos control some highly productive farmland, they are typically unable to obtain credit and are forced to rent or lease their lands to commercial farmers who have access to the credit needed to purchase equipment and agricultural supplies such as seed, fertilizer, and pesticides. Mayos seldom control social institutions in the region, which are also in the hands of mestizos. Even the Instituto Nacional Indigenista, supposedly an advocate for indigenous peoples, is suspected by many Mayos of being an institution used by mestizos for their own ends.

Mayos of Sinaloa, while implicated in many of the nineteenth-century rebellions involving their northern counterparts, faced different, almost bizarre social dynamics. In the mid-1880s an eccentric, idealistic North American, Albert Owen, founded a utopian community of colonists at Topolobampo on the Sinaloan coast in land that was commonly owned by several traditional Mayo communities. Owen's goal was to establish a communitarian society and connect it with the eastern United States by constructing a railroad through Chihuahua via Texas. Owen was granted an enormous chunk of land by President Porfirio Díaz on which to build his dream, and he enticed several hundred North American settlers to help fulfill his goal of constructing a socialistic community in which legal equality would be guaranteed and the productivity of the land shared on the basis of the vague term "equity."[51]

Owen's scheme fit in well with Porfirio Díaz's goal of developing the productive potential of the west coast of Mexico. Owen's ideological commitment was less important than his claim to have enough capital to found an agrarian community, establish a port at Topolobampo, and build the railroad. After being granted what he asked for, Owen directed the construction of canals and laid out the streets for a new city to arise from the thornscrub, but the project came to naught. The settlers soon abandoned the steamy lowlands infested with mosquitos and *jejenes* (no-see-ums). Owen apparently committed various fraudulent acts, and his company went bankrupt. Nevertheless, the enterprise established the agricultural potential of the Fuerte Valley, and growers were quick to take note and incorporate the considerable agronomic research resulting from the project's many studies. Owen's earthworks were taken over by hacendados whose descendants ultimately benefited from the construction of Miguel Hidalgo Dam on the Río Fuerte at El Máhone and the elaborate series of canals that connect with Josefa Ortiz Domínguez Dam on the Río Cuchujaqui and with the vast coastal plain of southern Sonora and northwest Sinaloa.[52] The Mayos of the Río Fuerte, on whose aboriginal lands Owen squatted, were the ultimate losers. Today their descendants work as impoverished day laborers on lands that once were theirs.

During the 1940s the federal and state governments undertook ambitious and aggressive programs to establish commercial agriculture in the fertile lands of the Mayo Delta. Fields were leveled, canals were dug and lined with concrete, Mocúzari Dam was built, and a network of paved roads was completed. The bulk of government subsidies found its way into the pockets of private investors rather than going to the original landowners, the Mayo communities that dotted the landscape. Some of the land remained under Mayo control, including nearly half of the irrigated lands (Germán et al. 1987:108), but the best lands—those nearest the large canals and key highway points—were allocated to non-Mayos. The capital resources necessary for commercial farming were lopsidedly allocated to private, non-Mayo investors, further increasing the

Mayo family, Yópori. In most Mayo villages, employment for young men is virtually nonexistent. They venture to cities and more populated areas in search of work.

Mayos' competitive disadvantage at developing the commercial potential of their lands. The clearing and scraping of the monte also cut Mayo residents off from their traditional gathering activities, destroying any possibility of continuing a subsistence way of life.

Huites Dam on the Río Fuerte was completed in 1995, providing the justification for clearing additional acreages south of the Mayo Delta. Thousands of hectares of monte formerly used for gathering plant materials incorporated into a subsistence way of life were cleared. They are now either irrigated or lie fallow awaiting promised irrigation water.

In the 1970s, the national and state government adopted a strategy of rural development that placed heavy emphasis on expansion of cattle grazing to complement the commercialization of agriculture. To increase beef production, officials promoted wholesale clearing of monte and the sowing of buffelgrass, an African import. Buffelgrass sowed in a cleared pasture can triple or quadruple beef production for the first few years (Yetman and Búrquez 1994), but productivity declines thereafter without aggressive and capital-intensive management. Vast areas (more than 1.5 million hectares in Sonora) have been thus transformed into monoculture grassland pastures, while an equal number have

Clearing for buffelgrass. Vast tracts of coastal thornscrub are being bulldozed to make way for pasture and commercial agriculture, jeopardizing the Mayos' traditional, semi-subsistence way of life.

occurred incidentally or on a haphazard basis. Thus the rural landscape in Mayo country continues to be transformed from highly diverse and ethnobotanically rich thornscrub to monocultural grassland and vast fields of commercial crops. The disappearance of native plants is reflected in the diminishing amounts of native plant materials to be found in the homes of Mayos, both rural and urban.

## MAYOS IN CONTEMPORARY SOCIETY

Traditionally Mayos have frowned upon the accumulation of money and financial competition. As a result, and because indigenous people in general lack social standing in northwest Mexico, Mayos have tended to be marginalized in the regional economy. Such a generalization, however, obscures some of the complexities of the Mayos' relations with local, national, and international economies. Most Sonoran Mayos—except on the coast, where they may be

Vicente Tajia contemplates the *pitahayal*. This botanical treasure yields tons of delectable fruit and other useful products.

fishermen—work part- or full-time as *jornaleros*, day laborers in fields, usually for agribusiness enterprises. Although most Mayos are members of comunidades or ejidos, for the most part their holdings are insufficient and the climate too dry for them to farm self-sufficiently, so they must gain income from other sources. On one Mayo ejido the lands are owned and worked collectively, but the remaining ejidos are divided into individual parcels and common lands. In some cases there are no common lands, only parcels. In a few instances the ejidatarios in the Distrito de Riego del Río Mayo farm their own irrigated parcels. In others, the parcels are leased to agribusiness producers for substantial sums, and the *parcelarios* are able to subsist on the income.

There are historical reasons to believe that Masiaca and the other comunidades were created as a labor reserve for the fields of the Mayo Delta (Yetman 1998). Within comunidades, contract leasing is not yet widespread, mostly due to the lack of reliable irrigation water. Cultural prohibitions against economic competition (Crumrine 1983) and a history of social marginalization of Indians (O'Connor 1989) have worked against the Mayos' participation in the mainstream of Mexican economic life.

Mayos living where monte is still available supplement their cash earnings with diverse materials gathered and gleaned near their homes. Not only is this resource of profound economic import (firewood alone is a major economic factor), but the familiarity with *juya ania* (the natural world) that such a semi-subsistence way of life necessitates results in a store of knowledge remarkable in its scope and intricacy.

# 4  Plant and Animal Life

VEGETATION

The plant communities found in the lands inhabited by Mayos are varied and complex, more so than in the lands of any other groups indigenous to northwest Mexico. In keeping with Martin et al. (1998) we have adopted the following categories for classifying vegetation: coastal vegetation, coastal thornscrub, foothills thornscrub, tropical deciduous forest, and oak woodland. The last category is represented only marginally in Mayo lands in specialized soil situations. Arroyos, canyons, and *cajones* (narrow canyons) also present special plant associations that add considerably to species diversity in the region. For example, species normally associated with tropical deciduous forest are found in arroyos adjacent to coastal thornscrub nearly to the coast, and the canyons and cajones of the Río Fuerte and Río Mayo are the northern limits of many tropical species. In the more arid thornscrub near Arroyo Cocoraque vegetation more typical of the Sonoran Desert reaches its southern limits, suggesting to Shreve (Shreve and Wiggins 1964) that this arroyo be designated the southern limit of that desert. Near Masiaca the *sahuaro* cactus (*saguo,* saguaro; *Carnegiea gigantea*) and *ejéa* (*palo fierro,* ironwood; *Olneya tesota,* Fabaceae), key species of the Sonoran Desert, reach their southernmost points.

## Coastal Vegetation

Coastal vegetation is divided into two subgroups: estuarine plants and dune and beach communities. Mayo lands contain several estuaries: Estero Tóbari, Estero Aquiropo, Bahía Yavaros, and the complex of estuaries known as Estero Agiabampo.[1] These are dominated by three species of mangroves: *ciali* (*mangle negro,* black mangrove; *Avicennia germinans*), *pasio tosa* (also called *moyet*) (*mangle,* white mangrove; *Laguncularia racemosa,* Combretaceae), and *canari* (*mangle rojo,* red mangrove; *Rhizophora mangle,* Rhizophoraceae). In

Mangrove estuary near Topolobampo, Sinaloa. Four species of mangrove are found here.

and near tidal flats a distinct community of salt-tolerant plants includes the mangrove-like *pasio* (*Maytenus phyllanthoides*, Celastraceae). The portion of this community lying above normal high tides is still subject to flooding during extreme high tides and hurricanes. Mangroves grow only where their roots can be permanently in saltwater or brackish water. The extensive mangroves of the Agiabampo complex form a maze of channels and small islands stabilized by the mangroves. South of Agiabampo in Sinaloa, the mangrove estuaries extend nearly to Bahía Topolobampo, encompassing the delta of the Río Fuerte.

Non-estuarine habitats extend along most of the coastline in Mayo country, a distance of more than 150 kilometers. Sand dunes and associated plant communities are found only near Agiabampo, Camahuiroa, Huatabampito, and Yavaros. Major Mayo villages, including Agiabampo, Camahuiroa, Las Bocas, and Yavaros, are located on the shore and are sustained by fishing and harvesting shellfish.[2] In Sinaloa, coastal villages are primarily mestizo, and Mayos live for the most part in inland villages.

Coastal thornscrub near Camahuiroa, an indigenous Mayo community and seaside resort. The influence of dew near the coast produces a denser, lusher growth than farther inland.

## Coastal Thornscrub

Immediately inland from the beach or estuary margin (including dense mangrove swamps) the vegetation assumes a predictable, if varied, aspect. Some uplands are depauperate, supporting only sparse plant life, while others nearby are heavily populated with small trees and cacti. The vegetation changes rapidly with variation in soil and soil horizons (McAuliffe 1995:108–13) and with distance from the water's edge. From northwest to southeast the coastal plain becomes slightly more moist, enough to produce a noticeably lusher aspect at the Sinaloan border. Inland locations tend to receive slightly more rainfall than those on the coast.

The most prominent feature of coastal thornscrub is the enormous numbers of *aaqui* (*pitahaya*, organpipe cactus; *Stenocereus thurberi*). Mayos refer to the thickest forests of these giants as *pitahayales*. They are found primarily on the flattest parts of the plain, flourishing in argillaceous soils. Also abundant is the arborescent *etcho* cactus (*Pachycereus pecten-aboriginum*), which is as tall as the pitahaya but never found in dense groves. A common, solitary tree is *jito* (*palo jito; Forchhammeria watsonii,* Capparaceae), which has a thick, straight bole (trunk) and symmetrical foliage. Its dark, persistent leafage makes it easy to spot during dry months, and it provides ample shade even in the heat of the spring drought. Equally prominent in the pitahayal is *báis cápora* (*saituna; Ziziphus amole,* Rhamnaceae), noticeable due to its roundish, bright-green leaves, and height of 13 m. Other trees include *choy* (*brea; Parkinsonia praecox,* Fabaceae), *júyaguo* (*guayacán; Guaiacum coulteri*), *murue* (*jaboncillo, ocotillo* or *palo pitillo; Fouquieria diguetii,*), *juupa* (*mezquite; Prosopis glandulosa,* Fabaceae), *sato'oro* (*papelío; Jatropha cordata,* Euphorbiaceae), *tásiro* (*sanjuanico; Jacquinia macrocarpa,* Theophrastaceae), and *to'oro* (*torote; Bursera fagaroides, B. microphylla*). Shrubs are also numerous, notably *ba'aco* (*Phaulothamnus spinescens,* Achatocarpaceae), *cósahui* (*tajuí,* Sonoran ratany; *Krameria sonorae,* Krameraceae), *jícamuchi* (*palo piojo; Caesalpinia palmeri,* Fabaceae), *juvavena* (*Capparis atamisquea,* Capparaceae), *ono jújugo* (*Adelia virgata,* Euphorbiaceae), *sa'apo* (*sangrengado; Jatropha cinerea,* Euphorbiaceae), *sigropo* (wolfberry; *Lycium andersonii,* Solanaceae), *huo'ótobo* (*vara prieta; Cordia parvifolia,* Boraginaceae), and the unusual *cantela oguo* (*candelilla; Pedilanthus macrocarpus,* Euphorbiaceae). Cacti include *ónore* (*biznaga; Ferocactus herrerae*); *choya* (chainfruit cholla; *Opuntia fulgida*); *navo* (*nopal,* prickly pear; *Opuntia wilcoxii*); the midsized columnar *musue* (*sinita; Lophocereus schottii*); *sibiri* (*choya; Opuntia thurberi*), often reaching 4 m in height; and *sina* (*Stenocereus alamosensis*). Also found are at least four species of *biznaguitas* (*Mammillaria* spp.), often abundant in the coastal thornscrub.

During las aguas a remarkable variety of climbers appear and grow with astonishing rapidity. Diverse vines are common, including the *choya huani* (*güerequi; Ibervillea sonorae,* Cucurbitaceae), whose lengthy vines emerge from a tuber submerged iceberg-like in the soft coastal dirt. The güerequi often resembles a gray jug half buried in the sand. The proliferation of vines as much as any other single factor contributes to the transformation of thornscrub when the summer rains arrive from a brown gray drab to a riot of greens presenting nearly a solid wall of rapidly growing plants.

In the gently undulating bajadas only a couple of meters above the flatter coastal plains coastal thornscrub merges into a semiforest of leguminous trees seldom more than 7 m high except along watercourses. In addition to mezquite, prominent legumes include *choy, caaro* (*palo verde,* blue palo verde; *Parkinsonia florida*), ejéa, *chírajo* (*güinolo* or *chírahui,* boat-spine acacia; *Acacia cochliacantha*), *jócona* (*Havardia sonorae*), *nésuquera* (*uña de gato,* catclaw; *Mimosa distachya*), and *cu'uca* (*vinorama,* sweet acacia; *Acacia farnesiana*). Other trees common in this leguminous belt but less common on the lower lands include *sire* (*granadilla; Malpighia emarginata,* Malpighiaceae) and *to'oro chucuri*

Coastal thornscrub near Jambiolobampo. Although *pitahayas* appear to dominate the landscape, other species are represented in greater numbers.

(*torote prieto*; *Bursera laxiflora*), which are shrubs on the flats but attain tree size on the rolling alluvium. During the dry season, jitos, saitunas, and etchos are easy to spot. The dark green jitos appear almost black, standing out as clearly as black sheep among white. It is easy to estimate the numbers of these great trees and their relative size by seeing them when the rest of their associates are dormant, appearing dead. The saitunas look ragged and craggy but often retain a few green leaves through the dry spring.

In late May and throughout June the pitahayas send out bulbous buds that gradually shade red. The white flowers open at night and remain available to pollinators till midmorning, when most of them shut forever. Pollinators abound in the evening and early morning hours. Lesser long-nosed bats (*Leptonycterus curasoae*) flock to the pitahayales at night, as do a host of moths. Bumblebees and other insects (including many beetles) lumber along taking in the last of the morning's nectar. Etcho fruits begin to ripen in May as well, attracting a different host of feeders, including fruit-eating bats and human beings. Natives used to begin collecting the etcho fruits in May, saving the liquid for syrup and the seeds as a source of oil and starch, but the practice has waned in recent decades.

For many natives May is their favorite time of year in spite of the lack of greenery and forage for livestock. The mornings are fresh and the nights cool. Woodcutters can walk easily through the bare monte unhindered by vines or large leaves. And etcho fruits are never far from hand.

Coastal thornscrub, with its remarkable pockets of pitahaya forest, should be considered the most endangered of all plant communities of northwest Mexico. While organpipes are common over a large area, the pitahayales occur only on the flat coastal plains that are, unfortunately, ideal for irrigated agriculture. We estimate that the original area of pitahaya-rich coastal thornscrub was roughly 600 km$^2$, only a quarter of which had dense growth of pitahayas. In the last three decades, about 25 percent of the area has been cleared, much of it sown with buffelgrass. Some of the finest stands of pitahaya—with up to several hundred individuals per hectare—have already been leveled near Agiabampo in anticipation of irrigation water from Huites Dam in Sinaloa. The owners of the ejido of Agiabampo, who lease out the fields to international agribusinesses, shrugged off the loss, saying that many pitahayas remain on the *faldas* (hillsides). Natives report that many more hectares are scheduled for clearing when issues surrounding water deliveries are cleared up. It would be a sad loss for native gatherers and fruit collectors if any more of the forests are felled. It would be an even sadder loss for the rest of the world, those who have not had the opportunity to amble through these million-armed jungles that teem with life tied to the abundance of the organpipes. Wandering through the pitahayales, especially a fortnight after a soaking rain, inspires a sense of wonder that only a few plant communities can match.

## *Foothills Thornscrub*

The vegetation of the coastal foothills, at elevations of 50 m up to more than 200 m, gradually changes into a different and more diverse mix of plants. Most coastal thornscrub species are represented, but in fewer numbers. The trees tend to be slightly taller and less dominated by leguminous species, even though the variety of legumes increases. Etcho cacti assume a role nearly equal to that of pitahayas. Representative trees include larger specimens of torote and torote prieto, *júchajco* (*brasil*; *Haematoxylum brasiletto*, Fabaceae), *aroyoguo* (*cacachila*; *Karwinskia humboldtiana*, Rhamnaceae), *tapichogua* (*copalquín*; *Hintonia latiflora*, Rubiaceae), *baihuío* (*guayavillo*; *Acacia coulteri*, Fabaceae), *mayo* (*mauto*; *Lysiloma divaricatum*, Fabaceae), *pómajo* (*palo de asta*; *Cordia sonorae*, Boraginaceae), the legumes *mapáo* (*palo colorado*; *Caesalpinia platyloba*), *ooseo suctu* (*palo fierro* or *palo pinto*; *Chloroleucon mangense*, the iron-

Foothills thornscrub, Cerro Terúcuchi. *Pitahayas, etchos,* and leguminous trees compose most of the vegetation here.

wood of the tropical deciduous forest), and *jícamuchi* (*palo piojo; Caesalpinia caladenia*), *jútuguo* (*palo santo,* tree morning glory; *Ipomoea arborescens,* Convolvulaceae), and *to'oro chutama* (*torote copal; Bursera lancifolia*). Flourishing at the transition to tropical deciduous forest is *joopo* (*palo blanco; Piscidia mollis,* Fabaceae), a tree uncommon in either fully developed thornscrub or tropical deciduous forest. Shrubs of foothills thornscrub include *candelillo* (*Euphorbia colletioides*), *siteporo* (*dais; Desmanthus covillei,* Fabaceae), *huoquihuo* (*huiloche; Diphysa occidentalis,* Fabaceae), *pisi* (*papache borracho; Randia thurberi,* Rubiaceae), *baijguo* (*palo dulce,* kidneywood; *Eysenhardtia orthocarpa,* Fabaceae), *samo* (*Coursetia glandulosa,* Fabaceae), *sitavaro* (*Vallesia glabra,* Apocynaceae), *torote lechosa* (*Euphorbia gentryi,* on Mesa Masiaca), and *júsairo* (*vara prieta; Croton flavescens,* Euphorbiaceae). These species occur infrequently in some areas, but often they are found in significant numbers.

Thornscrub of the Sinaloa foothills is subtly different from that of Sonora. It is host to a larger proportion of torote prieto and guayavillo, and the sa'apo on the slopes grows considerably larger. These combine to produce a more tropical aspect.

Foothills thornscrub, Mesa Masiaca. The *pitahayas* greatly surpass the surrounding vegetation in height. The basaltic rock absorbs more heat than the surrounding plains do, making for a more arid appearance.

Foothills thornscrub probably exhibits the sharpest contrast between the leafed-out state during las aguas and the mojino (gray) state in the drought of late spring. After several months of no rain, only the etcho cactus is reliably present and green. Torote prieto, growing in great numbers on some hillsides, shows up as a wispy maroon blotch. Cattle, desperately looking for food, compact the soil as they weave their way through the low forest. Collecting plants in May or June is a discouraging pastime. The forlorn landscape has almost nothing to yield to botanists. Still, as natives have long known, a great deal can be learned about bark textures and growth habits in this season. Each species has its signature when leafless, and our Mayo consultants have no more difficulty distinguishing trees in the dry season than they do in the wet season.

Large areas of foothills thornscrub have been cleared and replaced with artificial buffelgrass *praderas* (pastures) for Mexico's growing cattle industry. Still, huge amounts of the habitat cover the low ranges that rise from the coastal plain. Easy access can be found on cobbled roads leading to microwave towers on Cerro Prieto, a few miles east of Navojoa on the Alamos Highway, and

Tropical deciduous forest near Alamos in September. *Cuajilote, mauto* (*mayo* in Mayo; *Lysiloma divaricatum*), and *pochote* (*baogua* in Mayo; *Ceiba acuminata*) predominate. Note that the height of the canopy exceeds that of the *etchos*. (Photo by T. R. Van Devender)

on Mesa Masiaca, a basaltic flow near Mexico Highway 15, some 30 km south of Navojoa.

## Tropical Deciduous Forest

At higher elevations, rains are more generous and reliable, and plants respond with greater diversity and size, changing the structure and composition of the vegetation. Where the tops of trees overtake the tops of the etchos, ecologists arbitrarily (but not capriciously) posit the beginning of tropical deciduous forest (see Van Devender et al. 2000). The transition from thornscrub to tropical deciduous forest at about 200–400 m elevation in the northern part of the Mayo region is progressively lower to the south in northern Sinaloa. In fully developed tropical deciduous forest the canopy is complete when all plants are leafed out, meaning that at the height of rainy season growth, sunlight does not generally reach the ground. Etchos are common, scattered throughout the forest, but treelike pitahayas are few, restricted to drier sites such as rocky cliffs, canyon sides, and poor soils.[3] Most of the trees common in thornscrub grow

Mayo home, tropical deciduous forest, Los Capomos, Sinaloa. The greater rainfall, freedom from frost, and rich granitic soil give rise to more varied and dense vegetation than found in thornscrub. (Photo by T. R. Van Devender)

larger and taller in tropical deciduous forest but are less common, submerged in a large variety of other trees. Species restricted to tropical deciduous forest or of minor importance in thornscrub tend to have larger leaves and fewer spines. Some trees require the increased moisture and protection of the forest and cannot survive in foothills thornscrub or along arroyos.

Representative trees include mauto (the most common), *chi'ini* (*algodoncillo; Wimmeria mexicana,* Celastraceae), amapa, *batayaqui* (*Montanoa rosei,* Asteraceae), *tchuna* (*chalate; Ficus insipida*), nesco, *agia* (guásima; *Guazuma ulmifolia,* Sterculiaceae), *ciánori* (*palo barril,* buttercup tree; *Cochlospermum vitifolium,* Bixaceae), *joso* (*joso de la sierra; Conzattia multiflora,* Fabaceae), palo mulato, palo santo, *jupachumi* (*palo zorillo; Senna atomaria,* Fabaceae), *baogua* (*pochote,* kapok; *Ceiba acuminata,* Bombacaceae), sahuira, *samo baboso* (*Heliocarpus attenuatus,* Tiliaceae), *ca'ja* (*tempisque; Sideroxylon tepicense,* Sapotaceae), tepeguaje, *to'oro chutama* (*torote de incienso; Bursera penicillata*), *to'oro* (*torote copal; B. stenophylla*), *cuta tósari* (*vara blanca; Croton fantzianus*), and in Sinaloa, mora. Common shrubs include cumbro, *có'cori* (*chiltepín; Capsicum annuum,* Solanaceae), *huicori bísoro* (*flor de iguana; Senna pallida,* Fabaceae),

Transitional tropical deciduous forest near Yocogigua in March, when most trees are leafless. Dominant species are *mauto, nesco* (*Lonchocarpus hermannii*), and *palo santo* (*jútuguo* in Mayo; *Ipomoea arborescens*).

*mamoa* (*Erythroxylon mexicanum,* Erythroxylaceae), *limoncillo* (*Zanthoxylum fagara,* Rutaceae), *jósoina* (*papache; Randia echinocarpa,* Rubiaceae), sangrengado, *tatachinole* (*Tournefortia hartwegiana,* Boraginaceae), and *ta'aboaca* (*tavachín,* red bird-of-paradise; *Caesalpinia pulcherrima,* Fabaceae). In addition, dozens of species of vines and annuals appear during the rainy season, when the tropical deciduous forest becomes a steaming jungle, a nearly impenetrable wall of a hundred different shades of green.

In the dry season most of the color disappears. The etchos remain dark green, and a few species retain leaves, sometimes through the spring drought. Plants that tend to defy the spring drought include guásima (in low-lying areas), tepeguaje (on the hillsides, with bright green new growth in May), guayacán, sanjuanico, mezquite, *baisaguo* (*tescalama,* rock fig; *Ficus petiolaris*), and the riparian figs. The dormant mautos have a gray appearance, while the often numerous torote prietos of thornscrub tint the dry-season foothills a faint red. Thus, the mojino of the tropical deciduous forest has a subtly different texture and color than that of the foothills thornscrub.

Tropical deciduous forest in drought near Alamos. All the trees are leafless. *Etchos* (*Pachycereus pecten-aboriginum*), *palo piojos* (*Brongniartia alamosana*), and *palo santos* (trees with white bark) stand out in this photograph. People have been cutting wood on this hillside for centuries.

## Oak Woodland

Mayos live almost exclusively in thornscrub and tropical deciduous forest.[4] In Sinaloa and in a few areas in Sonora, their terrain includes small areas of oak woodland growing on hydrothermally altered soils; that is, red or white soils of high acidity that give rise to oak "islands." Although oaks are not often used by Mayos, the ecological uniqueness of these habitats merits discussion.

Few Mayos venture above 900–1,000 m elevation, where the transition from tropical deciduous forest to oak woodlands usually begins in the Mayo region. Where oaks occur at lower elevations on the scattered oak islands, soils are thin and support only a scattered plant community, a sharp contrast to the closed canopy of adjacent tropical deciduous forests. *Cusis* (encino; *Quercus albocincta*, Fagaceae) are found at 220 m on Cerros Colorados, not far from Piedras Verdes near Mocúzari Dam, and *encino* (*Q. chihuahuensis*) at 340–400 m near El Sabino in the Arroyo Güirocoba, at 340 m on the slopes of the Sierra de Alamos, and in the foothills east of the Río Fuerte. In these habitats we also expect to find *encino roble* (*Q. tuberculata*) but to date have not collected it. These oak species are not necessarily found at all locations, even though any of the three may occur. At Cerros Colorados, for example, only cusis are found, while near El Sabino only encinos are encountered. Certain other trees and shrubs are often associated with the oak islands, including *cascalosúchil* (*Plumeria rubra*, Apocynaceae), *chopo* (*Mimosa palmeri*, Fabaceae), *corcho* (*Diphysa suberosa*, Fabaceae), mamoa, *jupa'are* (*negrito*; *Vitex pyramidata*, Verbenaceae), *sapuchi* (*Randia laevigata*), tepeguaje, and *tornillo* (*Helicteres baruensis*, Sterculiaceae).

These islands of small, isolated oak woodlands are known to the Mayos but rarely visited. Francisco Valenzuela Nolasco lives only 20 km from Cerros Colorados, in El Rincón Viejo, but he had not been there before accompanying us one day. Nevertheless, he was familiar with all of the trees and shrubs from having seen them at higher elevations.

The oak woodlands of Sinaloa, where we hoped to gather a myriad of plants known to Mayos, are permanently off-limits to us because of the prevalence of marijuana and opium agriculture, and the zeal of growers to intercept and deter outsiders with hails of bullets.

## *Arroyos and Canyons*

Watercourses support different plant communities from those found only a few meters back from the damp riparian soils. In general they permit linear extensions of plants of the tropical deciduous forest downward into coastal thornscrub, often nearly to the edge of the sea, and also provide habitat for water-limited riparian species not found on slopes. The tropical deciduous forest trees are not as predictably found in foothills arroyos or coastal thornscrub as they are at higher elevations. They are probably established by rare flood events, and their seeds do not face good odds of propagating and surviving.

Examples of arroyo-limited species are *tupchi* (*abolillo*, soapberry; *Sapindus saponaria*, Sapindaceae), *chino* (*Havardia mexicana*), *joso* (*Albizia sinaloensis*, Fabaceae), *sabino* (Mexican bald cypress; *Taxodium distichum*, Taxodiaceae), *huata* (*saúz*, Bonpland willow; *Salix bonplandiana*; also Goodding willow; *S. gooddingii*, Salicaceae), and *júvare* (*igualama*; *Vitex mollis*). On the plains nearer the coast are found a few individuals of *aba'aso* (*álamo*, Mexican cottonwood; *Populus mexicana*, Salicaceae), amapa, nesco, and *caguorara* (*guayparín*, Sonoran persimmon; *Diospyros sonorae*, Ebenaceae). Guayparín is sufficiently represented in riparian habitats near the coast to constitute a reliable source of fruit and edible seeds.

The extension of tropical deciduous forest species down into arroyos makes these watercourses an enduring source of discovery. Arroyos and washes also have significantly more shade and are cooler than adjacent upland counterparts,

Tropical deciduous forest with *sabinos* (*Taxodium distichum*), Río Cuchujaqui. The great trees rooted in bedrock are able to survive floods. (Photo by T. R. Van Devender)

Río Cuchujaqui gorge near Alamos. Note the unbroken canopy, which is characteristic of tropical deciduous forest. (Photo by T. R. Van Devender)

This *ébano* (*Caesalpinia sclerocarpa*) on Arroyo Camahuiroa is well known in the region. Seedlings, although abundant, fail to reach maturity because of trampling by cattle.

and they constitute an apparent refugium for northern extensions of tropical trees. In arroyos near the coast, Mayos have shown us outliers of such tropical trees as *capúsari* (*Crataeva palmeri,* Capparaceae), *chaparacoba* (*Citharexylum scabrum,* Verbenaceae), *ébano* (*Caesalpinia sclerocarpa,* Fabaceae), *cuta béjori* (*palo cachora; Schoepfia shreveana,* Olacaceae), *palo limón* (*Amyris balsamifer,* Rutaceae), and a possibly undescribed *júchica* (*Sideroxylon* sp.). Along the coast the watercourses are also havens for livestock and woodcutters. Just above the coastal plains arroyos are beds of gravel, and within the plains, large deposits of sand, both of which are mined by hand shovel, an enterprise of villages within ejidos and comunidades. The mining of sand tends to degrade the bottom of watercourses, thus lowering the channel bottoms and increasing the distance that riparian trees must send their roots.

## ANIMALS

Snakes are feared and persecuted, with two notable and large exceptions: *la corua* (boa constrictor; *Boa constrictor,* Boidae), regarded throughout Sonora as a benevolent guardian of springs, and the *babatuco* (indigo snake; *Drymarchon corais*, Colubridae). We once caught a black and red *coralillo*, a magnificent nonpoisonous long-nosed snake (*Rhinocheilus lecontei* subsp. *antoni*, Colubridae), just outside a house in Teachive. Although the snake was completely docile, no one from the crowd of twenty people who gathered ventured within ten feet of us while we were holding it. We assured them that the snake was harmless and enjoyable to touch, and finally a small boy timidly inched forward, touched the snake, and ran. We released the creature outside of town to prevent it from being killed.

Vicente Tajia of Teachive reported that tortoises were once common in foothills thornscrub at 120 m elevation on the isolated volcanic Cerro Terúcuchi and are still occasionally seen in flatter coastal thornscrub toward the coast. In Teachive, he notes, tortoises are sometimes eaten in *birria* (meat stew). They are not kept as pets. The fact that during las aguas turtles emerge from their rock shelters and can be caught is common knowledge.

Skins of *sacahuí* (*escorpiones*, Mexican beaded lizard and Gila monster; *Heloderma horridum* and *H. suspectum*, Helodermatidae) and *coruas* are made into belts in leather shops in Masiaca. This is a mestizo, not a Mayo, custom but is now catching on among young Mayo men. During Holy Week a fariseo in Masiaca sported one proudly. Fine belts are also made from the skin of the spiny-tailed iguana (*Ctenosaura hemilopha*), and the craftsmen incorporate the lizard's head into the buckle. Local artisans occasionally fashion belts from the skin of Mexican green and western diamondback rattlesnakes (*Crotalus basiliscus* and *C. atrox*, Viperidae) for sale to tourists. Mayo men consider them ostentatious.

The region is rich in bird species, and Mayos are aware of this diversity. As with plants, several bird species reach their northern limit in the area, including the brown-backed solitaire (*Myadestes occidentalis*, Muscicapidae), lilac-crowned parrot (*Amazona finschi*, Psittacidae), purplish-backed jay (*Cyanocorax beecheii*, Corvidae), russet-crowned motmot (*Momotus mexicanus*, Momotidae), squirrel cuckoo (*Piaya cayana*, Cuculidae), tiger heron (*Tigrisoma mexicanum*, Ciconiiformes), and many others. A few birds are eaten (quail, doves, *chachalacas* [*Ortalis wagleri*, Cracidae]), and white-fronted parrots (*Amazona albifrons*) are captured young and kept in captivity. Boys tend to persecute hawks and owls, although one family kept a Harris hawk (*Parabuteo unicinctus*) as a pet. Teachivans enjoy hatching chachalaca eggs under a broody hen and raising the chicks as house pets. We have not conducted an extensive survey of birds identified by Mayos, but Vicente Tajia provided the Mayo names of more than 50 species found near his home in Teachive.

Most Mayo men who roam the monte are constantly searching for game, but Mexico's highly restrictive gun laws prevent most of them from owning firearms. Hence it is a rare occasion indeed when a deer is killed and venison eaten. On the other hand, several Mayo men have assured us that the meat of the skunk is palatable if the scent glands are immediately removed after the animal is killed. A resident of Masiaca is said to have made *albóndigas* (meatballs) of coyote, and some older Mayos, men and women, are said to snare rabbits.

These are only a few representative plant and animal species of the plant communities of the Mayo region. Mayos are familiar with more animals (especially bees) and many, many more plants. In the next chapter we describe how some of the most important plants fit into the Mayos' lives.

# 5  Eight Plants That Make Mayos Mayos

The plants described below have key roles in the life of Mayos, so much so that we suppose that without them the Mayo way of life would be quite different. We selected these species because of their variety of uses and the Mayos' general familiarity with them. They are presented in no particular order. While none of them is endemic to Mayo lands (the jito is endemic to the Cáhita region), their wide use indicates their importance. Most of these plants are well known to virtually all Sonoran Mayos, even though not all of them are found everywhere in Mayo lands. Where they do occur, they are of inestimable importance. In four cases (etcho, jito, *saya* [*Amoreuxia palmatifida* and *A. gonzalezii*], sitavaro) the Mayo name has become the Sonoran common name. For classification of plant uses and other plants used, see chapters 6 and 7.

Other plants of general importance—well-known species such as brasil, mauto, *mambia* (*chichiquelite,* nightshade; *Solanum americanum, S. nigrescens,* Solanaceae), and cósahui—are not included in this list. They are widely used by other native peoples in the region and throughout Mexico. Mauto, in particular, deserves more recognition for its ethnobotanical uses, but the Mayos of today live primarily in thornscrub below the mauto-dominated tropical deciduous forests. Prior to Spanish settlement, mauto was surely an important resource for the Mayos living from the Alamos area northward to the narrows of the Río Mayo, where Guarijíos are found today, and eastward well into Sinaloa. Mauto is a widespread tropical tree found from southern Baja California and southern Sonora south to Oaxaca and Veracruz. The Mayo name for *Lysiloma divaricatum* is not mauto, but mayo, for which the river and region are named.

*Cordia parvifolia* DC.                                  **huo'ótobo (vara prieta)**

Huo'ótobo is a scraggly, rather nondescript bush growing up to 3 m tall and 4 m wide. It is common on the coastal plain, rare in foothills thornscrub, and

*Pascola* mask carved from *torote prieto* by Francisco Gámez.
(Drawing © 2002 by Paul Mirocha)

absent in tropical deciduous forest. Following substantial rains it explodes with small leaves and delicate white blossoms 3–4 cm in diameter, with the texture and resilience of toilet paper. The flowers fall from the bush when barely touched. Huo'ótobo blossoms are a good indicator of recent rains or lack thereof, for the plant only blossoms in response to rain. The leaves are stiff, and the branches irregular but often arrow-straight. In very dry times the bushes shed nearly all their leaves.

Vicente Tajia notes that huo'ótobo is one of the most useful plants in the region. Indeed, it has cultural, technological, and medicinal significance. Smaller plants are preferred for medicine.

MEDICINE: The root of the plant is crushed and boiled into a tea that is administered to infants and children. It is said to help alleviate *resfríos* (colds) and the pimply rash that is associated with them in children.

ARTIFACTS: Straight sticks of the shrub, called huo'ótobos, are lopped off and scraped free of bark. They are then used to beat wool, cleaning it and rendering it more amenable to spinning and weaving. They are also used exclusively for the string heddle or shuttle, *hachomatua,* the part of the loom on which the yarn is wound, and which is passed through the warp to form a blanket. Huo'ótobos are also woven and tied together to make a *tarime* (bed), forming a springy mattress. *Zarzos* (racks or shelves suspended from the ceiling) made of woven huo'ótobo are widely used to protect cheese, drying meat, and other delectables from the depredations of dogs, livestock, and vermin.

CULTURE: Huo'ótobos are also incorporated at the beginning of the fiesta. The lead pascola holds a meter-long clean huo'ótobo in front of him and leads the other pascolas three times around the *ramadón* (festival ramada). The *alaguássim* (fiesta director) then takes the stick from the lead pascola and uses it as a probe, inserting it rudely into the baffle hole of the festival harp while the harpist looks on. The pascolas then, one by one, rather crudely sniff the end of the huo'ótobo, supposedly thus "measuring" the number of *sones,* or folk songs, available to sing. Onlookers find this ceremony, with its multiple layers of meaning, wildly funny.

In contrast to their earthy and playful uses in dances, the branches are harvested during Holy Week to be used in ceremonies by the alaguássim and pascolas to beat the ground as they call out "Gloria, Gloria" to the crucified Jesus.

Old-timers report that huo'ótobos were used for arrows and were especially effective when feathered with plumage from the Gila or other woodpeckers.

Women still use the pliable green branches scraped free of bark as the base of a wreath to which they attach flowers, fresh or plastic, to adorn tombs or homes.

Filemón Navarete, a singer for deer dancers and a resident of the ejido of Fernando Solís, a village of dire poverty, was kind enough to relate the following song for us:

| | |
|---|---|
| Yo huo'ótobo segua taca | From the wildlands blooms the huo'ótobo. |
| Seguátaca sanilócapo hueca | The branch is flowering. |
| Tósalise se segua | Along it burst the white flowers. |
| Yo huo'ótoboli taca | Near and far in the monte you see the huo'ótobo |
| Seguailo tátayowe bétana | From where the sun rises. |
| Amani huécali | There it is, the flowers opening. |

Filemón Navarete, Ejido Fernando Solís. Filemón is a well-known singer of deer songs.

| | |
|---|---|
| Sanílo hapo huécali | Out in the wild land blooms the huo'ótobo. |
| Cala liuliuti segua huéhueche | Now the flower falls. |
| Hili yo huo'ótoboli taca | Far out in the monte you see the flowers fall. |
| Cala liuliuti segua huéhueche | The flowers of the huo'ótobo are falling. |

***Stenocereus thurberi* (Engelm.) Buxb.**      **aaqui (pitahaya, organpipe cactus)**

A common columnar cactus in Sonora that changes life form from a multistemmed shrub branching from the ground in the Sonoran Desert to an 8–10 m tree in coastal and foothills thornscrub. It reaches southwestern Arizona and northwestern Sinaloa and is common in much of Baja California, but its largest populations are found in Sonora. In foothills thornscrub and tropical deciduous forest it grows as tall as 14 m, with some individuals developing hundreds of arms. Larger specimens possess a thick trunk extending a couple of meters above the ground. In general, though, the arms of pitahayas branch closer to

Vicente Tajia gathering *pitahaya* fruits near Coteco.

the ground than do those of etchos and sahuiras, which invariably have a discernible trunk. The arms of the pitahaya also tend to be thinner and diverge from the vertical axis at a greater angle than those of etcho and sahuira, whose branches tend to emerge horizontally from the bole for a short distance and then grow upward parallel to the main axis. The more numerous ribs of pitahayas give a smoother appearance to the branches than is the case with other columnar cacti.

A curiously crestate form of the cactus, called *aaqui nábera,* occurs sporadically throughout the region. It often yields no fruit. When it does, traditional Mayos warn, the fruits should not be eaten. The plant is bad, they say, and the fruit is thus contaminated. Felipe Yocupicio is adamant that the fruits should not be eaten: "Just look at the plant and you can see that it is a bad plant."

The pitahaya visually dominates coastal thornscrub and foothills thornscrub plant communities. In some areas it is also coverage dominant, but in others it is a codominant with leguminous trees. It occurs in enormous numbers and

*Aaqui nábera* (crestate *pitahaya*). Mayos believe the fruits of this unusual growth form should not be eaten.

in dense forests on the coastal plain south and east of Huatabampo. A walk through the greatest concentrations of pitahaya can only be compared with walking through some of the world's great forests. In places we counted more than six hundred mature plants per hectare. The individuals in these thick groves frequently exhibit a short trunk, but they do not attain the great stature of those in more mesic locales. This habitat, known locally as pitahayal (vegetation dominated by pitahayas), is a national treasure of Mexico, producing untold tons of commercially valuable fruits each year. Natives of the comunidad de Masiaca gather the fruits and sell them in local markets. Some are shipped to distant centers as well.

In spite of the commercial and aesthetic value of the pitahayal, the Mexican government is urging the Mayos to lease the lands to agricultural producers who will clear away the pitahayas and level the land for commercial agricultural production. In August 1999 a group of comuneros of Masiaca cleared a

*Pitahaya* fences such as this one near Bachoco afford privacy.

forty-hectare parcel near Coteco in order to plant buffelgrass. They left behind a few solitary cacti. Most of these were scorched by fires set to burn the slash. The remainder will surely perish due to full exposure to sunlight. In December 1999 almost all of Ejido Melchor Ocampo to the south of Camahuiroa was scraped and leveled in the hope that water for irrigation would arrive.

CONSTRUCTION: The ribs of the pitahaya are commonly used for fences (*chinami*), walls, ceilings, and some furniture. Along the coastal plain from Huatabampo south into Sinaloa the walls of many homes are made of woven pitahaya arms, often harvested green. The dried arms woven among strands of wire provide a dense fencing that is most pleasing to the eye and at the same time affords privacy and protection. Before the 1950s, when barbed wire became available and affordable in rural Sonora, large numbers of cuttings were planted in rows along with cuttings of etchos to produce living fences, many of which can still be seen. The lower wood of the trunk and arms is surprisingly sturdy and is used in many aspects of building, such as erecting crossbeams to bear heavy weights and creating the wooden form for saddles. The dried wood is used in great quantities as fuel for ovens, for baking bread, and in kilns for firing adobe bricks.

MEDICINE: The fleshy, moist stem is singed to remove spines, then applied

The entire pulp of the *pitahaya* fruit is eaten. Note the simple removal of the husk.

to the flesh directly for snake and insect bites, a remedy that we tested with positive results when an assassin bug bit Tom Van Devender. Francisco Valenzuela reported that the scorched peel of the fruit is applied directly to the anus for hemorrhoids, cautioning that it must be scorched enough to burn off the spines, which seemed prudent to us as well. Buenaventura Mendoza of Teachive reports that dried peels (*aaqui begua*) are boiled into a tea and taken for stomach problems and to stop hemorrhaging in women.

FOOD: The Mayos eat great numbers of the sweet, satisfying fruits (*aaqui tej'ua, aaqui buasi*) in late July, August, and September. They constitute an important component of the Mayo diet and a potentially valuable economic resource. Many people report that they eat more than fifty per day. Vicente Tajia says one never gets tired of eating them ("No se enfada uno") or overfull, and we share his opinion. Formerly the fruits were also dried and preserved or made into wine, but hardly anyone does that anymore.

One woman from Masiaca reportedly still uses the fruits to make tamales (*aaqui nójim*) commercially. At our request a woman in Teachive prepared some, boiling the pulp until it thickened and then pouring it into cornhusks to allow it to cool. We found them to be delectable. Reyes Baisegua says the best tamales are made from *poposahui* (nearly ripe but unopened fruits). These are laid on a cloth on top of a bed and squeezed to wring out the juice. Then they are boiled and the cooked pulp is poured into cornhusks.

Doña Gregoria Moroyoqui of Sirebampo demonstrated the preparation of *pitahaya seca* (dried organpipe fruits). She selected two dozen ripe fruits and placed the pulp in a skillet. She added a small amount of water and brought the mixture to a boil, stirring it for about five minutes. She then strained the boiled mixture (*beja buasic*) through a coarse piece of cloth to remove excess *miel* (syrup). The remaining mass, rather slimy, she spread out to dry, covering it with a screen to keep out flies and insects. Adequately dried, the pitahaya seca would last for several months. Francisco Gámez dries the fruits in a *tapanco*, a raised bed constructed from pitahaya slats laid side by side and close together, used to protect the fruits from dogs and pigs. The surface is ideal for drying pitahaya fruits, he says.

The pitahayas begin flowering in late spring, although some plants may flower earlier. By July the earliest fruits ripen, but serious collecting seldom begins before August, when prodigious numbers of ripe fruits appear and collecting is worthwhile. Immature fruits (*caboasi*—"it is still green") are avoided. *Poposahuim* ("between green and ripe") are collected and left to ripen in a bucket. Ripe fruits, called *buásim*, are often eaten then and there.

Serapio Gámez built for us an *aca'ari* (gathering bucket) used traditionally for pitahayas. It is a cylinder 45 cm tall and 30 cm in diameter built of pitahaya ribs lashed together with a woven bottom of deer hide. The top of the aca'ari is held rigid by a tightly lashed strip of guásima steamed into a circle. Attached to the bucket is a strap made of *tásic* (*ixtle*, fiber of *Agave vivipara*) worn over the shoulder. At one time all Mayos used aca'arim—large ones for men, smaller ones for women—but Mayos under forty years of age are unfamiliar with them.

Pitahaya gathering time is one of the happiest of the year in spite of the sultry heat. The best collecting begins early in the morning, before birds (doves, woodpeckers, mockingbirds, curve-billed thrashers, and white-fronted parrots), ants, bees, butterflies, and wasps have managed to wreak havoc among the ripening fruits and before the heat and gnats become overwhelming. The *paloma pitahayera* (white-winged dove; *Zenaida asiatica*) is especially attracted to the fruits.

Before gathering can begin, the collector (usually a man, although women collect as well) fashions a spear called *bacote* in Spanish and *jíabuia* in Mayo. The jíabuia is often made from a *quiote* (flowering stalk) of a tall *cu'u* (*mezcal; Agave vivipara*, Agavaceae) or, if that is not available, from the ribs of pitahayas or etchos lashed together. For lower-growing fruits, a long limb of jócona is

*Aca'ari* (*pitahaya* gathering basket) made by Serapio Gámez
(Drawing © 2002 by Paul Mirocha)

cut, leaving a fork at the tip in which the fruits can be wedged and wriggled off, thus preventing them from falling to the ground and getting bruised. For taller plants, a sharp point, usually made from *pisi* (*papache borracho*) is carefully lashed onto the end of the bacote. With the bacote balanced over a shoulder, the gatherer, often accompanied by children, walks briskly into the monte. He or she evaluates the egg-sized fruits and carefully impales those judged ready to be picked. These are gently wriggled from the cactus and lowered to the ground, where they are delicately removed from the spear point.

On good collecting days the bucket, or aca'ari, will be filled and left till late afternoon or overnight, during which time the spines soften, making peeling less hazardous. Usually a large number of fruits are consumed on the spot. An adult can easily down thirty fruits in a morning and as many in the afternoon. Most pickers dexterously peel the fruit without being pricked by the thorns.

*Pitahaya zarca.* While uncommon, these white fruits are well known and considered tastier than the more common reddish variety.

The pain inflicted varies from plant to plant, and the individual personalities of the cacti near villages are well known. Those whose thorns inflict the most painful punctures are treated delicately. Novices must practice with many fruits before becoming adept at peeling off the well-armed husks, avoiding both the long and short spines. By the time children are ten, they are usually dexterous in harvesting and will eat the first few dozen fruits they collect, staining their mouths, chins, and cheeks pink in the process.

The quality of the fruit varies from cactus to cactus, as does the color. The dominant color of the pulp is dark red, but the best fruits are often said to be those with purple red pulp. A few plants yield whitish pulp called *zarca* (Spanish) or *tótosi* (Mayo) with a more delicate flavor. Occasionally a pitahaya yields fruits with yellow pulp.

The number and size of fruits vary considerably from year to year. Mayos say that fewer and smaller fruits are produced in drought years; our observations over four years support that claim. In 1999, after an extraordinarily long dry spell, a few fruits ripened in early July. By early August, there were still enormous numbers of buds and unripe fruits, but hardly any ripe fruits. Vicente Tajia attributed the anomalous fruiting sequence to the occurrence of a lunar eclipse, which he believed caused the buds to fall off the cactus. After three consecutive days of rain, the fruits ripened and the full harvest began.

Filemón Navarete was kind enough to sing the following song for us. It is posed to the fiesta audience as a riddle, which they are expected to guess as the deer dancer performs. The correct answer is "¡Aaqui!"

| AAQUI | A RIDDLE OF THE PITAHAYA |
|---|---|
| Jita juya tácasuti | What is that plant growing there, out in the wildlands? |
| Juya ani hapo hueca | What is it growing there in the great wild fields? |
| Tos pólolote se sehua | The white flowers grow, white above white. |
| Ca'a sáhuaca hueca | It has no leaves. |
| Ili yo júyataca | It is a great plant without leaves. |
| Amanise sitepólopo hueca | It grows there, among the siteporos. |

The well-rehearsed deer dancer halts his dance abruptly with the last words as the enthusiastic audience continues shouting out guesses.

*Agave vivipara* L. [*A. angustifolia* Haw.]  cu'u (mezcal, maguey)

This agave is widespread in a variety of habitats throughout the lower elevations of the southern half of Sonora. Its leaves are straight and daggerlike, up to 1 m long. It flourishes in the clay soils of the coastal plain, in coarse bajada soils on alluvial slopes, and on rocky hillsides. It may grow to more than 1 m tall and 1.5 m wide, but it is usually harvested before reaching that size. It is the most common agave in the region.

FOOD: The flowering stalk (*cu'u varoa,* quiote) is lopped off with a machete shortly after it emerges from the leaves, and the plant is left undisturbed for a year while starches concentrate in the head. At the appointed time, the long, pointed leaves (*maiicua*) are chopped from the head (*cu'u coba, cabeza*), which is then pried from the ground. After the meager roots are stripped away, the cabeza is slow roasted (*tatemado*) in a *maya* (pit) for two to three days. The head is then exhumed and the leaf remainders on the roasted head (*cobata*) are consumed.

Most natives agree that this process must be followed closely for the best results. First a pit nearly a meter deep is dug and lined with stones. Firewood is placed on the stones with a few branches of torote prieto on top to add flavor. The cabeza, the starchy trunk of the agave, is placed in the hole. The firewood

*Cu'u* (*mezcal; Agave vivipara*). (Drawing © 2002 by Paul Mirocha)

is then ignited, and the burning wood is allowed to reduce to coals. It is then covered with a few branches of joopo (palo blanco). Dirt is placed on top of the whole mass. The head(s) are roasted for at least forty-eight hours before being removed and allowed to cool. The leaf stubs, resembling artichoke phyllaries, are chewed until the sweetness is gone. The quid is spat out. During the long period between the last pitahaya fruits in early fall and the first etcho fruits in late spring, cu'u was the only sweet food available and thus filled a void in the historic diet of the Mayos.

Cooked agave is very sweet, almost excessively so, with a strong molasses-like flavor. At first we could eat but a few, but the taste grew on us. Roasted agaves are relished by the natives, but they have become a rare treat as their numbers have dwindled in recent years. Elsewhere in the Sonoran range of *A. vivipara*, mestizos harvest the heads to brew into the local moonshine, called *bacanora*, *mezcal*, or *vino*, a fiery distillate. Diminishing numbers of agaves have interrupted this time-honored practice, immensely popular since distillation was introduced by the Spanish. *Maguey azul*, the tequila agave, is a southern cultivated variety of *A. vivipara*. Tequila and its wild variants constitute a common alternative to brews based on cane sugars.

MEDICINE: In addition to its food and fiber value, *A. vivipara* has medicinal properties. The pulpy head is chewed and the mildly purgative juice swallowed to "cleanse" the stomach. For scorpion stings, a leaf is scalded in a fire and then squeezed, the drops falling into hot water, which is drunk. The leaves are also used for veterinary purposes, according to several consultants. If a heifer will not bear a second calf, a leaf is roasted and squeezed, and the expressed juice is dripped into the young cow's mouth. Thus she is believed to be rendered more amenable to repeat fertility.

ARTIFACTS: The uncooked leaves are also scraped to make tásic (ixtle in Spanish), a fiber that is twisted into twine, especially for sewing the cocoons of ténaborim and making *morrales* (handbags) and rope. Doña María Teresa Zazueta of Choacalle still produces cord from agave leaves. She reports the best rope of all used to be made from ixtle, and that she and others still make it from time to time.

Don Reyes Baisegua of Sirebampo demonstrated the technique for making ixtle. First he donned a strange apron made from burlap and inner tube, which he said would protect his chest and arms from the agave's irritating juice. He then placed the *pencas* (long agave leaves) on a *burro*, a wooden tripod slightly more than a meter high, with the apical leg larger than the others. This largest leg was a curved mezquite log perhaps four feet long and four inches thick, slightly flat on the working surface. Reyes bent the agave leaf, secured it so it would not slip, and then held the other end flat on the surface of the working

Don Reyes Baisegua with a mass of *ixtle* (*tásic* in Mayo) ready to be woven into cord.

leg of the burro. He scraped the leaf with a knife, using a downward motion, until all the succulent green matter was removed and only the fibers remained. He then flipped the penca and repeated the process with the other end until only a long mass of white fibers remained. After he repeated the process with another penca, the fibers were dried thoroughly in the sun.

The production of the cord, the basic thread for weaving, begins with twisting the fibers into *mecate* (twine or rope) using a *malacate* (spindle) consisting of a small wagon wheel as a flywheel through which a smooth metal shaft about 40 centimeters long protrudes. Reyes's wife, Gregoria Moroyoqui, wound a piece of twine twice around the shaft and rapidly pulled it back and forth, twisting the attached fibers, while Reyes slowly backed away from the malacate, feeding fibers into the spinning mass of cord. When the mecate reached a length of roughly ten meters, they doubled the cord twice to make it thicker. Reyes then wove the cord into a morral, starting by stretching the cord between two pegs of guayavillo until the edging was strong enough to sup-

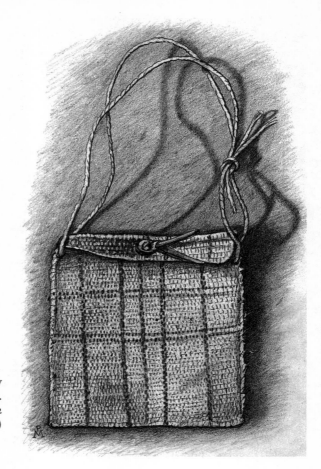

*Morral* woven by Don Reyes Baisegua. (Drawing © 2002 by Paul Mirocha)

port the weaving. He then transferred this warp to the loom, working in the shade of a giant mezquite.

The loom Reyes used was vertical, attached to a barbed-wire fence. He stood or sat while he strung the warp and wove the cord back and forth, tamping the woven material frequently with a *sasapayeca* (tamping stick) made from joopo. The resulting morral took about a day for him to finish and sold for about $5. With each morral requiring about fifty pencas, a healthy population of *A. vivipara* is crucial.

The procedures and tools used by Don Reyes vary little from those described by Beals (1945), who in 1930 noted widespread weaving of morrales in Masiaca, exclusively by men. Erasmus (1967) found considerable weaving of morrales in Las Bocas in the mid-twentieth century. Don Reyes made a couple of fine morrales de ixtle for us in 1995 and 1996 and continued to produce them through 1997. In 1997 men in Cucajaqui de Masiaca were once again making ixtle handbags and morrales. Until recently, several people in Teachive were making morrales, but in 1998, Reyes Baisegua was the only weaver in the region making morrales of historic high quality. They were a clean straw color with stripes of interwoven fibers dyed red and green. Sturdy and tightly woven, they would last for years. Poor people formerly used morrales to carry bundles and provisions on their journeys to and from the markets, the woven strap slung over a shoulder, but now they use cheaper plastic handbags.

Mayos have expressed some concern about declining populations of *A. vivipara*. If their heavy exploitation of wild populations continues, some cultivation will be necessary. Fortunately, most agaves respond well to cultivation and require little watering once established. Local Sonoran governments and individuals have begun experimenting with plantings.

*Pachycereus pecten-aboriginum* **(Engelm.) Britt. & Rose.** etcho

The etcho is nearly as important for Mayos as the pitahaya. It is scattered throughout coastal and foothills thornscrub but is much more common than the pitahaya in tropical deciduous forest. The Mayo name for this important cactus, etcho, has often been confused with the unrelated Spanish word *hecho* ("fact" or "deed").[1]

This stellar tree—which can exceed 10 m in height and have numerous arms—is greatly esteemed by Mayos, who named one of their major settlements, Etchojoa, after it. Unlike other columnars of the region, etchos may flower at any time, even though late winter into spring is the most common. The developing fruits appear to form large golden clusters on the arms, giving the giant cacti an attractive appearance. In the searing drought of the late spring, etchos are often the only trees on the hillsides retaining a vestige of dark green. (Pitahayas frequently turn a sickly yellow.)

ARTIFACTS: The fruits are covered with a thousand yellow spines, each about 5 cm long, that appear vicious but are relatively harmless and seldom penetrate the skin. These prickly fruits are said to have been used by Indian women to comb their hair, hence the species name *pecten-aboriginum*, which translates as "native comb." Vicente Tajia made one of the combs by scraping one side of the fruit clean, forming a handle, and then singeing the spines on the other side to remove the sharpest points. We theorize it would be passably useful in an emergency.

CONSTRUCTION: The strong lower portion of etcho ribs is used in house construction and in making looms and other household artifacts. The sturdy

96 / *Eight Plants That Make Mayos Mayos*

Vicente Tajia with *etcho* cactus near Camahuiroa. This specimen, well known in the region, is a fine fruit producer. Note the buds on the upper branches.

though light wood of the arms and the trunk is carefully sawed and shaped into boards for building benches and beds or for roof beams. María Soledad Moroyoqui of Teachive has used the same sasapayeca made of etcho wood in her loom for nearly thirty years. Cornelia Nieblas uses a tarime of etcho that she claims is more than a hundred years old.

MEDICINE: Juice from the flesh is squeezed from a section and drunk for *mal de orín* (bad urine) and as a remedy for *la prostatitis*. To stop bleeding, a few drops of juice are dripped onto the wound and a *gajo*—a piece from a cross

*Etcho* fruit. The spines, though daunting, are weak and avoidable.

section—is placed directly on the wound. The same method is said to work for insect stings and bites. A little piece of the flesh is chewed for toothache, according to Luciano Valenzuela of Los Muertos. Felipe Yocupicio of Camahuiroa, who has a wide reputation as an herbal healer, recommends squeezing a drop or two from the flesh into a glass of water to treat ulcers.

FOOD: In late May and early June, etcho fruits ripen in large numbers, revealing themselves as the spine-laden husk splits open, exposing the dark red interior. As the fruits mature, the spines gradually fall from the fruits, piling up menacingly (but harmlessly) at the base of the cactus.

The harvesting of etcho fruit requires a somewhat different technology from that used to harvest pitahaya. A bacote is fashioned from an agave stalk, and a spike of tough wood (pisi is preferred) is lashed to it. Lashed on both sides of the point is a wide piece of carved wood in the shape of a *w*. This secures the impaled fruit so that when the bacote is withdrawn, the fruit will not fall off.

The fruits (*jíconim*) are eaten raw or cooked and are made into wine and jelly. They can also be dried and preserved. The black seeds are roughly half as big as citrus seeds, far larger than those of the pitahaya. According to Doña María Gonzalez, born in 1905, the raw seeds were often ground into flour and

Living fence of *etchos, sahuiras,* and *pitahayas* near Güirocoba.

dissolved in water to make a gruel that was eaten as an *atol* (a thick mush) or drunk as *horchata* (a gruel-like sweetened drink). Both of these practices are now rare because the labor required to concentrate the seeds is considerable.

Many Mayos told us that etcho seeds mixed with a small amount of corn flour make excellent tortillas (*etcho tajcarim*). Older Mayos recalled regularly eating etcho seed tortillas as children. Fausto López of Nahuibampo said that he was raised on them. Doña María Teresa Moroyoqui of Teachive made us some tortillas from the seeds that her husband had collected and cleaned, and their flavor was reminiscent of sweet pancakes with the texture of buckwheat. At the same time she prepared atol that had the texture and flavor of cream of wheat. She also boiled the flesh remaining after the pulp is removed into *sitori* (miel, syrup). María Teresa said that her grandmother made cooking oil from the seeds, and that before there was lard, the old people would crush some seeds to oil a *comal* (griddle). They even squeezed oil from the seeds. She also said that two generations ago people regularly made wine from the fruit, but they have not for many decades.

Overgrazed pasture with *etchos*. The removal of other species has exposed these fine plants to heat stress.

Filemón Navarete claims that atol made from etcho seeds has aphrodisiac powers but must be eaten without sugar or other sweetening to be effective.

Small etchos, or perhaps stem cuts, are planted as fence rows and grow quickly (within ten years) into an impenetrable barrier, a practice that long predates the introduction of barbed wire into rural Sonora. Their strong trunks and spreading habit also make etchos valued as shade trees, an uncommon virtue for a cactus. Many times we have taken refuge from the fierce sun in the shade of an accommodating etcho, and one can get downright comfortable leaning back and resting on the thick, spineless trunk. At such times one can easily forget that the shade tree is a cactus.

***Vallesia glabra* (Cav.) Link.**                                                       **sitavaro**

A common, often puny shrub (sometimes a small tree) bearing tiny, whitish opal-like fruits the size of a pea, sitavaro is one of the best-known plants in the region, growing in abundance along arroyos. It ranges from southern Sonora to southern Mexico south to Argentina. It retains its green leaves through the spring drought, one of the few plants to do so. Under the right conditions sitavaro grows into a spreading tree nearly 6 m tall. Plants this size are known from the arroyos in the foothills of the Sierra de Alamos and from Sinaloa, where it is a shade tree in the yards of many homes. Nearly every Mayo is familiar with it, and a town in the Mayo Delta is named after the plant. It appears to be confined to silty soils in the vicinity of arroyos.

MEDICINE: The milky sap is highly recommended for cloudy or infected eyes. The sap exudes from broken leaves or branches and is carefully applied to the corner of the eye. When one of us had an eye infection, Vicente Tajia demonstrated its use and inquired afterward if it had helped, but all we could tell him was that it hadn't done any harm. Seferino Valencia of Las Bocas recalled that his father toasted the leaves over fire, pulverized them, and applied the powder to sores that would not heal. Thus prepared it is also said to cure infections rapidly. Francisco Valenzuela of El Rincón said that tea from sitavaro branches is used to wash sores or boils not only to help them heal, but also to prevent pathogens from the dead from entering through them into a living body. Certain vague diseases are thought to originate with death, and sitavaro is believed to be a powerful anti-infective against supernatural pathologies.

INDUSTRY: Doña Lidia Zazueta of Teachive wove a small woolen *cobija* (blanket) that bore figures dyed with sitavaro root, which is boiled to produce a mustardlike color.

CULTURE: Sitavaro is a favorite source of greenery for covering the fiesta ramadas during Holy Week, a tribute to its spiritual power as well as its evergreen decorative qualities. (If Holy Week occurs in April, most of the greenery is gone from the monte.) Large armloads of the branches are gathered and spread on roofs and ramadas as insulation. Sometimes the branches are covered with an additional layer of dirt.

The branches and trunks, though thin, produce a light and resilient wood, ideal for producing the penetrating thump of the drum. Balbina Nieblas of Teachive fashions sticks for this purpose.

On Good Friday, Vicente Tajia cut a branch and gently tapped each member of his extended family, saying, "May harm and sickness stay away from you." He explained that he uses sitavaro for this because the branches and leaves are soft and will not scratch or otherwise injure anyone during the ceremony. We asked him to scourge us as well.

### *Jatropha cordata* (Ort.) Müll. Arg.    sato'oro (papelío, torote panalero)

Sato'oro is a striking, irregularly branched, narrow tree reaching 8 m tall, with soft wood and a green to yellow to beige trunk from which peel large sheets of exfoliating tan bark with a yellow cast. In July the dark green leaves are accentuated by delicate bell-shaped pink flowers. It is common in thornscrub and tropical deciduous forest, where it is an aesthetic boon to the region, leafing

*Sato'oro* (*Jatropha cordata*) in thornscrub near Teachive. The tree is leafed out more than the surrounding *monte*.

out early and turning green whether or not it rains, bringing some measure of consolation to the natives in times of summer drought. The turning of its leaves to golden yellow marks the end of las aguas and the beginning of the tropical fall season.

Livestock do not touch the plant. Hence, it is very common, especially in those parts of coastal thornscrub where pitahayas occur only sporadically. The tree may be confused with torote, for both have succulent stems and exfoliating bark. Torote has compound leaves with small, narrow, pointed, dull leaflets, while those of sato'oro are cottonwood-like, simple, wide, shiny, and oval. Trunks of sato'oro tend to be straight, and the bark has etchings that encircle the tree. Torote trunks are craggy, and the bark has only random markings.

INDUSTRY: According to Mayos of Teachive, large rectangular sheets of bark were carefully cut from the live tree and used to wrap freshly gathered

Close-up of *sato'oro* showing shiny bark.

honeycomb (*panales*) from wild bees for transporting and storage. The bark is still said to preserve freshness in the honey as long as a year.

ARTIFACTS: Sato'oro is used on occasion as a living fence post but must be supported along the line by more durable posts. Mayos of the coastal plain use the wood to construct a tool called *chapa* for harvesting tunas, the fruits of the prickly pear cactus. A branch of sato'oro roughly 5 cm thick and 1.3 m long is cut, and the end is split with a machete for a length of about 20 cm. The crack made by the machete is wedged open with a short length (about 10 cm) of pisi bound into the crack with a length of güinolo bark. This leaves an opening about 3 cm wide in the end of the tool. The harvester works the tuna into this open crack and wiggles it about to detach the fruit from the prickly pear. The tuna remains gently wedged in the chapa until the harvester empties it into a container.

The sap is a powerful astringent and staining agent.

MEDICINE: A tea made from the bark is rubbed on the abdomen to relieve sore kidneys. It is also applied to bee stings and drunk at the same time, which is said to instantly alleviate the discomfort. Some maintain that if stung while in the monte, one should suck on a piece of the wood.

Filemón Navarete sang us the following song about the sato'oro.

### SATO'ORO

| | |
|---|---|
| Jita juya tácasu | There is a tree in the plain. |
| Sani laupu huécari | There it is, blooming in the monte. |
| Siquili si sesehua | The flowers, a soft red. |
| Amanisu sani laupu huécari | There it is, the only tree blooming |
| Sihueli léjajati sesehua | in the monte. |
| Amani seguauylo ta'ta amaong hueche bétana | The branches move, the red flowers tremble. Where the sun sets, there it is. |
| Amani lihueeca síqueli léjajati | It is moving, the soft red flowers moving too. |

*Amoreuxia gonzalezii* **Sprague & Riley**  saya, saya mome
*A. palmatifida* **Moç. & Sessé**

This little plant is an herbaceous perennial found in clay soils of thornscrub and more open soils in tropical deciduous forest. The plants emerge each year from a tuberous root the size of a carrot, sometimes larger. Certain locations are known to harbor numerous plants, but these may bear leaves one year and not the next. Throughout the dry months the storage root lies dormant. In mid to late summer, when the rains arrive, people and animals alike (especially *tayasu* [javelina, *jabalí*; *Tayassu tajacu*, Tayassidae]), rush to harvest the sayas, which are conspicuous by their showy yellow-orange flower, about 7 cm in diameter. Harvesting kills the plant, but the root system is complex, and apparently small tubers remain in the ground to sprout the following year. Saya is much sought after throughout the region. Although not rare, it is found only sporadically.

FOOD: Natives relish the *camote* (tuber), roasted or raw. Most prefer it roasted, however, because when raw it may taste bitter. The tuber's flavor is greatly enhanced with lime juice. The fruit capsule is also edible when it is green and tender. Some folks reportedly eat the entire plant, roots, flowers, and all. The black spherical seeds of *A. gonzalezii* (about 6 mm in diameter), well known to many natives, bounce off the ground when dropped. Seeds from both species dehisce from a capsule, and at that time they can be ground into a meal that produces a coffee substitute of purportedly excellent flavor.

Saya would appear to be an excellent candidate for domestication; it must have remarkable powers of propagation to endure the heavy harvesting that goes on in late summer.

*Saya (Amoreuxia palmatifida)*. The root is eaten raw or roasted, and the roasted seeds are ground and brewed into a beverage. The plant is invisible much of the year.

Filemón Navarete sang us the following riddle about the saya, often sung at fiestas by the musicians and singers who accompany the deer dancer. The audience calls out guesses throughout the song. The deer dancer follows the words and the rhythm with precision, teasing the audience.

### SAYA

| | |
|---|---|
| Jita juya tácasu | What greenery is that? |
| Saníloa pueca | There, out in the monte. |
| Síaleja jate je'eca | The wind blows through, making it green |
| Yldi yo júyataca | Through the fields and the forest. |
| Juya ániapo | The root shoots down with the first rain |
| Ca'a chuj nahua momayo | And the flower bursts forth. |
| Amani júbaute bulia yúcata | With the first rain, it bursts forth. |
| Huécheyo | The yellow flower |
| Júsale lijti se segua | The yellow flower, moving |
| Ylidi yo júyataca | In the field and the forest. |
| Jusali lijt se séhualo | The yellow flower moving. |

A male *jito* (*Forchhammeria watsonii*) tree in flower near Huebampo. Note the shape, which gave rise to its English name, lollipop tree.

***Forchhammeria watsonii* Rose**             **jito (palo jito; lollipop tree)**

This gnarled, strong-trunked, symmetrical tree grows up to 9 m tall, sometimes more, often resembling a lollipop in shape. It is endemic to southern Sonora and northern Sinaloa and is common in coastal and foothills thornscrub. It retains its dark green color throughout the year, leafing out anew in late spring a brighter green. As Gentry (1942:117) says, "The tree has a very individual appearance, suggestive of old olive trees in ancient Judea. In the burning days of late spring it is about the only tree that offers shade to weary beasts and man." The trees stand out clearly in the late spring, showing up as patches of vivid dark green, almost black, against the monotonous gray brown of the leafless forest.

Individual trees are recognized and respected. A remarkable example near Teachive is more than 10 m tall with a trunk of more than 1.5 m in circumference. Known as "El Jitón," it is widely renowned, and the shady ground beneath it is a resting place for foot travelers.

Vicente Tajia believes that jitos live to be a thousand years old. He has known

El Jitón nearly seventy years, and he said it has not grown appreciably in that time. Unfortunately, jitos apparently cannot be dated by tree-ring analysis (Thomas Harlan, pers. comm., 1995) because they are tropical trees without a distinct growing season: some years produce many rings, while other years may produce none. The dense foliage gives refuge to owls, which often perch on the upper branches, nearly invisible from below. Owls are often persecuted by Mayos, who attribute malevolent powers to them.

Recruitment of new individual jitos has been severely restricted due to heavy grazing, which results in the emerging seedlings being trampled. We have looked far and long for seedlings and found few, if any, more than a couple of years old and it is commonly said that "there are no young jitos."

Jito is not well known in Sinaloa, near the southern limit of its range.

FOOD: Jitos are dioecious. Each year the female trees yield several kilograms of fruits, which are boiled with sugar when tender and greatly enjoyed by natives. The fruits are boiled for about an hour to remove a bitter agent that rises to the surface of the boiling mixture as froth and is skimmed off.

These are only a few of the many plants Mayos use. A familiarity with them invariably leads (as it led us) to other plants and more uses. To these we now turn.

# PART TWO

Vicente Tajia beneath an unusually large *pitahaya* (*aaqui* in Mayo; *Stenocereus thurberi*). Giant specimens seldom yield much fruit.

# 6 Plant Uses

A traditional Mayo house constitutes a remarkable primer for initiating an ethnobotanical study. As the annotated plant list in chapter 7 indicates, a host of native plants are to be found in every such Mayo house, used for everything from building materials and home implements to foods and livestock management materials. Medicinal plants are stored between the vigas (beams) and *latas* (narrow poles laid side by side) of the ceiling, often near the portal. Many of the tools scattered around the home are fashioned from native woods. The beds (tarimes) and other furniture are usually fashioned from locally harvested materials. In traditional homes a house cross woven from willow, palm, *carrizo,* or pine needles is usually hung or nailed in a prominent place. It reveals which plants are symbols to the family. The woodpile, ever changing, yields important information about local plants. Cactus fruit of some sort is usually available during the hot months, and the fruits or husks are seen cast away on the ground. In other months berries or tree fruits ripen and can be seen in buckets or ollas.

We have classified the uses of plants according to various functions. We began doing this when we observed the remarkable variety of plants used in the homes of some consultants. These categories are not exhaustive or mutually exclusive, but they enable us to group plants in a manageable way.

AESTHETICS

In the aesthetics category we include plants that contribute to human enjoyment in their unmodified forms.

*Gloria* (*Tecoma stans* var. *stans,* Apocynaceae) appears as an ornamental in nearly every yard. Its leaves and yellow tubular flowers are also valued for their medicinal properties. Another valued ornamental is cascalosúchil, a small tree with brilliant white flowers that is also planted around homes. Horticultural varieties of this frangipani are often purchased as well. Other trees popular for their beauty include the rose-colored amapa and *huanaca* (*huisache*; *Caesalpinia cacalaco*, Fabaceae), which produces dense racemes of bright yellow flowers. Vicente Tajia and Juan Félix excavated shoots of a huisache growing near Jambiolobampo and planted them in Teachive as ornamental shade trees. The vine *bellísima* (*Solanum seaforthianum*, Solanaceae), naturalized in more mesic parts of the region, is welcomed as an ornamental as well. *Tavachín* is frequently planted as a hedge. Its beans may be eaten, but it is most valued in villages for its showy yellow, orange, and red flowers.

Plants are also valued for their pleasant aromas. Branches of *bíbino* (*salvia*, desert lavender; *Hyptis albida*, Lamiaceae) may be bundled together and ignited to give rooms a pleasant, fresh scent and also to help fumigate sickrooms. Palo limón (*Amyris balsamifer*, Rutaceae) is used in Bachomojaqui as a kindling wood, primarily because of the agreeable lemonlike fragrance given off by the burning wood.

During the hot season, which lasts from late April through October, people often seek refuge from the powerful sun, and some trees are renowned for the shade they provide. *Maco'otchini* (*guamúchil*; *Pithecellobium dulce*, Fabaceae), which may grow to a height of 25 m, is an excellent tree valued not only for its shade and beauty but also for its fruit and bark, which is used for tanning. Near watercourses, mezquites grow to 15 m in height and provide partial shade. In drier portions of the coastal thornscrub, the symmetrical jito is valued for its dense shade. At times it may be the only shade tree available. One tree near Teachive, known locally as El Jitón, is maintained free of undergrowth for the comfort of passersby. In thornscrub and occasionally in tropical deciduous forest large etchos and even an occasional aaqui provide respite from the pounding heat. In the watercourse of Baimena, Sinaloa, a lengthy grove of giant and ancient sabinos is cherished, with individual trees familiar as landmarks. During the hottest days, shade can be found at all hours in the grove. Along arroyos in hilly or mountainous country, fig trees, primarily tchuna and nacapuli, afford dense shade, and at certain times of year a mist visible in the sunlight, part of the trees' evapotranspiration, showers from the canopy, functioning as a natural evaporative cooler.[1]

Some plants are valued because they are attractive oddities. Examples are *choca'ari* (*ayal*; *Crescentia alata*, Bignoniaceae), which is planted as a conversation tree (owners love to talk about them) as well as for its ayales (gourds) (see chapter 7), and *túbucti* (*brincador*, Mexican jumping bean; *Sebastiania pavoniana*, Euphorbiaceae), whose wriggling seeds are viewed as toys enjoyed by young and old alike.

The *criollo* (creole) cow is a mixed breed that is drought tolerant and willing to endure on the varied fare of the *monte*.

## LIVESTOCK

Some plants are used to feed livestock, and others as materials for managing livestock via fencing, enclosures, and corrals. Although livestock were introduced by Spaniards and were not part of pre-Conquest life, the sale of livestock and related products, especially cheese, is now the single most important source of land-based income for the Mayos. Because deer and wild bees are of great importance, we include them in this category.

A sophisticated knowledge of plants consumable by livestock and toxic to them is essential to the Mayos' enterprise. Furthermore, deer (though wild) are monitored and killed whenever possible, and the venison adds important protein to a diet deficient in animal protein. During drought periods, many Mayos collect fodder in the monte and carry it to their homes or pastures to supplement the meager feed available there. Among the wide variety of herbs, shrubs, and trees that produce fodder are leguminous trees such as mezquite, *brea*, ejéa, and jupachumi. Their edible pods and branchlets fall to the ground in dispersed

112 / Plant Uses

Corral of *brasil* trunks (*júchajco* in Mayo; *Haematoxylum brasiletto*).
(Photo by T. R. Van Devender)

sequences and constitute important sources of time-released fodder. The fallen corollas from the flowers of some trees (for example, guayacán and palo santo) are important feed sources for livestock and deer. The bark and inner wood of palo santo and güinolo are eaten by burros in times of drought. The foliage of many legumes (including brea, jócona, and mezquite) nourishes goats, which generally eat all species consumed by cattle plus many others, especially shrubs. Several species (algodoncillo, chapote [*Casimiroa edulis*, Rutaceae], mezquite, palo zorillo) provide pollen and nectar for bees (Mayos are devotees of honey and are constantly on the lookout for *panales* [wild hives]). In addition to grasses and the trees mentioned above, a multitude of herbs, including several members of the family Acanthaceae are recognized as important sources of fodder.

Although most cows and goats forage on common lands, animal husbandry is nearly as important as animal feeding

Chickens are raised in individual homes, and methods of providing them with shelter and fencing are vitally important. Short trees with bare boles and dense foliage (ejéa, guayacán, and mezquite) provide roosts for chickens where they are safe from predators, including dogs. Ladders (notched poles) of mauto or palo colorado are provided for the chickens to make access to the roosts easier.

The wood of other trees (algodoncillo, palo colorado) are valued for fence posts, while some (chilicote, etcho, jaboncillo, torotes [*Bursera* spp.]) are used to create living fences, and others for building corrals (often constructed from brasil, mezquite, and tepeguaje). Etcho cacti, jitos, and palo santos are valued for the shade protection they afford. Water and feed troughs are constructed by hollowing out logs of *torote copal* (in this case, *Bursera stenophylla*) or tepeguaje.

CONSTRUCTION

Nearly all Mayo homes built more than two decades ago were constructed entirely, or almost entirely, from native materials. Within the past twenty years, burnt clay bricks have become popular and have to some extent replaced sun-dried adobe, which is considered poor people's material. Nevertheless, many houses are still constructed out of materials that reflect an ancient tradition and centuries of experimentation with local resources.

The typical house consists of a frame of vigas (roof beams) and *horcones* (posts), walls of adobe or wattle and daub (interwoven branches plastered with mud), and a flat roof of cross-hatched latas laid upon the vigas and covered with a thick layer of dried grass upon which several inches of mud are placed. After several years, plants sprout from the dirt roofs, a fascinating sight for those unaccustomed to living roofs.

Certain hardwoods—amapa, cedro, mauto, mora, joopo, and palo colorado—are especially valued for their durability and resistance to damage by water and insects. These qualities make them desirable as vigas and horcones, and also for constructing *tinajeras* (tripods supporting water jugs), which are found in every home. Other woods—batayaqui, pitahaya, and vara blanca—though less durable, are easily cut and resilient, making them useful for cross-hatching in roofs and for wattle. Grasses, including giant canes such as carrizo and *otate* (bamboo; *Otatea acuminata*), may be used to construct walls. The walls of homes near the coast are constructed primarily of woven pitahaya ribs, with wood of various trees used as ceiling vigas. Inland homes are more often constructed of adobe with vigas of amapa, mezquite, palo colorado, or even copalquín, with roofs of dirt. Decades ago, many homes in tropical deciduous forest had roofs of woven *palma* (*Brahea aculeata*, Arecaceae) leaves. Burnt adobe bricks and poured concrete roofs are a relatively new innovation and represent a mestizo, not Mayo, form. Houses of dressed (carved) stone with dirt or palm roofs still stand at El Paso.

Aurora Moroyoqui weaving a *cobija*. Women in ten or so villages continue to weave fine woolen blankets despite dwindling markets. Aurora uses only natural colors.

ARTIFACTS

The artifacts category includes plants and plant parts used in making tools, implements, furniture, utensils, and so on. Given the Mayos' relative poverty in a poor country, they produce many items that would be purchased in urban areas or in more developed economies. Hence a wide range of plant materials are identified that solve particular needs.

Palo de asta and palo dulce are especially noted for their durability in tool handles. Guásima is the chosen wood for light chairs and tables, while sabino, laboriously sawed into slabs, is ideal for making fine furniture and cribs. Tough posts of guayacán are used in building looms. Mauto is the preferred wood for tinajeras. Mezquite, palo chino, torotes, and *teso* (*Acacia occidentalis,* Fabaceae) are carved into bowls and ollas (containers). Álamo is cut into thin, wide planks for tabletops. Rods, staffs, and poles of huo'ótobo and sina are indispensable in weaving. Chilicote is carved into cork stoppers for *bulis* (gourd canteens). *Jíchiqiam* (brooms) are fashioned from various types of Malvaceae (genera *Abutilon* and *Sida*) and from dais. An array of commonplace and ingenious uses add to this list, which we find to be ever-expanding.

INDUSTRY

The industry category includes plant products consumed in producing other materials.

Chemicals necessary for producing arts and crafts, for construction, and for products required in everyday activities are derived from a wide variety of plant sources. Blanket weavers produce various shades of wool with dyes derived from brasil, chiju (*añil*, indigo; *Indigofera suffruticosa*, Fabaceae), mezquite, palo dulce, and sitavaro. A mordant for dyes comes from the root of sa'apo. The barks of guamúchil, joso, mauto, and palo chino are used to tan leather, each one producing a different color. Soap is manufactured by pulverizing the fruits of tásiro, by collecting the liquid derived from pounding the long leaves of *aimuri* (amole; *Agave vilmoriniana*, Agavaceae), and by mashing the bark of jaboncillo. The roots of brea were formerly burned to produce ashes used in making soap. Samo baboso is added to lime to make it adhere to plaster or adobe. The bark of *papelío* is used to keep honey fresh in the comb. Tempisque bark or sap is used to coagulate milk in making cheese. Numerous woods are used as firewood, the selection often dependent on the temperature desired.

MEDICINE

While nearly all Mayos consult with Westernized practitioners for health problems, most also continue to use traditional remedies. Nearly all the Mayos we interviewed relied on doctors and nurses for their most pressing health care needs but professed confidence in traditional medicine and native curers as well. This probably reflects the limited availability of modern health care combined with a pervasive conservative conviction that the native pharmacology is capable of addressing most infirmities. While manufactured pharmaceuticals are more concentrated and thus quicker acting, traditional remedies may in the long run be equally efficacious, or so many believe. *Curanderos* (healers) charge very little and are usually willing to accept barter for services. Herbal and traditional remedies are also available for free, or nearly so, and using them allows Mayos to avoid contact with the urban doctors and the associated health system, which many feel treat them poorly because they are poor Indians. Few are the households that do not include an array of dried herbs collected from the nearby monte or from houseplants, even some from distant places, to treat common ailments. These are prepared and administered for common ailments. One woman asked us to bring her a supply of *gobernadora* (creosotebush; *Larrea divaricata*, Zygophyllaceae), a common Sonoran Desert shrub, about which she had heard much, but which is absent from the Mayo region.

Pre-Columbian medical practices were rooted out by the Jesuits, so we can only speculate on what they might have been. Mayos call upon a traditional Iberian physiological paradigm to explain the etiology of disease: A healthy body is one in which forces of hot and cold are balanced, and these forces are easily thrown off by alien influences and must be restored before the body can return to health. For example, a young woman from Teachive was believed to be incapable of conceiving because she had eaten ice cream, drunk cold sodas, and eaten an orange (at a critical time) when she was young. Only the proper herbs and potions could restore her fertility. Similarly, some infant illnesses, especially diarrhea, are attributed to a noxious imbalance brought on by their mother's working too near a stove or drinking something icy.

In addition to the "hot" and "cold" paradigm, Mayos incorporate a widely held conviction that spiritual balance can easily be disturbed by exotic forces, resulting in considerable, even permanent, physical harm. A curandero informed one of our consultants that he lost his eye to the spine of a jaboncillo because an enemy had placed a hex upon him. In general, chronic diseases are thought to be the result of hot/cold imbalance, while behavioral disorders and injuries due to accidents are deemed brought on by spiritual forces. At times, however, no boundary between physical and spiritual infirmities is observed. For example, *susto* is the name for a bodily disease that may be caused by fright brought on by appearances of supernatural beings (discussed further below).

In spite of frequent nonphysical diagnoses, Mayos are sophisticated in identifying organic disorders and diagnosing infestations of endoparasites.[2] Whenever possible, they are also likely to consult with Western practitioners for curing acute infectious diseases (especially pulmonary ills) rather than resort to local curanderos. Native medicine seems ineffective at dealing with the common cold and its complications.

Aside from common colds and flu, the most common ailments are kidney problems (kidney pain and urinary disorders, including prostatitis), digestive maladies (often related to infections with endoparasites), diabetes, female disorders, and snake and insect bites and stings. Diabetes is reportedly increasingly common and is often correlated with increased alcohol and sugar consumption.[3]

Many plants in the annotated plant list (chapter 7) are used to treat urinary and digestive tract difficulties. Kidney problems are predictable in the region due to the torrid climate and resultant frequent dehydration, which is not generally recognized for the health hazards it poses. Many digestive infirmities are related to general problems of sanitation, including contamination of drinking water, lack of adequate domestic waste disposal, lack of running water and hot water, swarms of houseflies and cockroaches, and a general lack of awareness surrounding sanitary food preparation.[4]

Stings and bites are common, especially from arthropods and reptiles. Several species of venomous scorpions, centipedes, and spiders are common in the region, and most homes provide ideal habitats for them. A wide variety of stinging bees and wasps plus an intimidating number of biting ants populate all habitats. While stings and bites are a nuisance and often painful, they are seldom fatal; however, the arrival in the 1990s of Africanized bees, which are especially aggressive and will attack without provocation, has added a new and potentially lethal danger.

Among the several species of venomous reptiles are the Mexican green rattlesnake, the western diamondback, and the tiger rattlesnake (*Crotalus tigris*), which are always killed on sight, as are most other snakes. Rattlesnake bites are not unusual and can be extremely painful, but victims seldom die. One reason that bites are common is that rattlesnakes are viewed as a marketable resource whose skins are made into belts. Also, Mayos eat the meat and use the rendered fat and extracted bile as medicinal remedies. The skeletons and dried tissue are ground into a powder believed to bestow recuperative powers when sprinkled on food.

Other venomous reptiles include the elusive *pichicuate*, or cantil (*Agkistrodon bilineatus*)—a relative of the cottonmouth moccasin that is rarely seen but widely feared—and two species of coralillos (coral snakes), relatives of the cobra. The Arizona coral snake (*Micruroides euryxanthus*) of coastal areas is venomous but so small that it is unlikely to inflict a serious bite. The large tropical coral snake (*Micrurus distans*), uncommon but found along arroyos and in the foothills forest, is capable of inflicting a fatal bite. Mayo country is also home to the world's only venomous lizards, the escorpiones: the uncommon Gila monster, found in coastal thornscrub from Masiaca and Yocogigua north, and the Mexican beaded lizard, a large (up to nearly a meter in length), black and yellow, lumbering lizard common in foothills thornscrub and tropical deciduous forest. The latter is respected but not much feared because its potentially dangerous bite is easy to avoid.

Two harmless geckos (Gekkonidae), locally called *salamanquesa*, are nocturnal predators of small insects. The western banded gecko (*Coleonyx varie-*

*gatus*) is a common ground lizard in thornscrub. Another gecko (*Phyllodactylus homolepidurus*) is common in tropical deciduous forests, where it is often seen on the walls and ceilings of houses. Both species are viewed with terror by many residents, who believe the hapless and innocuous creatures are deadly and insidious. Francisco Valenzuela related to us an anecdote about a child who reportedly died in transit to the hospital as a result of a bite from a gecko.

Animal products are often combined with plant parts in producing remedies. Especially important is the *manteca* (fat) gathered from various fauna, including rattlesnakes, lizards (especially spiny-tailed iguanas [*Ctenosaura hemilopha*]), *cholugos* (coatimundis; *Nasua narica*), and *torim* (packrats; *Neotoma* spp.). Lest practitioners of Western medicine dismiss the natives' insistence on the use of fats, it is important to recall that many important plant chemicals are fat soluble and may be rendered available to the human system only if so dissolved. Burro milk, reportedly high in fat, is often a recommended medium for administering herbs and is also considered beneficial on its own. Mayos who administer herbal remedies emphasize the importance of including all recommended ingredients. The omission of any will alter the effects of the remedy.

Curanderos are more common among mestizos than among Mayos (only three Mayo curers are known in the region, in Camahuiroa, Huebampo, and Buayums).[5] Mayo healers, called *móriac* (hechicero, sorcerer), were systematically hunted down and exterminated by the conquering Spaniards, so little is known about their methods of curing.[6] Most Mayos, however, know the whereabouts of curanderos and do not hesitate to consult with them. The curers usually invoke mystical and occult practices, incorporating herbs and other artifacts into their cures. Typically, they warn of spells cast on victims by other people or by animals, especially owls. For a cure they will sometimes administer massage, sometimes prescribe herbal potions, and sometimes suggest rituals that may include manipulation of animal parts as well.

Mariano Dórame Buitimea of Buayums, near Huatabampo, is one of three Mayos recognized as a curandero by the Instituto Nacional Indigenista. Mariano uses many plants of the rich thornscrub surrounding Buayums along with minerals, spells, massages, gestures, prayers, and animal parts and products to perform cures. The wide range of services he offers include removing spells from one's vehicle, chasing evil spirits from one's house, and curing a myriad of infective diseases. He says his specialty is curing *malpuesto,* an evil spell or hex. Mayos and mestizos alike visit him at his home to seek his services. In spite of his reputation as a curandero, Mariano lives in great poverty and in fear of the increasing violence and crime that plague the Mayo region. Alejo Bacasegua of Camahuiroa is also a curandero, though his practice is less formal than Mariano's. His cures include remedies for rabies, hexes, and venereal diseases.

It is widely assumed among Mayos that *brujos* (witches) and, to a lesser extent, hechiceros frequent the region. They are thought to be difficult to detect, but curanderos often prescribe herbs to ward off or counteract their evil activities. Sitavaro, particularly, is thought to be effective against ambient evils—evil spirits that move through the air—especially spiritually infective agents emanating from the dead.

Another common ailment is susto, a result of being traumatically frightened, often by a spiritual encounter. One child was said to have seen a ghost and consequently came down with a high fever, and an acquaintance of ours was inflicted with susto after he encountered a malevolent owl while he was walking alone in the monte. The symptoms are similar to those of hysteria, once a familiar malady diagnosed by psychotherapists, but now as uncommonly diagnosed as possession by demons. Curanderos usually cure susto with a combination of herbal decoctions, potions of animal parts, massage, and spiritual manipulations.

The ubiquity of remedies for intestinal disorders indicates a high incidence of parasitism, just as the wide variety of remedies for insect bites and stings indicates an environment rich in venomous invertebrates. The remedies suggest that several plants are natural insecticides, miticides, antihelmetids, and antiprotozoans.

Although curanderos are not common, each village has at least one person who prepares a *tónico*, a mixture of teas from a variety of herbal sources. This tonic is taken daily to improve energy and ward off disease. A tea brewed from *chicura* (canyon ragweed; *Ambrosia ambrosioides,* Asteraceae) is routinely administered to women during and after childbirth. Those tending toward diabetes are encouraged to drink a tea from copalquín, which may also have antimalarial properties. Its family (Rubiaceae) includes the genus *Chinchona,* from which quinine is derived.

Remedies are administered in many forms. Often the plant or plant part is used in its natural form. In other cases, a part of the plant must be cooked, roasted, ground, pulverized, or boiled. In a few cases herbal parts are preserved in alcohol, and the mixture is inhaled. Often a combination of these preparations is deemed more effective than any one used singly. The most common method of administering native medicine is through a tea brewed by boiling the bark, root, branch, leaf, flower, or a combination of these. Patients usually drink the tea, all at once or as a substitute for drinking water, but in some cases it is massaged into the skin or onto the affected area or used to bathe the patient.

Many herbal remedies used by Mayos arise from their deep, shared knowledge of plants, including growth forms, habits, and ecological requirements. Part of Mayo socialization includes detailed and repeated discussions of the efficacy of various remedies for the multitude of common afflictions. Mayo herbal medicine is founded on endless exchange of information among mothers and grandmothers, neighbors and visitors, relatives and strangers over which plant can cure which illness, how each remedy should be administered, and how much is a proper dosage. This dynamic cultural data bank gives rise to an ongoing repository of medical knowledge and lore that evolves parallel to the formal tenets of medicine practiced by formally trained practitioners. But the monte is essential for the constant reinforcement and renewal of this knowledge—this ongoing experience of plants in their vital environment, the richness of herbal remedies, and the creativity of ongoing socio-medical experimentation. Without the monte, the Mayo-ness of medicinal plant use will fade. Curanderos, often secretive and jealous of their techniques and formulas, will survive in the cities, but the Mayo social pharmacology will not.

Below is a list of common ailments and examples of various Mayo strategies for curing them.

- Cancer. The juice from the flesh of a five-ribbed (*not* six-ribbed!) sinita is widely viewed as having cancer-curing properties. Even more common is the use of the large tuber of güerequi. The root is pounded, dried, and ground into a powder that is then poured into capsules and taken regularly for any sort of cancer. Another remedy is a tea brewed from the bare stems of *jutuqui* (graythorn; *Ziziphus obtusifolia,* Rhamnaceae).
- Circulatory problems. The leaves of gloria are brewed into a tea widely drunk for lowering blood pressure and alleviating heart problems. The bright red tea of the cambium of brasil is thought to produce similar results, with the added benefit of shrinking varicose veins. A tonic made by steeping flowers of sanjuanico is also thought to "strengthen" the heart. The raw flesh of biznaga is often recommended to lower blood pressure.
- Colds, grippe, and upper respiratory ailments. Mayos are no more successful in fending off or curing the common cold than any other people. For a cough, they usually recommend tea brewed from the bark of torote prieto or copalquín. Along the coast a tea is brewed from the leaves of *mariola* (*Aloysia sonorensis,* Verbenaceae), while in the foothills a similar shrub also called mariola (*Lippia graveolens,* Verbenaceae) is similarly decocted. Some herbalists produce

Dried flowers of *sanjuanico* (*tásiro* in Mayo; *Jacquinia macrocarpa*). A tea brewed from the corollas is thought to strengthen the heart. The flowers are also used to produce a mustard-colored dye. (Photo by T. R. Van Devender)

an inhalant by steeping the leaves of *aroyoguo* (*cacachila*; *Karwinskia humboldtiana*, Rhamnaceae) in alcohol and storing the resulting liquid in a vial. A similar potion for alleviating asthma is made from the root of *jupachumi* (*Petiveria alliacea*, Phytolaccaceae), steeped and inhaled. Another root considered effective for bronchitis and asthma is chanate pusi (rosary bean; *Rhynchosia precatoria*, Fabaceae), which is brewed into a tea and drunk. Chewing the sweet sap of brea is also widely believed to improve pulmonary function.
- Digestive problems. For hemorrhoids, two remedies stand out. The leaves of *tebue* (*toloache*, sacred datura; *Datura lanosa*, Solanaceae) are scalded and carefully rubbed on the inflamed area. Alternately, the powdered ashes of güerequi are flung at the hemorrhoid from a distance, supposedly with excellent curative results. For intestinal parasites, papaches eaten on an empty stomach are the easiest and most popular remedy. For more persistent cases, a strong tea of *lipazote* (*epazote*; *Chenopodium ambrosioides*, Chenopodiaceae) is

widely recommended. Equally effective is said to be a tea from *chíchibo* (*estafiate*; *Ambrosia confertiflora*, Asteraceae). This tea is also prescribed for relief of diarrhea, as are the cooked roots of choya and the fruits of sibiri and a number of other plants.

For bad stomach, the scorched husks of pitahayas, the chewed leaves of pasio, and a tea brewed from dried leaves of júsairo are generally believed to be effective. Goma Sonorae, the commercially marketed lac found on branches of samo, enjoys a wide reputation for its ability to alleviate all manner of digestive problems.

- Ear and eye problems. For deafness, a concentrated tea from the leaves of *paros pusi* (*sacamanteca*; *Solanum azureum*, *S. tridynamum*, Solanaceae) is poured directly into the ear. For earache, the flesh of a roasted biznaguita is squeezed, and the drops are directed into the ear. For foreign objects in the eye, a seed of *comba'ari* (chani; *Hyptis suaveolens*, Lamiaceae) is placed in the eye, reportedly causing the foreign body to adhere to it. Eye infections are treated with the milky sap from a leaf of sitavaro. The sap of sa'apo is frequently used to remove foreign objects and to clear up "cloudy" eyes. The similar sap from sangrengado is also said to clear red eyes.
- Fatigue, lassitude, or chronic lack of energy. This condition is commonly diagnosed as a result of "weak blood." The goal of treatment is to strengthen the blood, for which a tea from copalquín bark—or, more accurately, the bark plus a thin layer of the underlying heartwood—is administered. Copalquín, a common small tree throughout the region, produces a bitter tea and is often flavored with cinnamon or honey. Equally effective, according to many, is a far more palatable tea brewed from the bark of palo mulato, but the tree is limited to the moister portions of the region so is less readily available.

    Fatigue may also be the result of malpuesto, a hex. To counteract this evil spell, a curandero must be consulted and the appropriate measures taken.
- Folk diseases. We use this description not to deny the reality of these conditions or to disparage their diagnosis, but because our Western classification of pathologies does not include such illnesses as malpuesto and imbalances of hot and cold. To treat malpuesto, the root of cacachila is boiled into a potion applied by curers. The concentrated tea from boiled brasil wood is believed to alleviate

*tiricia* (sadness or depression). Tea from tepeguaje bark is widely believed to restore the proper balance of hot and cold so that a woman may conceive. A wash from steeped leaves of sitavaro may help fight off the malign influence of the dead. Use of pounded root of *na'atoria* (*baiburilla*; *Dorstenia drakena*, Moraceae) in the proper setting is believed to heal *alferecía* (boca torcida, perhaps Bell's palsy).

- Infectious diseases: The sharp-pointed leaves of sanjuanico are brewed into tea for *tos ferina* (whooping cough). The entire plants of various *mochis* (*Boerhavia* spp., Nyctaginaceae, and *Tidestromia lanuginosa*, Amaranthaceae) are brewed and administered for combating measles. A tea composed of a mixture of *cordoncillo* (scalystem; *Elytraria imbricata*, Acanthaceae) and *nata'ari* (*ortiga*, noseburn; *Tragia* sp., Euphorbiaceae) and a tea steeped from copalquín bark are both considered effective against malaria. Remedies for rabies include the fruits from a particular Sinaloan cacachila tree or the powdered bark of ébano, which is sprinkled on dogs and humans, a remedy said to be both preventive and therapeutic.
- Kidney and urinary tract problems: The liquid in which *teta segua* (*flor de piedra*, resurrection plant; *Selaginella novoleonensis*, Selaginellaceae) has been steeped for a day is thought to dissolve kidney stones. This remedy is said to be enhanced when a few branches of cósahui are steeped along with the teta segua. Among numerous remedies for aching kidneys is a tea brewed from the leaves of *miona* (*Commicarpus scandens*, Nyctaginaceae), which is drunk, and a tea from torote panalero that is rubbed on the kidneys. A tea in which the root of tatachinole has been steeped is often prescribed as well. Several other herbs are also touted as an aid to healthy kidneys.
- Reproductive problems: For women who have trouble conceiving, a hot tea from the pungent *cuchu pusi* (*hierba de venado*, deer weed; *Porophyllum gracile*, Asteraceae) is thought to promote conception as well as relief from menstrual cramping. The ground-up seeds of an unidentified species of *Ipomoea* (*trompillo*, morning glory; Convolvulaceae) are thought by some Mayos of Los Capomos to promote fertility. For impotence, teas from copalquín and palo mulato are believed effective. For venereal diseases, the fruit of a particular choya cactus (*Opuntia* sp. X) is sometimes touted as a cure, a

remedy that may go back hundreds of years. Another remedy for love's mischances is a tea brewed from the branches of *jutuqui,* another unidentified choya species. Prostatitis is a common malady among older men, and numerous teas and potions are recommended, including a decoction from the spines of güinolo and sibiri. A few drops of etcho juice taken directly are recommended for alleviating male urinary problems.

- Musculoskeletal problems. The most popular remedy for bad bruises, sprains, dislocations, and broken bones is the tescalama sap and bark. Bone breaks, sprains, and arthritis are also treated with direct application of leaves of *sanarogua* (*mataneni; Callaeum macropterum,* Malpighiaceae). Most villages have someone knowledgeable in setting broken limbs, but more frequently this job is left to urban doctors.
- Skin conditions. A decoction of sitavaro is applied to pimples and boils. The milky sap of *cuépari* (golondrina spurge; *Euphorbia* spp.) is sometimes applied to skin that lacks pigmentation, as is the crushed root of *toto'ora* (*cresta de gallo; Plumbago scandens*). Ashes from the root of guayacán are thought to aid in healing the crusts and flakes brought on by psoriasis. For scabies, a decoction of the leaves of *mala mujer* (*Heliotropium curassavicum,* Boraginaceae) is used as a body wash (Bañuelos 1999:107). Even more commonly used is a wash brewed from the bark of nesco or from the pounded and pulverized branches of juvavena.
- Tonsillitis: Mayos have two remedies for alleviating this malady. The leaves of two plants, each named *mo'oso* (*Bebbia juncea* [sweetbush] and *Machaeranthera tagetina,* Asteraceae) are brewed into a tea used as a gargle. The bark of cumbro is steeped and drunk for the same condition.
- Diabetes: Remedies for diabetes abound in Mayo pharmacology. Most prominent is powdered güerequi poured into gelatin capsules. Also highly recommended is a tea brewed from the combined roots of cósahui and sa'apo. The steeped leaves of *chíchel* (*toji; Struthanthus palmeri,* Loranthaceae), a common mistletoe, are frequently mentioned as combating diabetes. Many Mayos recommend a tea brewed from copalquín bark sweetened with cinnamon to overcome the bark's bitterness.

We have not found any report of a remedy for tetanus. One of Masiaca's finest musicians died from this dreadful disease after having a foot crushed by a cow.

## FOOD

Mayos were agriculturalists at the time of the first contact with Europeans, but their diets also included abundant wild foods gathered from the monte. Over the intervening centuries, these plants have constituted an irreplaceable supplement to cultivated foods, but the practice of gathering is becoming less common. In towns and villages of the Mayo Delta, such as El Júpare or Buaysiacobe, none of the original monte remains, so only a small percentage of wild food is consumed. In villages surrounded by monte, however, most families include at least some wild plants in their diets. In villages such as Baimena and Los Capomos, where monte is abundant and rainfall is usually sufficient for raising corn, beans, and squash, much of the food consumed by households is raised or gathered locally.

Many wild plants can be eaten without cooking. Cactus fruits, especially etcho, pitahayas, and tunas, are eaten directly from the plant. Seeds from several cactus fruits (etcho, biznaga) are mixed with ground corn to make tortillas or are boiled into atol (porridge). The fruits from trees such as miscellaneous figs (*Ficus* spp.), guayparín, igualama, saituna, and tempisque are eaten raw or stewed. Bush fruits such as sire and cumbro are plucked and eaten on the spot. The thick husk of papaches is removed and the pulpy contents eaten. Roots such as saya and *masasari* (*sanmiguelito,* queen's wreath; *Antigonon leptopus,* Polygonaceae) are eaten raw, while those of pochote and *chócola* (*Jarilla chocola,* Caricaceae) are eaten cooked.

Mezcal heads are roasted and the leaf bases eaten in nearly all villages. Wild greens such as chichiquelite and *hué* (*bledo,* pigweed; *Amaranthus palmeri,* Amaranthaceae) are eaten eagerly following rainy periods. Spice plants, including *orégano* (*Lippia palmeri,* Verbenaceae) and chiltepín, contribute to the cuisine of nearly every household. The bark of palo mulato is brewed into a refreshing tea. The sweet pulp of guamúchil pods is eaten raw or toasted.

While miscellaneous seeds do not form an important component of the Mayo diet, a number of edible seeds are nibbled on by those who work in the monte. Examples are *aguaro* (*Martynia annua* and *Proboscidea altheaefolia,* Martyniaceae), comba'ari, and the seeds of pochote fruits.

## CULTURE

The culture category includes plants and plant parts that have cultural or spiritual significance in Mayo culture. While most of the plants in this category are shrubs and herbaceous plants, some trees are used as well. Abolillo, sigropo, and palo dulce are used to make beads for rosaries, necklaces, and crucifixes. The points for the lances carried by fariseos in Lenten processions are made of brasil wood. Sitavaro is used to counter spells or infections from the dead; it is taken as a tea to prevent ethereal pathogens invading through boils and sores. Álamo branches are used to deck ramadas and procession routes during Holy Week. Brea flowers are sprinkled on fallen house crosses on Good Friday. Specific herbs may be described by a curandero to combat a spell placed on one person by another. Huo'ótobo is used by the lead pascola to start off fiesta ceremonies.

This list and the specific uses included in chapter 7 reflect the pervasiveness of native plants in Mayo culture. Although most of our consultants were older men, older women also appear to be highly sophisticated in plant identification, but because they spend a disproportionate amount of time in the home, they have far less opportunity to become familiar with plants in their native habitat. It is usually the women, however, who process the plants that the men collect in the monte.

We have encountered few younger Mayos with sophisticated field knowledge or intimate applied knowledge of plants and their uses. Benjamín Valenzuela of Las Rastras, born in 1973, is the only younger person we found willing to guide us into the monte and describe plants. This is not to deny that other such young people exist, for our experience has taught us that many rural Sonoran youths have immense knowledge of plants and their uses, but fewer are involved in the study of plants and their traditional Mayo uses. Moreover, most young men from small villages are forced by economic necessity to seek work elsewhere. Once they become assimilated into city life, they lose their close identification with the land and its products. In many places, traditional Mayo knowledge of plants is not being passed on to younger people, who when they begin school at the age of five are expected to speak Spanish instead of Mayo. The more subtle uses of plants for food and medicinal remedies may be lost in a generation or so, replaced by food from the *tienda* and medicines from the *farmacia*. In contrast, plant uses for construction, livestock, and other things shared with the mestizo culture will persist longer than the Mayo language. One of our goals in this study has been to assure the Mayos that their knowledge of plants is valuable and should be passed on to future generations.

In chapter 7 we present the plants known to and used by Mayos. The remarkable spectrum of uses and diversity of plants used will afford the reader an opportunity to ponder the depth of Mayos' familiarity with the natural world they find in their lands.

# 7 An Annotated List of Plants

INTRODUCTION

In gathering the following data on the Mayos' knowledge and use of plants, we visited the region and consulted with Mayos for nearly ten years, beginning in the early 1990s and concluding in 2000. We also familiarized ourselves to the maximum possible extent with Mayo history, the geography and geology of the region, the region's fauna and flora, and the state of contemporary Mayo culture. We have doubtless omitted important data and considerations, but we hope that others will complement our work and continue to encourage Mayos, young and old, to carry on the old traditions which our many acquaintances have taught to us. We especially hope that the day will soon arrive when researchers can safely explore the mountainous areas adjacent to the Río Fuerte accompanied by Mayos. Drug traffic and the associated violence now preclude such investigations.

*Methods*

During this study we made more than fifty four- to ten-day visits to the Mayo region. We selected Mayo consultants from a variety of villages: Nahuibampo on the middle Río Mayo, the many villages in the delta of the Río Mayo, Alamos, villages in the southern foothills and coastal plains of Sonora, and villages in the basin of the Río Fuerte in Sinaloa. These individuals were known locally to possess superior knowledge of plants, and none of the consultants knew those from other villages. All were paid at a rate higher than the prevailing day wage. In every case, we carefully explained the nature of the project, stressing that we considered the Mayos' knowledge to be of great importance, and asked them to teach us about their identifications and uses of plants, and to share information about the significance of plants in the Mayo view of the world.

The consultants accompanied us into the field, occasionally to more distant areas but usually near their home village, which for most of them is also where they were born. Patricio Estrella, Vicente Tajia, and Francisco Valenzuela regularly accompanied us on trips to locations far from their homes. Vicente Tajia journeyed to a wide variety of other villages and proved exceptionally adept at making new acquaintances and explaining the nature of our work.

In Spanish, we asked all consultants to identify plants, to state both the Spanish and Mayo names, and to list and explain the uses of the plants. We asked the following questions:

- What is this plant called?
- What is it good for?
- How do people use it now?
- Have you used it? How did you prepare it?
- What does the name [if given in Mayo] mean in Spanish?

We also encouraged the consultants to use free association and describe what plants meant to them. Whenever possible, consultants' reports were verified by asking identical questions of other consultants in another area. Vouchers for nearly all of the ethnobotanical records were collected by Tom Van Devender and deposited into the herbaria at the University of Arizona, the Universidad de Sonora, Arizona State University, and others. Plants were collected under permits to Van Devender and Alejandro E. Castellanos V., Departamento de Investigaciones Científicas y Tecnológicas de la Universidad de Sonora, from the Secretaría de Medio Ambiente of the Recursos Naturales y Pesca in Mexico City. Specimens were identified using comparative collections and literature in the University of Arizona Herbarium, and with the assistance of many taxonomic specialists at other institutions. Collection data, scientific and common names, and ethnobotanical information have been entered into a computer database. Our collections, added to those made by our colleagues, provide an informative ecological narrative of the Mayo region.

The list below includes all plants for which uses, however humble, have been noted by Mayo consultants. Mayos refer to the sum of all these plants as *juya ania;* that is, everything that lives and moves in the monte. Our list cites the most current authority and notes synonyms where confusion might arise. Spanish and English common names are included in parentheses. When the significance of a Mayo common name has been ascertained, it is translated into English in quotation marks. Families are listed alphabetically, as are the species within them.

In many cases consultants could not recall a specific name or use. Often they explained that they had once known but had forgotten, or that they had heard of uses or names of a plant but could not recall them. We interpret this to indicate a decline in intimacy with the juya ania as Mayo culture has succumbed to the pressure of outside forces.

We caution that Mayos, like all of us, are prone to make hasty generalizations about the efficacy of plant remedies. News travels rapidly when someone recovers or is healed after ingesting a certain remedy or receiving a certain treatment, and the remedy involved may assume an unjustified glory, at least for a while. Also, curanderos are sometimes protective of their formulas and may attribute misleading properties to them. We have tried to note when uses were noted by one individual or merely a few, and when they are more widely disseminated.

This list contains some 370 species from 82 families. The largest families represented are Fabaceae (Leguminosae) with 53 species, Euphorbiaceae with 26 species, Cactaceae with 25 species, and Asteraceae (Compositae) with 23 species. Of these, approximately 135 (36 percent) are trees, 124 (33 percent) are shrubs, and 82 (21 percent) are herbs. We note 21 vines and 15 grasses. In several cases our consultants lumped more than one species under the same name. We believe this reflects the great floristic diversity in the region and confirms that the number of plants used by Mayos today and historically has been greater than the number used by indigenous groups to the north who live in less diverse habitats. In all probability, indigenous people to the south, where diversity is even greater, were familiar with a larger variety still. It is also likely that more herbaceous species were used in the past. The ubiquitous trampling of the ground by cattle, although it usually leaves the canopy little affected, has heavily affected the understory that would otherwise constitute a complex and varied component of ungrazed natural monte.

## Mayo Taxonomy and Classification of Vegetation

We have not succeeded in deciphering any Mayo taxonomy from the information we have gathered over the years. We initially hoped to detect grounds for a folk taxonomy, believing, along with Berlin et al. (1974) that such a "deep" system would offer "major implications for a better understanding of human cognition" (xv). Our quest for underlying patterns was rendered more difficult by consultants' varying nomenclature. Similarly, translation of Mayo plant names into Spanish yielded virtually no insight into any underlying systematics. Indeed, some translations only deepened the mystery of some Mayo names. We now believe that such a system is not to be found among Mayos.

This apparent lack of an underlying classification scheme does not imply that our consultants did not make connections, for they did indeed acknowledge some phylogenetic relationships. They recognize, among others, cuépari (spurges), *mochis* (*Boerhavia* spp.), and to'oro (*Bursera* spp.), but they do not apparently use any "deep" systematic scheme of classification. They adopt Spanish classifications for plants. The families Malvaceae (*Abutilon* spp., *Anoda* spp., *Bastardiastrum cinctum, Pseudabutilon scabrum, Sida* spp., *Wissadula hernandioides*) and Poaceae (nearly all grasses), for which they have few, if any, Mayo names, are grouped using the Spanish names *malva* and *zacate*. The *confiturillas* (Spanish) are a mixed group of shrubs with similar flowers. *Confiturilla amarilla* is *Lagacea decipiens* in the Asteraceae, while *confiturilla blanca* (*Lantana achyranthifolia, L. hispida*) and *confiturilla negra* (*L. camara*) are in the Verbenaceae. Our consultants also recognized various members of the Acanthaceae (*Dicliptera resupinata, Henrya insularis,* and *Tetramerium abditum*) as *rama del toro,* even though the growth forms range from herbs to shrubs, and flowers are purple or red.

Several consultants recognized subtle relationships; for example, two of them quickly associated an undescribed species of *Sideroxylon* with other members of the genus, but each with a different species. Their treatment of mangroves is instructive. Each species has a separate Mayo name, with no implied or stated relationship. In Spanish, on the other hand, all are grouped together as *mangles* (mangroves), including pasio, a strictly terrestrial plant. The plants universally called mangroves—including black (*Avicennia germinans*), red (*Rhizophora mangle*), and white (*Laguncularia racemosa*), as well as mangle (button mangrove, *Conocarpus erecta,* Combretaceae)—are a functional, not taxonomic, group. In other cases our experience was similar to that of Berlin et al. (1974:53), who found among the Tzeltal of Chiapas, Mexico, that

> general botanical collecting quickly revealed that the Tzeltal lacked legitimate plant names for much of the local flora. [But] . . . when presented with a particular plant specimen, informants rarely responded that the specimen had no name. Instead, they would systematically attempt to classify or relate the specimen under observation to one of the categories in the named taxonomy.

We suspect that Berlin and his associates may have demonstrated an understandable enthusiasm for names and connections that was quickly perceived by their informants. Eager to please their employers, they felt free to use creative means of naming and relating whenever their local knowledge ran out. Lest we be perceived as observing Mayos from a patronizingly scientific pedestal, we would like to emphasize that not only were we repeatedly corrected by local consultants when we incorrectly identified plants, but we became involved in vigorous arguments with Mayos as to whether two plants were different species or merely variations within a single species. Our winning percentage was not as good as that of a decent baseball pitcher.

We agree with Berlin (1992:8) that "human beings everywhere are constrained in essentially the same ways—by nature's basic plan—in their conceptual recognition of the biological diversity of their natural environment." Mayos responded historically to their environment as all people respond, by seeking to order, classify, and demystify it as well as to render it available for exploitation. We speculate that Mayos may once have commonly spoken of broader and more intricate relationships among plants, and that they probably once noted a greater variety of natural "kinds" of plants and grouped them systematically. Their history of dispersion over the last century and their contemporary isolation in small communities with little opportunity for pan-Mayo communication have undoubtedly extinguished much of any "deep" system of classifying plants. Furthermore, the disappearance of monte near the largest population centers has left most younger Mayos with no laboratory for studying wild plants and incorporating their similarities and differences into their community plant repertoire.

In contrast, several consultants clearly distinguished among plant communities, elaborating on the generic term *monte*, which is universally used to denote "the bush" or "the field." Thus we heard consultants refer to *manglares* (mangroves), pitahayales (coastal plains dominated by *Stenocereus thurberi*), monte mojino (vegetation, usually foothills thornscrub or tropical deciduous forest, of a characteristic brown gray color in the dry season), *monte grande* (probably better developed tropical deciduous forest), *monte de arroyo* (arroyo vegetation), *monte verde* (semideciduous tropical forest found in deeper canyons), *encinales* (oak woodland), and *pinales* (pine forests). For an elaboration on this folk classification, see Yetman et al. 1995:334–36. Missing is any terminology for designating thornscrub (*matorral espinoso* in Spanish), perhaps because what is remarkable to foreign ecologists is everyday backyard scenery to natives of the region.

Our collecting efforts were hampered to some degree by the calamitous drought that affected northwest Mexico during most of our period of visitation. In general, equipatas failed between 1994 and 2000, and through much of the region las aguas were only marginally sufficient. As a result plant growth was hampered, especially during the winters, and our ability to find winter-responsive plants was curtailed.

Finally, we present Mayo basic terminology for plant anatomy. Bark—*begua;* flower—*sehua;* fruit (general)—*taca;* glochid—*sebua;* spine—*huitcha;* leaf—*sahua;* leaf (of agave)—*maiicua;* pad (of a cactus)—*berua;* plant—*oguo;* root—*nahua;* seed—*batchia;* trunk, stem—*cuta;* tuber—*nahua;* vegetation—*juya ania.* Some important plants have individual names for their fruits.

## Mayo Names and Exotic Species

Of the 739 species of the Río Cuchujaqui southeast of Alamos, in the heart of Mayo country, 6.2 percent are introduced species (Van Devender et al. 2000). Mayos do not live in these tropical deciduous forests today but probably did in recent historical or earlier times. Thus 46 species of plants arrived with or after the Spanish. Many of these are widespread species that our modern Mayo consultants recognize. Mayos have Mayo names for most native species that they use and find distinctive. They do not have a general term for weeds (*malezas* in Spanish), referring to unwanted plants. They are aware that seeds of exotic plants, such as *Malva parviflora* (malva; Malvaceae), *Melilotus indica* (*trebolín,* sour clover; Fabaceae), and *Sisymbrium irio* (*pamitón,* London rocket; Brassicaceae) are carried inland from fields in the nearby river deltas via cattle feed, especially alfalfa (*Medicago sativa,* Fabaceae). These Eurasian annuals are called by their Spanish names.

In a few cases, native species, such as *Euphorbia colletioides,* are referred to only by their generic Spanish name, possibly because the plants are not useful or conspicuous. The massive, conspicuous *Taxodium distichum* is an anomaly because it bears only the Aztec (*ahuehuete*) and Spanish (sabino) names. Most introduced grasses (*Dactyloctenium aegyptium, Dichanthium annulatum, Eleusine indica, Eragrostis cilianensis,* etc.) are casually included with the Mayo *ba'asso* and Spanish *zacate.* The giant cane *Arundo donax* (carrizo; Poaceae)—used extensively for house walls by Mayos and Yaquis on the coastal plain—likely assumed the name *ba'aca nagua* from the native *Phragmites communis* that it has almost completely replaced.

Most introduced species have no Mayo names. *Ricinus communis* (castor bean) is an interesting example. This common, conspicuous plant apparently entered the Mayo region after Gentry's 1942 *Río Mayo Plants.* Mayos and mestizos call it *higuerilla* (little fig). Vicente Tajia referred to *Nicotiana glauca* (tree tobacco) as *marihuana* at first, then as *cornetón.* Vicente is a brilliant Mayo botanist, but considering that *Cannabis sativa* is called *mota* rather than marihuana in rural southern Sonora, and that *N. glauca* is called *juanloco* in urban Sonora, he clearly was not familiar with this South American immigrant.

In some cases when Mayos have a name for a plant, it indicates a long familiarity with it. This is especially interesting in regard to widespread New

World tropical species whose pre-Spanish distributions may never be known. Examples (for details, see species accounts below) include large trees prized for shade and fruit (caguorara [*Diospyros sonorae*]; maco'otchini [*Pithecellobium dulce*]), showy ornamentals (ta'aboaca [*Caesalpinia pulcherrima*]), useful shrubs (chiju [*Indigofera suffruticosa*]), and disturbance plants (guacaporo [*Parkinsonia aculeata*]; yete ogua [*Mimosa asperata*]). These species are likely native tropical species near the northern limits of their natural range.

We recognize that scientists are constantly discovering new associations among plants. The use of DNA as a taxonomic aid will doubtlessly result in some changing of taxonomic grouping. The identifications below represent the most up-to-date plant classifications available. The Mayo and Spanish common names may provide the permanent basis for many taxonomic decisions.

---

### ACANTHACEAE

*Carlowrightia arizonica* A. Gray  ánima ogua "tree of death" (palo blanco)

This is a common, white-flowered shrub of thornscrub and tropical deciduous forest that leafs out after summer rains. MEDICINE: The leaves are brewed into a tea taken for fevers. It is considered more effective when combined with leaves and inflorescence of *Elytraria imbricata*. It constitutes important forage for LIVESTOCK as well. Its Mayo name is a mystery, since it is 0.5–2.0 m tall and not a tree and is used to cure illness.

*Dicliptera resupinata* (Vahl) Juss.  (alfalfilla; rama del toro)

A common perennial herb with small purple flowers enclosed in twin bracts that turn white with age and become conspicuous, it abounds in upper foothills thornscrub, along arroyos, and in tropical deciduous forest. LIVESTOCK: It is an excellent food for cattle, which seem to thrive on it.

*Elytraria imbricata* (Vahl) Pers.  (cordoncillo; scalystem)

Cordoncillo is a common and well-known blue-flowered perennial herb of moist thornscrub, tropical deciduous forest, and oak woodland. MEDICINE: Its upright, tail-like inflorescence and leaves are boiled and taken as a tea for fever. Most Mayos believe it more effective when mixed with other herbs. It is often brewed with *Carlowrightia arizonica*. The presence of the two seems to have a synergistic effect. Others recommend that it be brewed with an unidentified plant called *"ortiga"* (perhaps *Tragia* sp.). Such a tea is effective for malaria, according to Francisco Valenzuela. Felipe Yocupicio uses a tea from the whole plant to treat diarrhea.

## 134 / Acanthaceae

***Henrya insularis*** **Nees**  (rama del toro)

This common yellow-flowered perennial herb or subshrub, along with other similar Acanthaceae, constitutes a critically important component of LIVESTOCK feed in the tropical deciduous forest. Francisco Valenzuela pointed out that many goats and cows depend on such plants for survival following the regional rains. The lack of a Mayo common name is probably due to the fact that its use as livestock feed is a post-Contact use.

***Justicia californica*** **(Benth.) D.N. Gibson**  semalucu "hummingbird flower" (chuparosa)

An uncommon shrub with bright red flowers, occasionally found in coastal thornscrub, sometimes growing into bushes nearly 2 m high. Valued as forage for LIVESTOCK. As the name indicates, hummingbirds swarm around the flowers.

***Justicia candicans*** **(Nees) L.D. Benson**  ma'aso o'ota "deer bone" (chuparosa, muicle cimarrón, rama de venado, ciática)

A 0.5–1.5 m shrub more commonly found in tropical deciduous forest than in thornscrub. Flowering nearly year-round, its brilliant red flowers make it easy to identify. MEDICINE: Francisco Valenzuela believes the leaves are effective against malaria when brewed into a tea and drunk. Cinnamon is added to make the tea more palatable. The plant is also valued as a forage crop for LIVESTOCK and for deer.

***Ruellia nudiflora*** **(Engelm. & A. Gray) Urban**  (papachili, rama del toro, tronador)

This common perennial herb of coastal thornscrub grows to 1 m high with large purple flowers. It is especially common along coastal arroyos. MEDICINE: The leaves are brewed into a tea taken to alleviate fever. CULTURE: Children play a game with the fruits, soaking them in the mouth and then spitting them into their hands. The saliva is absorbed rapidly, causing the seeds to explode, much to the delight of the onlookers. *R. intermedia* Leonard is a very similar species that is widely distributed in tropical deciduous forest. The same uses as for *R. nudiflora* are attributed to it.

***Siphonoglossa mexicana*** **Hilsenb.**  (cordoncillo)

This is a locally common small shrub found only in tropical deciduous forest. It is a poorly known taxon that should be transferred to *Justicia*, as were the other *Siphonoglossa* (Thomas F. Daniel, pers. comm., 1994). The only Sonoran population is located at El Rincón Viejo north of Alamos. Similar purple-flowered acanths that are more widely distributed in tropical deciduous forest are *Justicia caudata* A. Gray and *J. phlebodes* Leonard & A. Gray, while *J. masiaca* T.F. Daniel is locally common in coastal thornscrub near Masiaca. Uses for these species were not mentioned by our consultants. MEDICINE: A tea is made from the branches that can also be used as a wash for rheumatism, according to Francisco Valenzuela. This and other members of the genus are important LIVESTOCK feed.

***Tetramerium abditum*** **(Brandegee) T.F. Daniel** (rama del toro)

This is a red-flowered shrub similar to *Justicia candicans* but with viscid leaves; it is common in tropical deciduous forest, where it constitutes a LIVESTOCK food.

## ACHATOCARPACEAE

***Phaulothamnus spinescens*** **A. Gray** ba'aco "snake weed"

This large shrub is common in both foothills and coastal thornscrub and grows into the transition to tropical deciduous forest as well. The translucent fruits are rumored to be poisonous but are gobbled up by birds, especially *cenzontle* (mockingbird; *Mimus polyglottos,* Mimidae), of whose lengthy and varied songs the Mayos are particularly fond. LIVESTOCK: The leaves are heartily eaten by goats. ARTIFACTS: Vicente Tajia used a 20 cm length of a branch sharpened at each end and pressed into the tip of an agave stalk to collect pitahaya fruits.

## AGAVACEAE

***Agave aktites*** **Gentry** huítbori (lechuguilla)

We did not find specimens of this medium-sized agave, whose habitat is confined to stabilized coastal dunes, until Vicente Tajia located them for us. Gentry found huítbori to be common along the coast of the Mayo region and noted that it was widely collected and roasted. His collection was from Yavaros, as is ours. It was not reported in collections between the time Gentry collected it in the 1950s and the specimen found by Friedman near Agiabampo in 1996. The plant has narrow leaves, much shorter than those of *A. vivipara.* The leaves are too short for the plant to be a source of usable ixtle (agave cord). FOOD: It pups liberally, which is fortunate, since its roasted starch is considered by Mayos as the finest of all agaves. A maya (roasting pit) is dug, and the procedure for roasting (tatemando) is the same as that for *A. vivipara.* Its numbers appear to have declined substantially, probably a reflection of heavy local consumption.

136 / *Agavaceae*

*Baogoa* (*mezcal*; *Agave* cf. *colorata*) from Sibiricahui. Note the pocketknife on the leaf in the right foreground.

### *Agave* cf. *colorata* Gentry      baogoa (mezcal)

This wide-leafed agave with jagged spines on the leaf margins grows only on the steep slopes of the volcanic mountains in the coastal region. Its leaves tend to be constricted near the base. It has blue gray leaves with whitish blotches and often emerges from apparently solid rock cliffs. It is fairly common on Cerro Sibiricahui, north of Masiaca. Its common name is the same as that of *Ceiba acuminata*. FOOD: According to Juan Félix of Teachive, the heads are delicious when roasted.

### *Agave* cf. *jaiboli* Gentry      aiboli (mezcal)

The aiboli grows in the same habitat as *A. colorata*. It is a larger plant, with longer, narrower leaves and less spectacular spines. The leaves are not constricted at the base. It also has a somewhat greener hue than *A. colorata*. In proper soil it grows to nearly a meter in height and equally wide. FOOD: Vicente Tajia notes that it is not as tasty when roasted as either *A. vivipara* or *A.* cf. *colorata*.

ARTIFACTS: Although ixtle can be made from its leaves, specimens with adequate leaf size are uncommon and are distant from human habitations, where the common long-leafed *A. vivipara* is widely used for fiber.

*Agave vilmoriniana* A. Berger                    aimuri (amole; octopus agave)

Amole is a large agave with broad, flexible leaves to 1 m long that rise weakly from the head and sprawl quickly and grotesquely, making the plants resemble huge spiders. They abound on sheer, shady volcanic rock faces throughout the Mayo region, often growing in great numbers and turning the dark brown vertical surface into a tableau from science fiction. They are known only locally in the land of the Mayos, though they are common in canyon country. On the steep cliffs of Arroyo de la Culebra near Baimena they are found in large numbers. Natives warn that the plant cannot be used to produce mezcal. INDUSTRY: The leaves are pounded and the oozing liquid collected and used as soap. Natives maintain that hair shampooed with amole will not turn gray. Gentry (1982:84) notes that soap made from dead leaves was found in markets along the west coast of Mexico.

*Agave vivipara* L.                    cu'u, juya cu'u (maguey, mezcal)

Juya cu'u is the most common agave in the region. It is roasted, its leaves are made into rope and fiber, it is used as a MEDICINE, and its stalk is used to gather cactus fruits. For a more complete description, see pages 90–95.

## AIZOACEAE

*Trianthema portulacastrum* L.                    mochis (verdolaga; horse-purslane)

This is a sprawling succulent annual with purplish flowers found growing on disturbed, often salty, soils in the driest parts of coastal thornscrub, a typical *mochis* (see *Boerhavia* spp.). In general, the name "mochis" refers to the genus *Boerhavia*. It is a good LIVESTOCK food, and several consultants believe it was eaten by humans long ago.

## AMARANTHACEAE

*Amaranthus palmeri* S. Watson                    hué (bledo, quelite; pigweed)

These fast-growing erect annuals explode from the bare ground, appearing in enormous numbers during las aguas, the summer rains. FOOD: The young plants are gathered while still tender and eagerly consumed by adults, children, and LIVESTOCK alike, all racing to see who can get to them first. Well they might, for they are delectable, tastier than any spinach. As soon as their inflorescence develops, the plants become largely inedible. The dry fruiting stalks become covered with tiny spines that produce allergic reactions in many people. Wild and cultivated amaranths are raised for grain by Guarijíos to the north, but we have not yet recorded such a use among the Mayos of the region.

*Tidestromia lanuginosa* (Nutt.) Standl.              mochis (hierba lanuda)

This prostrate annual is widespread along arroyos, in thornscrub, and in tropical deciduous forest. MEDICINE: A tea brewed from the leaves and drunk is said to alleviate the pain of ant bites. It is also taken to help cure measles in children.

## AMARYLLIDACEAE

*Hymenocallis sonorensis* Standl.            guoy cebolla "coyote onion"
(cebolla de coyote; spider lily)

A semi-aquatic lily with a delicate, showy white flower, it is common along watercourses. It is a good indicator of more or less permanent water. MEDICINE: The tuber is roasted and applied directly to snake or insect bites, according to Don Reyes Baisegua of Sirebampo.

## APOCYNACEAE

*Plumeria rubra* L. f.                           (cascalosúchil; frangipani)

A small, irregularly shaped tree with thick, candlelike branches that explode into exquisite white blossoms in late May, June, and July. Although flowers often appear when the plant is leafless, leaves will appear with early rains. The brilliance of the blossoms contrasts handsomely with the green of the leafage. AESTHETICS: Mayos recognize the tree for its beauty and are said to have planted it near their houses. Inexplicably, we have found no one familiar with a Mayo name for the plant. MEDICINE: Rosario "Chalo" Valenzuela of Baimena reports that the clear sap that drips from a cut branch helps heal persistent sores.

*Stemmadenia tomentosa* Greenm. var. *palmeri*      berraco (huevos del toro)
(Rose & Standl.) Woodson

A small tree with large leaves and milky sap. The paired fruits, the size of walnuts, open in senescence to expose a bright red interior, hence the common name "huevos del toro," or bull's testes. When they are thus in fruit they are rather embarrassingly clinical in their resemblance. MEDICINE: Rosario "Chalo" Valenzuela believes the sap might be used to heal pinkeye but was not certain.

*Vallesia glabra* (Cav.) Link.                                                sitavaro

A common shrub bearing tiny, whitish fruits the size of a pea, it can grow into a quickly spreading tree nearly 6 m tall in the foothills of the Sierra de Alamos. It retains its green leaves through the spring drought, one of the few plants to do so. In Sinaloa it is a shade tree in the yards of many homes. It is one of the best-known plants in the region. See pages 99–100 for a more complete description.

ARECACEAE

*Brahea aculeata* (Brandegee) Moore         ta'aco (palma, palmilla; fan palm)

This is a relatively small palm with trunks 2–9 m tall, with teeth on the leaf petioles. ARTIFACTS: Its leaves, harvested when quite green, are used for making hats and baskets, though they are considered inferior to *Sabal uresana* for that use. The pencas (frond strips) are woven into strong ropes that serve a variety of purposes, remaining flexible even when dried out. CONSTRUCTION: Ta'aco leaves are a standard roofing material for homes, providing unmatched protection from rain and heat, and quiet during heavy rains. Few building materials present a more agreeable appearance than the palm thatching used in many older Mayo homes in the tropical deciduous forest. Roofs thatched with ta'aco are said to last twenty-five years. Unfortunately, it is not found in the lower parts of the region and hence not used there. In Sinaloa upstream from El Fuerte it is still widely used for roofing.

*Sabal uresana* Trel.                         ta'aco (palma, palma real; fan palm)

This ta'aco is rather taller and more majestic than the more common *B. aculeata*. It may reach 15 m in height and has unarmed petioles. It occurs in eastern and west-central Sonora south into Sinaloa (see Felger and Joyal 1999). ARTIFACTS: It is the preferred palm for weaving. Several elderly women in Los Capomos weave hats and baskets from *S. uresana* for their livelihood. They recall that in the old days they gathered the *cogollos* (young fronds) themselves in nearby hills, but now usually purchase them from woodcutters who bring in a few at a time. The closest palms have been destroyed and populations are more distant. Artemisa Palomares, a widow of Los Capomitos, notes that the cogollos must be cut in the spring or the pencas (individual strips) will not weave as well. The greater humidity in the Río Fuerte area makes the pencas more malleable there than farther north in Sonora, where the mestizo and Mountain Pima women must construct special submerged huts (*juquis*) to maintain the proper humidity for weaving. The hats are broad-brimmed and

Doña Artemisa Palomares of Los Capomitos, Sinaloa. She is well known for weaving baskets and hats of *ta'aco* (*palma; Sabal uresana*).

double-woven. They are renowned for keeping the head cool. On extremely hot days wearers often dip the hats in water and enjoy a few moments of air conditioning. Most men now prefer mass-produced commercial cowboy hats, however, for they are more stylish and less likely to identify the wearer as an Indian. The finely woven baskets are in considerable demand in the region. Mayo women from coastal villages covet the baskets from Sinaloa for storing tortillas and trinkets and are an important market for the wares.

### ARISTOLOCHIACEAE

*Aristolochia watsonii* Woot. & Standl.  guasana jibuari "Indian weed"
(hierba del indio; pipevine)

An uncommon prostrate herb with sagittate leaves up to 8 cm long. It is usually found on loose soils near water. Vicente Tajia found it in the Arroyo Masiaca not far from San Antonio and on a moist arroyo bank near Camahuiroa.

MEDICINE: The leaves are boiled and the resultant tea is used to expel worms and cure stomachache. It is also used as a wash for sores. Gentry (1942) reported that the roots are roasted and brewed into a tea for kidney and stomach ailments. *A. quercetorum* Standl. is a similar species in tropical deciduous forest.

### ASCLEPIADACEAE

*Asclepias subulata* Decne.    ma'aso pipi "deer teats" (tetas del venado)

A common white-flowered, shrubby milkweed consisting of leafless stalks growing to around 0.5 m tall in coastal thornscrub. It flourishes on overgrazed soils near Yópori, north of Masiaca. It also ranges well into tropical deciduous forest and into the Sonoran Desert in Arizona. *A. subaphylla* Woodson, endemic to coastal dunes at Huatabampito, is very similar. LIVESTOCK: Vicente Tajia observes that goats will eat it when they are very hungry.

*Marsdenia edulis* S. Watson    mabe (tonchi)

A common liana in the milkweed family, it is frequently seen climbing in trees and large shrubs throughout the region. The dangling pods resemble small mangoes. When they are scratched, milky sap quickly oozes from the cut (see p. 142). FOOD: Children gather the young pods and bring them home, where they are eaten with great relish. If left to ripen, they soon become hard and full of puffy seeds that float away in the wind when the dry pod splits. After the pod dehisces, the dried sections resembling the husks of small coconuts fall to the ground. Domingo Ibarra of Huasaguari demonstrated to us that the vine, when stripped of bark, exudes a milky latex that gives off a sweet flavor when it is gently chewed on. He and his family routinely suck on the stems when walking through the monte. MEDICINE: The roots and stems (which take on trunklike proportions) are boiled in water. The liquid is used to soak compresses that are applied to help reduce varicose veins.

*Matelea altatensis* (Brandegee) Woodson    jo'osi pipi "devil's tit"
(teta del diablo; milkweed vine)

A vine that produces pods up to 10 cm long, tapered at the ends, with harmless, blunt thorns protruding from the sides. Resembling a green Christmas ornament, it is found in coastal thornscrub. LIVESTOCK: The plant is good forage for goats but is not consumed by humans.

*Sarcostemma cynanchoides* Decne.    huichori (milkweed vine)
subsp. *hartwegii* (Vail) R.W. Holm

A common vine with pretty purplish white flowers. It is found throughout the region and well to the north and south. MEDICINE: Vicente Tajia reports

142 / Asphodelaceae

Mabis, fruit of *mabe* (*tonchi; Marsdenia edulis*). The fruits are widely eaten when young and tender. Note the droplets of milky sap.

that the herbage boiled in water produces a wash effective in alleviating scorpion stings. The milky sap can be applied to foot calluses and corns to remove them, he says.

### ASPHODELACEAE

*Aloe barbadensis* Mill. [*A. vera* (L.) L. ex Webb, not *A. vera* Mill.] (sábila; aloe vera)

This introduced yellow-flowered aloe is widely used and praised. It is locally established in coastal thornscrub near Bachoco and can be seen growing naturalized in other areas. It is universally known and used in the region. MEDICINE: The juice is routinely applied to burns, cuts, bruises, and rashes. It is taken by the spoonful for digestive problems as well. CULTURE: Vicente Tajia reports that if the sábila is blessed by a priest and placed on a corner of a house, it will bring good fortune and help ward off diseases.

## ASTERACEAE

*Alvordia congesta*                         cochibachomo "pig seepwillow"
(Rose ex O. Hoffman) B.L. Turner

A bushy composite with willowlike leaves and small yellow flowers, it is locally a dominant in coastal thornscrub, flowering in the fall. LIVESTOCK: It is said to be excellent goat forage.

*Ambrosia ambrosioides* (Cav.) Payne        jiogo (chicura; canyon ragweed)

This ubiquitous weak shrub, one of the most common plants in the region, grows very tall—up to 4 m—and thick in arroyo bottoms and on disturbed soils, sometimes forming virtual jungles. At times the stem resembles the trunk of a small tree. Its presence in large numbers is usually an indicator of severe overgrazing, for cows refuse to eat it. Its large lanceolate leaves are covered with a viscid substance that makes them less than succulent. LIVESTOCK: In spite of its coarse leaves, it is eaten with relish by beasts of burden, while spurned by cattle and goats. Pack animals (burros and mules) will often stop for a brief mouthful and tear off half a bush while plodding toward their destinations. MEDICINE: Francisco Valenzuela reports that a tea is made for women after giving birth to cleanse them and help them return to a normal menstrual cycle, and to coax the stomach into returning to its normal size. It is also used in general as a remedy for menstrual problems. These uses are common throughout northwest Mexico. A novel use was mentioned by Leonarda Estrella of Los Capomos. For people with bladder problems (those who need to urinate frequently at night) a good remedy is to form an ersatz diaper from the large, sticky leaves to wear at night.

*Ambrosia confertiflora* DC.            chíchibo (estafiate; slimleaf bursage)

A common perennial herb of disturbed soils. The plant has grayish, deeply lobed leaves. MEDICINE: Estafiate is widely viewed as a valuable medicinal plant. The roots are made into a tea to cure diarrhea and to expel parasites. Some Mayos report that the entire plant is effective for this purpose. Alejo Bacasegua commonly uses a tea made from the leaves to alleviate the pain of rheumatism. INDUSTRY: In 1997 a Japanese social worker working with weavers from Teachive discovered that estafiate, when mashed and steeped in water, produces a powerful green dye suitable for wool. After her departure, weavers had yet to incorporate the green into their color scheme. CONSTRUCTION: Chíchibo, as well as sitavaro and chicurilla, is widely used in constructing roofs. A thick layer of the green plants is laid on top of the latas (roof slats). Dirt is then packed on top. Properly done, a roof so built will never leak, according to the Mayos.

*Ambrosia cordifolia* (A. Gray) Payne       cau nachuqui "guachaporo of the hills"
                                              (chicurilla; Sonoran bursage)

This tough perennial shrub covers many hundreds of acres of overgrazed, rocky pastures. Prior to the introduction of *Pennisetum ciliare* (buffelgrass), it was an important plant in the succession from milpa (rain-fed cornfield) back to tropical deciduous forest. LIVESTOCK: While it is often viewed as a trash weed and spurned by cattle, Luis Valenzuela of Las Rastras noted that it is good fodder for goats and burros. Mules often seem to prefer it to any other forage, passing up apparently delectable grasses in favor of the tough weed.

*Baccharis salicifolia* (Ruiz & Pav.) Pers.       bachomo (batamote; seepwillow)

The leaves on this common seepwillow resemble those of a willow. It is a large shrub sometimes exceeding 3 m in height and growing in large thickets along watercourses. It gives off a not unpleasant odor common to arroyos. INDUSTRY: Many Mayos say batamote produces a soap that will neutralize body odor (a problem gringos have, according to certain blunt and crass Mayos). A few leaves are commonly slipped into shoes under the feet to eliminate sweaty feet and the resultant foot odor. One of us asked Don Vicente Tajia for some, and he presented us with a handful; the treatment seemed to be effective. ARTIFACTS: Seferino Valencia notes that his ancestors fashioned arrows from the straight branches. They needed little paring or straightening to be effective in hunting.

*Baccharis sarothroides* A. Gray       (hierba de pasmo, romerillo; desert broom)

A common plant of the drier parts of the region, it proliferates on disturbed soils, especially on roadsides and near and in dry washes. MEDICINE: Amelia Verdugo Rojo of Basiroa reported that the branches and leaves are boiled, and the tea is applied to sores or itches.

*Bebbia juncea* (Benth.) Greene       mo'oso (sweetbush)

A common low shrub with numerous bare branches and terminal yellow flowers. It is widely distributed in drier thornscrub. MEDICINE: A tea made from the branches is used as a gargle to alleviate the symptoms of tonsillitis.

*Brickellia brandegei* B.L. Rob.       juya chibu (rama amarga; brickel-bush)

This shrub grows on exposed rock faces and rocky hillsides in tropical deciduous forests. It is excellent forage for LIVESTOCK, according to Francisco Valenzuela.

*Brickellia coulteri* A. Gray        bacot tami "snake's teeth" (diente de culebra; brickel-bush)

A common subshrub, the brickel-bush of the Chihuahuan and Sonoran Deserts. It is well known to Mayos, who are generally at a loss to think what it is used for. LIVESTOCK: We noticed that goats eat it, and that is important in the frequent droughts that plague the region.

*Encelia farinosa* A. Gray        choyoguo "brea bush" (rama blanca, incienso; brittlebush)

Extremely common in the Sonoran Desert, this shrub has limited distribution in coastal thornscrub as far south as northern Sinaloa. MEDICINE: Alejo Bacasegua reports that a tea made from the leaves is commonly used in Camahuiroa to relieve rheumatism.

*Encelia halimifolia* Cav.        choyoguo "brea bush"

This subshrub, common near the coast, has yellow ray flowers and distinctively russet-colored disk flowers. Vicente Tajia knew of it and thought it had a use but could not recall what it was.

*Eupatorium quadrangulare* DC.        cocolmeca (tepozana)

This large perennial herb develops a straight stalk growing up to 3 m tall, assuming treelike proportions. It grows in moist, shady habitats in tropical deciduous forest such as the Arroyo Las Bebelamas north of El Rincón near Alamos and in Arroyo de la Culebra near Baimena. The tough stalk is square in cross section, hence the scientific name. It produces a dense display of white disk flowers that dry and remain attached to the plant for many months after it has died back. MEDICINE: Rosario "Chalo" Valenzuela of Baimena reports that he brews the flowers into a tea to expel amoebas. Some older people applied the leaves directly to the forehead to alleviate pulsing headaches.

*Helenium laciniatum* A. Gray        júptera "for sniffing" (rocía; sneeze weed)

A prolific annual with bright yellow blossoms appearing near the southern coast after las aguas. The basis for the Mayo name was a mystery to us until we discovered the English name for the plant. When we asked Vicente Tajia about the Mayo name, he demonstrated by crushing a dried flower. The dust it gave off produced prodigious sneezing. MEDICINE: The leaves, flowers, and stems are boiled with salt. The liquid is applied to rashes and allergic itching.

## *Hymenoclea monogyra* Torrey & A. Gray

jeco (hierba de pasmo, romerillo; cheesebush)

This common shrub with conifer-like leaves grows up to 3 m tall, achieving nearly tree size. It abounds in arroyos, where it serves the valuable function of retarding the velocity of flash floods. Large shrubs typically lean in the flow direction of the floods. The bush survives because it is unpalatable to livestock. ARTIFACTS: Vicente Tajia reports that the long branches are cut and stripped of bark and made into a mat called a *zarzo*. This is suspended from the ceiling of houses where *panelas* (chunks of pressed cottage cheese) are placed to cure, safe from animal thieves. MEDICINE: Francisco Valenzuela reported that a wash made from leaves is used to bathe mange on animals or itchy skin on humans. Alejo Bacasegua swears by it as a remedy for colds, sores, fever, and the hot and cold discomfort experienced by women.

## *Machaeranthera tagetina* Greene

mo'oso

This pretty purple aster can be an annual or an herbaceous perennial growing to nearly a meter high in thornscrub and tropical deciduous forest. It thrives on disturbed roadsides and in overgrazed bottomlands and waste places, its apparent bad taste providing some degree of protection from grazers. MEDICINE: Vicente Tajia touts mo'oso as a good remedy for inflamed tonsils. The leaves are brewed into a tea that is gargled.

## *Montanoa rosei* Rose ex B.L. Rob. & Greenm.

batayaqui "dressed as a Yaqui" (batayaqui)

A tall, flexible shrub or small tree about 5 m tall with slender, straight, and smooth stems, growing in extensive, often impenetrable thickets in tropical deciduous forest and moist foothills thornscrub. In January the trees produce a profusion of dense white blooms that permeate the region with a most agreeable perfume. Groves of the plant at this time are abuzz with bees, whose honey is harvested wild or domestic. Batayaqui grows quickly in canyon bottoms and is in abundant supply despite heavy use by natives. Even so, in 1999 natives reported that it was being overharvested. The only tree composite widely recognized in the area, batayaqui is a most versatile, widely used tree. A mountain range in the region and a village are named after the plant. ARTIFACTS: Beds (tarimes, Spanish; *tapestes,* Mayo) are made by lashing lengths of poles together, often in a frame of etcho wood. Such a bed was demonstrated to us at El Rincón and proved remarkably comfortable. The poles each give way to a protuberance of the body, much like a mattress of many springs. The wood is widely used to make backs for chairs, animal cages, and crates for shipping. Arrows are still made from the straight stalks. CONSTRUCTION: The springy

Woven wall of *batayaqui* (*Montanoa rosei*) poles. Such frameworks are often covered with mud or lime plaster. (Photo by T. R. Van Devender)

poles are woven to produce a sturdy wall or roof that may be plastered or not. FOOD: Honey is harvested from the bees attracted to its flowers. MEDICINE: Gum from the trunk is applied to sores and is said to be a healing agent.

*Palafoxia linearis* (Cav.) Lag. var. *linearis*  bahue cuépari (golondrina del mar; Spanish needles)

This low-lying shrub covers many small dunes or tussocks above the high-water mark along the Sea of Cortés. Its pink purple flowers apparently bloom in response to rain and can be found any month. MEDICINE: Juan Félix reports that the flowers and leaves brewed into a strong tea make an effective wash for sores and cuts. This remedy has been used in his family for many years, he says.

*Parthenium hysterophorus* L.  chíchibo (estafiate)

A native annual or weak perennial herb common in disturbed soil and waste places in many areas, it is thought to cause allergies. It bears the same common name as *Ambrosia confertiflora,* another gray herbaceous perennial that it superficially resembles. In Arizona and New Mexico "estafiate" refers to *Artemisia ludoviciana,* an aromatic, gray-leaved herb in mountain habitats.

LIVESTOCK: Chíchibo makes good forage for goats. In Teachive a pile of the herb had been left for goats to eat during the spring drought.

*Pectis coulteri* (Harv. & A. Gray) ex A. Gray    goy sisi "coyote piss" (chinchweed)

An abundant yellow herb that carpets many fields after the summer rains. LIVESTOCK: The fast-growing and drought-tolerant plants supply important forage. They have a strong urinelike odor but are eagerly eaten by cattle and goats.

*Perityle cordifolia* (Rydb.) S.F. Blake    oseu nácata "lion's ear" (rock daisy)

A bright green herb with large, viscid leaves and showy yellow flowers, it is found on volcanic rocks high on the hillsides, especially on Cerro Terúcuchi and Ayajcahui near Masiaca, and in similar rocky regions in thornscrub throughout the region. It flowers in late winter and early spring. MEDICINE: Juan Félix of Teachive swears that the following happened. His grandfather had told him that a tea brewed from the whole plant would put the drinker to sleep. He and a friend were curious but reluctant to try it themselves, so they boiled some leaves into a tea and poured it in a dog's water dish. The animal drank the liquid and directly fell into a deep slumber from which it could barely be aroused.

*Porophyllum gracile* Benth.    cuchu pusi "fish eye" (hierba del venado)

This aromatic shrub occasionally reaches more than a meter in height in the region. It often can be detected from some distance by its pungent though not unpleasant odor as the crushed herbage releases fumes redolent with terpenes. It produces many white to purplish flowers in response to rainfall. Its Spanish name may refer to its use as a spice to cook venison, as was reported for a related species (*Porophyllum macrocephalum* DC.) in Curea, just north of the Mayo region. MEDICINE: Cuchu pusi is renowned for its uses as a remedy, so much so that it is known to natives in the Mayo region who live outside the coastal plain where it is most frequently found. Although it is common throughout the Sonoran Desert well into Arizona, it is uncommon in foothills thornscrub and tropical deciduous forest. Near Las Bocas it is nearly universally used. Paulino Buichílame Josaino of Las Bocas says that cuchu pusi is brewed into a tea to alleviate menstrual cramps and muscle cramps. For women who are unable to conceive (*tienen matriz fría*), the hot tea will make them "hot" and enable conception. On the other hand, for women who are pregnant, drinking the tea may cause miscarriage. Vicente Tajia also reports that the herb is a good remedy in general. Felger and Moser (1985:287) quote a Seri woman who said of *P. gracile,* "The thing it is not good for does not exist."

*Pseudognaphalium leucocephalum* (A. Gray) Anderb. [*Gnaphalium leucocephalum* A. Gray]    talcampacate (gordolobo; white cudweed)

The gordolobo is a common annual up to a meter tall with cottony white flower heads. It is frequently found growing in disturbed soils and along overgrazed arroyos, where it is collected by natives. We list the name as Mayo, but it is probably of Aztec derivation. MEDICINE: Many Mayos and mestizos alike prescribe a tea made from the flowers to alleviate aches and fever. Turibio Estrella of Los Capomos claims it is also effective on mange and scaly skin when brewed into a wash. Talcampacate is often marketed commercially in Mexico, usually found in bunches in the produce displays. *P. leucocephalum* and *P. canescens* (DC.) Anderb., *P. stramineum* (H.B.K.) Anderb., and *P. viscosum* (H.B.K.) Anderb., recently segregated out of the genus *Gnaphalium*, likely are the aromatic gordolobos used medicinally rather than the smaller species relegated to *Gamochaeta*. The commercial gordolobo widely distributed in Mexico likely represents several species of *Pseudognaphalium*.

*Tagetes filifolia* Lag.    (yerbanís)

A tiny annual with narrow leaves and a pleasant licorice smell. Our specimen was flowering in December, quite late for the species, and was collected very low, at 340 m. MEDICINE: Francisco Valenzuela reports it is brewed into a tea that is good for the blood; that is, good for what ails you. The larger *Tagetes lucida* Cav., the common *yerbanis* of the Sierra Madre near Yécora, is widely used as a tea by mountain people and city dwellers alike.

*Xylothamia diffusa* (Benth.) G.L. Nesom    jeco

INDUSTRY: A scraggly composite with yellow flowers found in sandy soils near the coast, it is used to poison tilapia (an introduced freshwater fish), according to Balbaneda "Nelo" López of Yavaros.

## BIGNONIACEAE

*Crescentia alata* H.B.K.    choca'ari (ayal)

A curious-looking tree to 10 m tall, with compact, spindly branches that appear to be mostly naked. The leaves and gourdlike fruits develop directly from the trunk and main branches, producing a most odd appearance. This tree is widespread in the New World tropics. Don Patricio Estrella of Los Capomos and Marcos Valenzuela of Baimena report that the trees that grow near the mouth of Arroyo de la Culebra near Baimena are native. Elsewhere to the north it appears to have been introduced and possibly naturalized. For example, specimens growing near Etchojoa and near Güirocoba, flourishing on the ruins of an abandoned hacienda, are surely naturalized cultivars. Farther south they

## 150 / Bignoniaceae

*Sonajas* (rattles) made of *ayales* (fruits of *Crescentia alata*), Los Capomos. (Photo by T. R. Van Devender)

become common around human habitation. CULTURE: The gourds are made into *ayales* or *sonajas* (rattles) used by deer dancers in Mayo fiestas. MEDICINE: For pneumonia, bruises, sprains, and lack of energy, wine or mezcal is poured into the gourds and allowed to sit for a few minutes before being drunk. We are unaware of the success rate of this fine remedy but endorse it on general principle.

**Tabebuia chrysantha (Jacq.) Nichols**     to'bo saguali (amapa amarilla)

A strong-boled tree similar in configuration and leaf structure to the amapa. The trees are scarcer and appear to be generally smaller than *T. impetiginosa*. The tree is nowhere common, limited to scattered individuals on slopes in the tropical deciduous forest, usually distant from human habitation, suggesting a history of heavy human exploitation. This latter is unfortunate indeed, for in March and April the tree bursts into cascades of yellow blooms whose brilliance is unmatched in the world of flowers. The ethereal glow can be seen for many miles, a vibrant transcendence on the mojino of the leafless canopy. To stand beneath an amapa amarilla in full flower is to be transported to an unworldly realm of splendor. This sensational flowering has not benefited the tree in modern times, for it gives away the tree's presence to woodcutters. In South America it is commonly referred to as *el beso de la muerte* (the kiss of death), a reference to its suicidal blooming. When not in flower it can be distinguished from the pink amapa by its lighter bark and rather rougher leaves. CONSTRUCTION: Mayos say that the lumber is excellent for vigas and horcones, just as the wood of amapa is (see below), except that it is reputed to be even more durable. This is difficult to verify, however, for it seems doubtful that the tree ever occurred in sufficiently great numbers to constitute a significant source of lumber. Still, all express an admiration for the tree and identify locations where it grows.

*Tabebuia impetiginosa* (Mart.) Standl.                                    to'bo (amapa)

A strong tree up to 12 m tall (shorter in thornscrub, taller in moist canyons in the tropical deciduous forest) with a spreading crown. The straight trunk is covered with dark, rough bark. In winter and early spring the bare branches explode into a riot of pink blooms, enhancing the forest in all directions. The profuse and colorful blossoms, a blessing to man, have been a curse to the species, marking the trees as easy targets for woodcutters and at one point pushing the species to near extinction in some areas. It grows south through Mexico to Argentina. The Mexican government has wisely chosen to protect the tree, and midsized individuals have appeared once again. With the cessation of underground mining, it has again become common. The wet winter-spring of 1998 brought on a particularly showy display of the flowers. Near Nahuibampo the hillsides were brushed with bright pink in all directions. One large (rather unexpected) individual was in full bloom near the Sea of Cortés in Arroyo Camahuiroa in early December 1997 and was visible from a distance of 2 km. In contrast, hardly any flowered in the dry winter-spring of 1999. Thus amapas are known from near the coast to the oak woodland. Vicente Tajia of Teachive planted an amapa in his backyard. After seven years, it bloomed for the first time. The soil in his yard, high in clay, is apparently not to the amapa's liking. CONSTRUCTION: Untold millions of these trees were cut for roof beams for buildings and for mine timbers propping up the hundreds of miles of tunnels from which the silver of Alamos was extracted. The stout, heavy bole of dense wood is the best stock for posts and beams for homes, resisting rot and infestation by boring insects. Most older homes in its habitat feature vigas and horcones (forked posts that support vigas) of amapa. Beams from abandoned houses, often more than a hundred years old, are eagerly collected for use in new buildings. INDUSTRY: Ashes of the wood are used like lime to soften corn for grinding into flour.

*Tecoma stans* (L.) H.B.K.                 (gloria, lluvia de oro; yellow trumpet bush)
var. *angustata* Rehd.

Mostly a large shrub or small tree with numerous brilliant yellow tubular flowers that contrast most agreeably with its dense green foliage of narrow green leaves. In the lower oak woodlands it occasionally grows to 5 m. The native *T. stans* var. *angustata* Rehd. is widespread across the southwestern United States and northern Mexico. It is uncommon in its wild state in the Mayo region, and we found only one wild specimen during our fieldwork. We have been unable to determine its Mayo name, which suggests that it is not well known in the wild. MEDICINE: The leaves brewed into a tea or directly applied to the head are widely believed to have remarkable medicinal powers, especially for lowering blood pressure and alleviating heart problems.

*Tecoma stans* (L) H.B.K. var. *stans*      (gloria, lluvia de oro; yellow trumpet bush)

In contrast with *T. stans* var. *angustata*, *T. stans* var. *stans* is a tropical shrub, often a tree, with broad leaves. It occurs natively from Baja California Sur to Argentina. AESTHETICS: It is a popular ornamental throughout tropical Mexico, common in all Mayo villages, where it is also valued for shade and as MEDICINE for lowering blood pressure and treating heart problems. Nearly every house has at least one gloria plant. Our Mayo consultants only casually differentiate between the two varieties.

## BIXACEAE

*Amoreuxia gonzalezii* Sprague & Riley                      saya, saya mome
*A. palmatifida* Moç. & Sessé

This important little plant is an herbaceous perennial of thornscrub that emerges each year from a tuberous root the size of a carrot, sometimes larger. FOOD: The roots are eaten raw or cooked, as are the seeds. The taxonomic differences between the two species are subtle, requiring laboratory instruments for determination. For a more complete description, see pages 103–4.

*Cochlospermum vitifolium* (Willd.) Spreng.    ciánori "yellow tree" (palo barril, palo barriga; buttercup tree)

A most imposing tree with a thick trunk covered with silvery gray bark that resembles elephant skin. The smooth, symmetrical bole rises straight and tall, frequently in excess of 10 m. In late winter, when the tree is leafless, brilliant yellow cup-shaped flowers up to 10 cm in diameter explode from the tips of the branches. Seen from a distance, they resemble yellow golden ornaments. In late fall the broad sycamore-like leaves turn shades of red, yellow, and gold reminiscent of more northern autumns. Palo barril is common in tropical deciduous forests and disturbed secondary growth to the south. Reports describe it as reaching its maximum height in the northern limits of its distribution. Specimens shown to us by Francisco Valenzuela in Arroyo de las Bebelamas near El Rincón are the only records from the Río Mayo basin, but it becomes abundant farther south. MEDICINE: The wood is carved into a basin and filled with water, which is allowed to stand overnight. Women drink the water for varicose veins. ARTIFACTS: Among Sinaloan Mayos the easily worked wood is used to make musical instruments, including harps and violins.

## BOMBACACEAE

*Ceiba acuminata* (S. Watson) Rose      baogua "it is cooking" (pochote; kapok)

The pochote grows to more than 15 m tall, its branches often developing in linear zigzags, and its large palmately compound leaves illuminate the landscape with a light green, turning yellow in early fall. Young trees become covered with large laminated thorns that are innocuous and become pyramidal, softening with age and lending an air of strangeness to the great trees; they often render the young trees unclimbable (see p. 154). The pendant fruits usually develop in spring, bursting open in fall in the form of softball-sized puffs of cotton, conspicuous from great distances on the leafless trees across the leafless monte. MEDICINE: A vessel of the wood is filled with water. Women drink the liquid as a cure for varicose veins. FOOD: In times of scarcity the roots are eaten with great enthusiasm by native people. The seeds, eaten in Sinaloa, have a peanut-like flavor and consistency. CONSTRUCTION: In Los Capomos the wood is used for vigas and pronounced solid as long as it is protected from water. ARTIFACTS: The wood is used for making cots. The puffy fibers are used to stuff pillows and as cushioning in cribs. Recently, the women of Aduana, near Alamos, have begun using the thorny portions of the bark as stock for carving miniature landscapes of the type introduced many decades ago in southern Mexico and still common there. They also make necklaces from the dry black seeds.

*Pseudobombax palmeri* (S. Watson) Dugand      (cuajilote)

A gnarly, contorted tree seldom more than 7 m tall with ruby wine-colored bark peeling into concave sections that produce an attractive mottled red coloring. It flowers when leafless in April. The buds swell on the tips of branches and then burst open, presenting pollinators with large, sensuous upright blooms

Young *pochote* (*Ceiba acuminata*) near Yocogigua. Note the many thorns, which are harmless. The pods burst open when dry and are used for stuffing pillows and quilts.

with a multitude of stamens that spill lasciviously over the edge of the brilliant white blossom. The flowers open at night and are visited immediately by nectar-eating bats and moths. The large leaves turn gold in the fall, retaining their vigor and presenting a sharp outline of color until they fall to the ground, still strong and entire. Our Sonoran Mayo consultants were familiar with the tree, even though they live outside its range. It is common and well known among Sinaloan Mayos. MEDICINE: The wood is carved into a bowl into which water is poured. Women drink it to cure varicose veins. FOOD: The tender fruits are eaten.

### BORAGINACEAE
*Cordia curassavica* (Jacq.) **Roem. & Schult.**                                    jusa'iro

A small, scraggly bush with small, nondescript greenish white flowers, it apparently reaches its northern limit in coastal thornscrub just north of the Mayo

*Boraginaceae* / 155

*Cuajilote (Pseudobombax palmeri)* in dry season near Güirocoba. The tree flowers in April. The Mayos of Los Capomos eat the fruits.

region. ARTIFACTS: Felipe Yocupicio says the stems bundled together make good brooms.

*Cordia parvifolia* DC.          huo'ótobo (vara prieta)

A scraggly bush becoming quite large. Following substantial rains it explodes with leaves and delicate white blossoms 3–4 cm in diameter resembling toilet paper. Branches are used for cleaning wool and in loom technology. Straight sticks are widely used in fiesta ceremonies. See pages 79–82 for a more complete description.

*Cordia sonorae* Rose          pómajo (palo de asta)

A slender, clean-barked tree up to 10 m tall with a smooth, silvery gray trunk, leafing and flowering toward the tips of the branches. It flowers a delicate white in late winter and early spring, the individual blossoms visible from afar as white points in the forest. The flowers turn a distinctive coffee color and persist on the trees for days. The Spanish name means "flagpole," suggesting the rapid, straight growth of the young trees. The tree is common in hilly thornscrub and tropical deciduous forest. It grows well on disturbed soils. CONSTRUCTION: The hard, vigorous wood is widely used for posts, beams, and vigas in houses. ARTIFACTS: The wood is widely used to make handles for tools and implements, for which it is chosen because it is tough and resilient. Few Mayo homes are without an implement with a handle of pómajo. Scraped and polished it provides a smooth surface, hard and free of splinters.

*Heliotropium angiospermum* Murray — laven co'oba (cabeza de violín; violin head)

A perennial herb growing up to 1 m tall, with tiny white flowers. Francisco Valenzuela believes it had some use but could not recall it.

*Heliotropium curassavicum* L. — (mala mujer; alkali heliotrope)

We have not collected this plant, but Bañuelos (1999) reports that it is collected near the coast in the municipio of Huatabampo. MEDICINE: According to Bañuelos, the plant is boiled and used to treat mange or scabies.

*Tournefortia hartwegiana* Steudal — tatachinole "hot permanent [hairdo]"

A small shrub that produces a white, curling inflorescence that in turn develops into black berries the size of BBs. The flowers are quite noticeable and strange. The plant is uncommon outside tropical deciduous forest, but within the forest, when it flowers from December to April, it is quite common. In lower tropical deciduous forest it is abundant on disturbed soils. MEDICINE: The root is used for a wash to soothe the pains of rheumatism. A tea is made from the raw root as well, said to be good for kidneys. Vicente Tajia touts the application of a leaf on a sore that will not heal. We tried it and concurred. FOOD: The ripe berries are eaten, although we found them unappealing.

## BRASSICACEAE

*Descurainia pinnata* (Walt.) Britt. subsp. *halictorum* (Cockerell) Detling — (pamita; tansy mustard)

A common herb with lacy leaves and tiny yellow flowers, it proliferates after rains. FOOD: The seeds are mixed into water to make a refreshing drink. When tender, the greens are eaten as a vegetable. The stems are cooked as well. MEDICINE: According to Francisco Valenzuela, a tea made from the leaves is taken to alleviate dysentery.

*Dryopetalon runcinatum* A. Gray　　　　　　　　　　　ma'aca (mostaza)

This annual appears in moist soils after soaking winter rains. The lacy white flowers turn lavender with age. FOOD: The young plants are eaten uncooked. Lus Valenzuela of Las Rastras pronounced them to be tasty. We tried them and agreed. They taste as fresh as the finest watercress.

## BROMELIACEAE

*Bromelia alsodes* H. St. John　　　　　chó'ocora "very salty" (aguama)

A pineapple-like terrestrial bromeliad often growing in large, impenetrable clusters. The leaves are armed with sharp barblike spines. The flowers and fruit are brilliant red. Although the plants are found in a few scattered thornscrub and drier tropical deciduous forest locations near Alamos, Mayos of the Río Mayo appear to be unfamiliar with them. FOOD: According to Patricio Estrella of Los Capomos, the fruits are regularly eaten even though they are a bit sour.

*Hechtia montana* Brandegee　　　　huítbori (aguamita, mescalito; false agave)

An agave-like succulent reaching a height of 0.5 m that sends out a flowering stalk as tall as 4 m. It grows at somewhat higher elevations than most Sonoran Mayos frequent. Francisco Valenzuela was familiar with it and thought it had a use but could not remember. FOOD: Guarijíos eat the heart of the plant (Yetman and Felger, unpublished data), so Francisco's memory is probably correct. Vicente Tajia had heard that people from the south eat the hearts. LIVESTOCK: Vicente Tajia noted that goats do well on the plant.

*Tillandsia exserta* Fern.

See *T. recurvata*, below, from which it is not distinguished by Mayos even though *T. exserta* is uniformly larger.

*Tillandsia recurvata* L.　　　　huírivis cu'u (mescalito de huitlacoche; ballmoss)

This epiphytic plant is abundant in locations with high humidity. It grows in a more or less spherical habit, seldom more than 20 cm across, sending out slender shoots of purple flowers in late summer. On the coastal plain, within a few kilometers of the water, it grows in great numbers, hitching on to many different plants, sometimes obscuring their natural form. *Ziziphus amole* is a particularly gracious host. In spite of its abundance, Mayos identified its uses only as a food for LIVESTOCK, for which it is excellent. During times of drought (nearly every spring) cowboys will pull the *Tillandsia* from branches and feed it to cows or goats. Huírivis is also the name of the curve-billed thrasher (*huitlacoche*; *Toxostoma curvirostre*, Mimidae).

## BURSERACEAE

***Bursera fagaroides*** **(H.B.K.) Engl.**     to'oro, to'oro sahuali (torote, torote de vaca)
**var. *elongata* McVaugh & Rzed.**

A prominent tree during the leafless season, growing to 8 m tall. The light-colored bark ranges from sandy to light green and is usually colored with layers of wispy exfoliations that whistle in the wind. When it is leafless during the spring drought, the trunks stand out in the monte as yellow white. When leafless it is easily confused with *Jatropha cordata* but has a more angular trunk and grows into a far more spreading habit. Birds love the dried fruits, which cling to the apparently dead branches for many months. ARTIFACTS: The wood is used for fence posts and living fences. Some individuals carve it into dishes and utensils, but according to others, it is inferior for this purpose. INDUSTRY: At Teachive, the dead wood is thrown into the kiln as firewood for baking bricks. At Los Capomos it is the favored wood for firing clay pots, according to Leonarda Estrella, producing just the right temperature for firing. MEDICINE: At Nahuibampo the sap is boiled down until thick and applied to sprains.

***Bursera grandifolia*** **(Schldtl.) Engl.**     to'oro mulato (palo mulato)

Palo mulato is a handsome spreading tree much esteemed by natives, up to 10 m tall with unpredictable branching. Its bark is dark bluish green to silvery gray to red, with tan exfoliations that give it an exotic appearance. It is a good indicator of well-developed tropical deciduous forest, but occurs as a small tree in thornscrub on Cerro Sibiricahui. Except for the rare (in Sonora) *B. simaruba*, *B. grandifolia* is the most tropical Sonoran *Bursera*. The meaning of the Mayo *mulato* is obscure. One source asserted that it means "tea maker," but others have reported the name merely to be a very old Cáhita word with no obvious meaning, certainly not that of "mulato" in English and Spanish. ARTIFACTS: The white wood is used to carve masks, utensils, and bowls. FOOD: The thick, reddish inner bark is relished as a tea and as a coffee substitute. In Nahuibampo it was the only beverage available to the Mayos, who had no money to purchase coffee. MEDICINE: Tea from the bark of palo mulato is widely used as a tonic, said to restore energy by thickening the blood. The redder the bark of the tree, the more effective the remedy. Certain individual trees have heavily scarred trunks, indicating they are known to have especially potent bark. Many natives store a package of the dried bark in their homes, and so do we.

***Bursera lancifolia*** **(Schldtl.) Engl.**     to'oro chutama "resinous to'oro";
    to'oro síquiri "red to'oro" (torote copal)

A large, spreading tree up to 10 m tall distinguished from *B. fagaroides* by its larger, lance-shaped leaflets usually 6–8 cm long. Its gray green exfoliating bark

A leafless *palo mulato* (*Bursera grandifolia*). The bark of this tree, which is found only in tropical deciduous forest, is widely used as a coffee substitute. (Photo by T. R. Van Devender)

exudes a hint of incense when freshly cut, hence its Spanish name. *B. fagaroides*, from which it is difficult to distinguish when leafless, has no noticeable aroma. During the dry season the bark of torote copal usually turns grayish. Its fruits are not persistent, unlike those of *B. fagaroides*. ARTIFACTS: The sap is used as a resin for violin bows. MEDICINE: The sap is chewed for toothache and coughs. The fragrant bark is steeped in hot water to make a tea also used to treat coughs. Smoke from the wood is said to cure headaches.

*Bursera laxiflora* S. Watson     to'oro chucuri "black to'oro" (**torote prieto**)

In tropical deciduous forest, torote prieto is a large spreading tree with smooth, nonexfoliating dark gray bark and tiny lacy leaves. The ends of the branches are elongated and bear a reddish hue which renders the tree readily distinguishable when leafless in the dry season. The bole often becomes quite thick—up to a meter—with age. *B. laxiflora* grows as a large shrub in the Sonoran Desert and in thornscrub. Only in tropical deciduous forest does it reach tree size. It is also the only *Bursera* in the region that leafs out opportunistically following substantial rain. This perhaps explains why it thrives (though in much smaller habit) in the highly variable climate of the Sonoran Desert to the north. Opportunistic leafing also occurs in the genera *Fouquieria* and *Jatropha*. ARTIFACTS: The fragrant wood is carved into masks, spoons, and trays. The late Benigno Acosta of Teachive pronounced it the best for this purpose, rivaled only by *B. grandifolia*, which is unavailable there. The plants' natural incense permeated the workshop area of his house when he carved the wood. INDUSTRY: At Teachive branches of the tree are placed at the bottom of a fire pit when agaves are roasted. They are said to impart a pleasant aroma and add sweetness to its flavor. MEDICINE: Throughout the region the bark is brewed into a tea and drunk to cure coughs.

*Bursera microphylla* A. Gray                                    to'oro (torote)

This small *Bursera* is very rare in the Mayo region, confined to steep hills on the dry coast. It is common in the Sonoran Desert to the north. Seri Indians have multiple uses for it (Felger and Moser 1985:241–44), but we found no Mayos acquainted with it. Vicente Tajia observed that it was a torote that had some uses, but said it was more important for the Yaquis than for the Mayos.

*Bursera penicillata* ( Sessé & Moç.        to'oro chutama (torote, torote acensio
ex DC.) Engl.                               [= de incienso], torote puntagruesa)

The blunt, rough ends of the branches of this *Bursera* help distinguish it in the dry season. The tree's serrate leaves, 5–8 cm long, and its bark give forth an aniselike fragrance detectable from many yards away. We have often had the experience of noting the fragrant smell before seeing the tree. The tree's bark is a nonexfoliating light gray. *B. penicillata* is said to be the largest *Bursera* in the region, with specimens often exceeding 15 m. It can also be distinguished by the thickness of its foliage compared to the rather sparse leafage of other Sonoran species. In late September or October the senescent leaves turn bright yellow and illuminate the hillsides of the tropical deciduous forest. In some parts of the forest *B. penicillata* is the most common torote. ARTIFACTS: The wood is excellent for carving into bowls and utensils. FOOD: A cut in the bark soon begins to ooze aromatic gummy sap that is sometimes chewed like gum. CULTURE: Elsewhere in the region, the sap is collected and burned as incense.

*Bursera simaruba* (L.) Sarg.     to'oro mulato (gumbo limbo)

Mayos do not distinguish this tree from *B. grandifolia*. *B. simaruba* grows only in the semideciduous forests of moist, shady box canyons that drain tropical deciduous forest slopes, in the company of such evergreen trees as *Aphananthe monoica* (guasimilla; Celtidaceae) and *Drypetes gentryi* (palo verde; Euphorbiaceae), hence it is unknown in Mayo country except to natives of Baimena. It is well known throughout dry tropical forests of the New World. It grows into a very tall, stately tree with dark green exfoliating bark at times showing red beneath, so much so that the bark appears to be blood red. When the outer bark is cut, the inner bark is white or pinkish, unlike that of *B. grandifolia*, which is blood red. It also retains its leaves throughout most of the year, dropping them only when the water source dries up. The specimen we examined was 20 m tall. Baimena natives report that it grows much taller in canyons upstream, but we dared not visit because of the hazards posed by cultivators of poppies. The uses of *B. simaruba* are said to be identical to those of *B. grandifolia*.

*Bursera stenophylla* Sprague & Riley     to'oro, to'oro sajo (torote copal)

The small, dainty leaflets formed into lacy leaves help distinguish this large *Bursera* from other species in the genus. Its bark does not exfoliate. Individual trees grow to great, sprawling size, the trunks sometimes reaching 75 cm in diameter. It is uncommon and was unknown to most of our Mayo consultants. ARTIFACTS: The sap is collected and burned as incense. The wood is carved into bowls and utensils. MEDICINE: The fragrant sap is chewed as a gum and is said to fight off colds.

CACTACEAE

*Acanthocereus occidentalis* Britt. & Rose     tasajo

This remarkable cactus (possibly introduced) is strictly epiphytic, though it superficially resembles *Lophocereus schottii*. It produces straight, three-ribbed arms more than a meter long with shiny, smooth skin. It grows only in trees, requiring a substantial crotch to support its weight. Near Baimena one grows 6 m from the ground in the crotch of a mezquite. The showy white flowers open only at night, suggesting bat pollination. FOOD: The fruits become large and sweet, according to Rosario "Chalo" Valenzuela, who says that harvesting is difficult but worth the effort.

*Carnegiea gigantea* (Engelm. ex Emory) Britt. & Rose     saguo (sahuaro)

This great cactus is found in large numbers on the basaltic slopes of Mesa Masiaca, only a few kilometers from Teachive. This is the southernmost popula-

*Sahuaro* (*saguo* in Mayo; *Carnegiea gigantea*) on Mesa Masiaca, where the southernmost population of these great cacti flourishes.

tion of the species. Other populations are found on basalts on Cerro Prieto east of Navojoa, and on Cerro Yópari, some 10 km north of Teachive. In spite of the nearness of the population to several villages and the importance of the sahuaro in the lives of other indigenous groups, our consultants said they knew of no important purpose for the plant and expressed no particular interest in it. FOOD: The fruits ripen in July and are recognized as edible; however, given the hot, rocky habitat and inconvenient access, as well as the fact that natives pronounce the fruit to be inferior to those of other local cacti, they deem it hardly worth collecting. The ribs, used widely by native peoples to the north, are weaker and shorter-lived than those of etcho or pitahaya.

***Ferocactus herrerae* J.G. Ortega**  ónore (biznaga; barrel cactus)

This barrel cactus grows to nearly 2 m in height and half a meter in diameter. It is confined to the thornscrub of the coastal plain. Its ribs, covered with stout spines, exhibit a most agreeable spiral growth form, reminiscent of a barber's pole. The egg-sized yellow fruits cluster thickly at the apex. FOOD: When the slightly sour fruits are pulled from the plant, the tiny black seeds spill out the bottom. According to Vicente Tajia, when in the past the supply of white corn was exhausted, old people gathered *onobachia* (the seeds) and made tortillas from them, preferring the cactus-seed tortillas to those made from yellow corn, which they viewed as more suitable for pigs. They also hack off the waxy epidermis and spines with a machete and roast the inside flesh with sugar, pronouncing the cooked starch to be delicious. We verified their pronouncements. LIVESTOCK: Unfortunately, these often large cacti, though not rare, are never common and grow slowly. No one will ever be saved by drinking the water contained therein. Mayos seem to be inclined to respect their right to live. Cows and goats, however, have no such scruples. During prolonged dry spells, cattle have learned to bowl large specimens over to gain access to the fruits. Goats simultaneously nibble at the roots. Although the specimens die slowly, the uprooting is usually fatal. Throughout coastal thornscrub entire populations—hundreds upon hundreds of plants—of this noble cactus have been knocked over and uprooted, endangering the coastal population of a valuable plant. MEDICINE: Florentino "Queriño" Piña of Las Bocas eats the raw flesh of ónore to control high blood pressure. He collects a medium-sized specimen weighing perhaps forty kilograms (many of which have been knocked over by cows), leaves it in his yard, and from time to time extracts a plug from the thick cactus as though he were testing a watermelon.

### *Lophocereus schottii* (Engelm.) Britt. & Rose var. *tenuis* G. Lindsay          musue (sinita)

A columnar cactus with many upright stems originating from a common base usually a maximum of 2.5 m in height. A large specimen near Bachomojaqui reaches more than 5 m in height, however. With age the upper needles turn white, giving the top portion a hoary appearance, hence the English name "old man cactus." Specimens grow as far north as Organ Pipe Cactus National Monument in Arizona. *L. schottii* var. *tenuis* differs from the typical Sonoran Desert varieties in having more slender stems with more ribs. The white flowers open at night and are pollinated by a moth that occurs only on it (T. Fleming, pers. comm., 1998). FOOD: The small, round, tasty fruits are difficult to collect due to competition with birds and insects. MEDICINE: The flesh and skin of the cactus are boiled into a tea that is drunk to relieve the pain of arthritis. Alejo Bacasegua promotes the flesh as the source of a tea excellent for bathing the body. The material must come from a specimen with five ribs, he warns, not from one with six or seven ribs. Vicente Tajia pronounces the five-ribbed variety, which is uncommon but abundant in certain localities, to have cancer-curing properties as well. Some plants have both five- and six-ribbed arms.

*Mammillaria grahamii* Engelm.   chicul ónore "rat barrel cactus"; to'oro bichu
*M. majatlanensis* K. Schum.   (biznaguita, pitahayita; fishhook pincushion)

These are small cacti with cylindrical stems, fully armed with fishhooklike spines, seldom growing more than 15 cm tall. The cacti are quite common but variable throughout the region, appearing in clumps of several to many individuals. FOOD: The small, elongated red fruits are quite tasty, though hardly substantial, resembling a small jellybean. MEDICINE: For earache, a head is collected and scalded to burn off the numerous spines. The flesh is squeezed until a drop appears. It is dripped directly into the affected ear. Rosario "Chalo" Valenzuela of Baimena says that applying a small section of the flesh to a sore that will not heal speeds the healing process.

*Mammillaria mainiae* K. Brandegee   (fishhook pincushion)

Specimens of this apparently rare cactus, difficult to distinguish from *M. grahamii* (above), were identified near Yocogigua by Braulio López. Its names and prescribed uses are identical with those of *M. grahamii*.

*Mammillaria yaquensis* Craig

This dwarf, multiheaded cactus was found growing abundantly near Tierra y Libertad and identified and described with the same name and uses as *M. grahamii*. It is endemic to coastal thornscrub.

*Opuntia fulgida* Engelm.   choya (choya; chainfruit cholla)

In waste places throughout the coastal plain and elsewhere, this powerfully armed cactus proliferates and is sometimes nearly the only plant found over many square meters of overgrazed land. While the cactus "invades" many village spaces, it is not condemned, for all know it constitutes emergency rations. Farther north in Sonora, the cactus has proliferated enormously in overgrazed pastures, replacing native grasses and even the introduced buffelgrass. The cactus flowers a deep rose color in late spring and summer. With flowers and new growth on the cactus, the plant becomes most attractive. Vicente Tajia says that when a couple without much physical beauty gives birth to a beautiful child, people say, "How is it possible that such an ugly couple could have such a pretty baby?" Others answer, "Well, how is it possible that a plant as ugly as a choya could have such a pretty flower?" FOOD: The fruits are eaten, though with a good deal of caution, for they are covered with potent, tiny spines called sebua (Mayo) or *aguates* (Spanish) (glochids). The husks must be removed, usually with a machete, and are often reserved for pigs. The sectioned sour fruits are eaten with lime juice. MEDICINE: The fruits are often prescribed for stomachaches. Vicente Tajia says that the fruits are heated and eaten to control high blood pressure. Francisco Valenzuela says he and others use the fruits and roots for *mal de orín* (when urine is bloody or urination is difficult). Alejo Bacasegua of Camahuiroa verified this use, and said he also uses the roots to make a tea that is in general good for the kidneys and, he added, looking around to make sure no women were within earshot, "When your *thing* is sick." Elsewhere in Sonora the fruits are a remedy for the common cold. LIVESTOCK: In spite of its intimidating spines, *O. fulgida* is eaten by cows and goats in times of drought. Some ranchers will torch the plants to remove the spines, but cows, especially the rangy old breeds (referred to as *criollo* or *meco*), seem unfazed by the long, sharp thorns and munch away contentedly at the potent mass. It is not at all unusual to see cattle with one or more sections of choya cactus stuck on the head, even around the mouth.

***Opuntia gosseliniana* F.A.C. Weber**  navo (nopal, duraznilla; Sonoran purple prickly pear)

This cactus grows robustly among the basaltic rocks of Mesa Masiaca. As Vicente Tajia pointed out, the rounded pads are gray, turning purplish in winter, contrasting with *O. wilcoxii*, whose pads are green. The spines are of variable length. Often they are exceptionally long and menacing. This prickly pear often has a short trunk covered with long, flexible spines. FOOD: The fruits and new pads are edible and tasty despite the nasty thorns.

***Opuntia leptocaulis* DC. var. *brittonii* (J. G. Ortega) Bravo**  jijí'ica "it is cooked" (choya; desert Christmas cactus)

This pencil-thin choya is common in the coastal lowlands, where it grows in thickets. Rabbits seem to feast on the red fruits. Although *O. leptocaulis* is widespread across the southwestern United States and northern Mexico, variety *brittonii* is endemic to the coastal thornscrub from Sinaloa to Sonora as far north as the Empalme area. It differs from other populations in its elongate (not spherical) fruits splashed with scarlet and the new stems growing from them. FOOD: The fruits can be eaten, according to Vicente Tajia, but it is hardly worth the effort, for they are covered with legions of glochids.

***Opuntia leptocaulis* DC. × *O. thurberi* Engelm.**  sibiri (?) (hybrid choya)

*Opuntia leptocaulis* is one of the more promiscuous cacti, hybridizing with virtually all other nearby choyas. This hybrid was found in coastal thornscrub.

These are "instant" taxa (rapidly emerging hybrids), intermediate in size, flowers, fruits, and spines. Their uses parallel those of each parent species.

**Opuntia pubescens H. Wendl. ex Pfeiff.**   to'otori huita "hen dung" (choya)

A rather plump, spiny, brittle choya, it is a common understory plant in the tropical deciduous forest. It was pointed out by Braulio López and Santiago Valenzuela of Yocogigua on the slopes of El Bareste in transitional tropical deciduous forest. Several Mayos mentioned that the choya could be eaten by LIVESTOCK during a drought, but they knew of no other uses.

**Opuntia thurberi Engelm.**   sibiri (choya)

This often nondescript, rambling choya cactus grows to tree size, often with a trunk hardly distinguishable from those of surrounding trees. Individuals may exceed 4 m in height and sport a trunk nearly a foot in diameter. It is a rather spindly, lightly spined choya, with branches protruding irregularly from the main stem or trunk. The flowers and fruits are a greenish yellow. Sibiricahui, which means "hill of choya," is the name of a mountain north of Masiaca. FOOD: The green, tender young spines and fleshy leaves are sometimes eaten as a snack. In times of severe spring drought, LIVESTOCK, especially cattle, will eat the cactus. MEDICINE: The young spines and leaves are eaten before they become dry and sharp as a cure for diarrhea and *pujos* (bloody stools). Francisco Valenzuela pronounced the root to be a good remedy for diabetes when brewed into a tea. Ramón Morales of Los Muertos used the boiled root for kidney pain.

**Opuntia wilcoxii Britt. & Rose**   navo (nopal; prickly pear)

Lowland Sonora's largest prickly pear, growing into a 5 m tree in tropical deciduous forest, it has bright yellow spines that make the pads seem to sparkle. Although they are locally common, they are relatively scarce in thornscrub and tropical deciduous forests in Sonora but increase in abundance south as far as Jalisco. There are several varieties of navos with varying properties. Navojoa means "place of the prickly pears." FOOD: Humans eat the tender pads (*nopalitos*) and the fruits (*návoim* [Mayo]; *tunas* [Span.]). Tacos made from nopalitos are most appealing. The fruits are refreshingly juicy and, once one gets used to chewing the seeds, substantial and capable of quenching the most powerful thirst. They are widely eaten, although due to relative scarceness of the plants, never in the numbers of pitahayas. One must take great care when harvesting the fruits to avoid the bothersome and painful aguates (glochids). The tender young pads, a bright green, are reported to be an important source of vitamins A and C, a boon for the Mayos' diet that appears to be lacking in fresh vegetables much of the year. Nopales have been well domesticated in Mexico and produce delectable fruits. O. wilcoxii and other prickly pears are valuable for LIVESTOCK, which in times of drought eat the mature pads.

*Opuntia* sp.                                                                                                                        jutuqui (bachata)

A choya cactus with rather vicious spines, it is abundant in the monte near Teachive. MEDICINE: The root is brewed into a tea used for prostate problems. For venereal disease in men, it is said to be applied directly to lesions (chancres). Some believe the tea to be effective against cancer.

*Pachycereus pecten-aboriginum* (Engelm.) Britt. & Rose           etcho

The Mayo name (etcho) for this important cactus has often been confused with the unrelated Spanish word *hecho*. This stellar tree is greatly esteemed by Mayos, who named one of their major settlements, Etchojoa, after it. FOOD: The fruits are edible and yield seeds ground into flour for tortillas. The seeds are a source of oil for seasoning griddles. ARTIFACTS: The lumber is widely used for furniture and doors. See pages 95–99 for a more complete description.

*Peniocereus marianus* (Gentry) Sánchez-Mej.                  nómom (sina)

An uncommon slender cactus similar to *P. striatus* but with strongly four-sided stems. Vicente Tajia collected this strange cactus on Cerro Terúcuchi and on Sibiricahui to the north. FOOD: The fruits are edible (see p. 168).

*Peniocereus striatus* (Brandegee) Buxb.            nómom (sina, sacamatraca)

A pencil-thin cactus plant often resembling a vine more than a cactus, it grows among thick thornscrub. It is common near the coast in coastal thornscrub and ranges in the Sonoran Desert as far north as southern Arizona. AESTHETICS: Some women along the coast keep specimens in flower pots for the beauty of their pink-tinted flowers. FOOD: The fruits are eaten.

*Pereskiopsis porteri* (Weber) Britt. & Rose                           jejeri

The only cactus in the region bearing fully developed leaves. It is strange in appearance, often growing vinelike into shrubs and trees or snaking along the ground. It may grow 5 m high in a tree and then send out many reptilian branches. The powerful spines and glochids can be hazardous to the inexperienced, who may spy the leaves and be misled into thinking it is not a cactus. FOOD: The orange fruits are eaten only in emergency because they are sour and heavily laced with glochids. As in *Opuntia leptocaulis*, new stems grow directly out of the fruits.

*Nómom* (*Peniocereus striatus*) cactus. Note the fruits next to the pocketknife.

*Pilosocereus alensis* Weber                    aaqui jímsera "bearded pitahaya" (pitahaya barbona)

A medium-sized (to 7 m tall) columnar cactus growing on steep slopes in tropical deciduous forest. On the upper portions of the branches the spines grow into dense mats of gray white hairs, hence the common name "bearded pitahaya." It has a decidedly individual appearance. In Mayo country it is found only at Baimena, where it is common and known for its delectable fruits. Francisco Valenzuela of El Rincón (in tropical deciduous forest) was unfamiliar with it, although it is found in the nearby Sierra de Alamos. FOOD: Marcos Valenzuela of Baimena highly recommended the fruits, saying they were as tasty as those of sahuira (*Stenocereus montanus*) yet bore no spines.

*Selenicereus vagans* (K. Brandegee) Britt. & Rose        sina cuenoji "tamal cactus" (sina volador)

This most unusual cactus occurs only in the southern portions of Mayo lands.

Domingo Ibarra with *sina cuenoji* (*Selenicereus vagans*) on *palo chino* near Huasaguari. The cactus has engulfed the tree and probably caused its death.

Even there it is scarce. It typically grows in the crotch of a large tree, often a mezquite, sometimes moving up thick branches hydralike, becoming massive and engulfing a sizeable portion of the tree. The many arms droop octopuslike into the open air. Several specimens were growing in mezquites near Huasaguari north of Masiaca, including one of immense proportions that resembled nothing more than a thirty-foot-long mass of writhing snakes. The flowers are white. FOOD: The cactus produces edible fruits variable in size, some larger than those of pitahaya. Domingo Ibarra says they are tasty, if not sweet. Fruits collected from a specimen in the Mayo village of El Paso on the Río Cuchujaqui were said to be delectable.

***Stenocereus alamosensis* (J.M. Coult.) Gibson & Horak**     **sina (galloping cactus)**
[***Rathbunia alamosensis* (J.M. Coult.) Britt. & Rose**]

Sina is a small, rambling columnar cactus. The arms, arising from thickets often more than 10 m across, seldom exceed 2 m in height. The bright red tubu-

Francisco Matus and *sinaaqui,* an apparent hybrid between *pitahaya* and *sina,* near Piedra Baya. The fruits and wood are different from those of either parent.

lar flowers are diurnal and are visited by hummingbirds. The red fruits are the size of ping-pong balls and ripen throughout the summer. The large expanse of cactus sometimes renders the ripened fruits accessible only to animals, which seem usually to harvest the fruits before humans can gather them. FOOD: The small fruits are especially relished. MEDICINE: In addition to their fine taste, the fruits are also reported to help ease the pains of rheumatism and to alleviate the pain and swelling of insect stings. Vicente Tajia collected the husks of ripe fruit to boil into a tea for soothing his granddaughter's heat rash. ARTIFACTS: Weavers use a length of carved stem to elongate the *na'abuia* (shed shuttle) of their *telares* (looms) to accommodate wider blankets. The wood is carved into a cylinder with a tapered end and inserted into the hollow end of the shuttle made of carrizo (*Arundo donax*). The width of the blanket can thus be extended indefinitely. The sina is strong enough to withstand the tamping pressure of the sasapayeca and the moving warp.

***Stenocereus alamosensis*** × **(J.M. Coult.)**                                                     sinaaqui
**Gibson & Horak** × ***S. thurberi*** **(Engelm.) Buxb.**

This uncommon cactus appears to be a hybrid between the pitahaya and the sina. It occurs only as isolated individuals and assumes the form of a small pitahaya, reaching nearly 5 m in height in a columnar growth form with 10 ribs, but its branches have the characteristic shape of the sina. According to Vicente Tajia, the flowers are bright red, as are those of the sina, but the fruits are smaller. ARTIFACTS: The wood is much harder than either pitahaya or sina, and the stems are hollow. It is sought after and used to make swings and cribs for children. The hollow stems are cut into lengths and threaded, thus making a flexible side or support for a swing that will prevent a child from falling through but will not cause injury. The cactus is well recognized in the region. Vicente pointed out specimens in several different localities.

***Stenocereus montanus*** **(Britt. & Rose) F. Buxb.**                   sahuira (pitahaya)

An uncommon, stately (10 m tall, 13 m rarely) columnar cactus, typically found in higher, well-developed tropical deciduous forest of southern Sonora, usually with several large arms. It superficially resembles the etcho in that it has a well-developed trunk and arms that issue from the trunk well above the ground. Its needles are brown to black, unlike those of the etcho, which are white. The spines of the fruits are far fewer, but sharper and stronger than those of the etcho. In extreme southern Sonora and in tropical deciduous forest along the Río Fuerte the sahuira becomes common, outnumbering etchos and the dwindling numbers of *S. thurberi*. Individual sahuiras there grow to more than 13 m tall. In granitic soil near Baca an enormous individual has grown in the form of a candelabra, with more than 50 upright arms. It casts a considerable and comfortable shadow. As is the case with pitahayas, however, the most prolific fruit producers are not the largest trees. In Sinaloa sahuira supplants pitahaya in many domestic uses, providing lumber and ample fruits for which the gathering technique and consumption are similar to those of pitahaya. FOOD: The fruits, locally called pitahayas, are eagerly eaten and said to be superior to the delectable fruits of *S. thurberi*. CONSTRUCTION: The wood is widely used for lumber and is useful for vigas and horcones. It is a valuable wood, said to be more durable than that of etcho or pitahaya. Due to its small numbers in Sonora, however, its use is limited. Some Mayos expressed a desire to cultivate the plant. Near El Chinal and

An exceptionally large *sahuira* (*Stenocereus montanus*) near Baca, Sinaloa. Note the many fruits on the branches. Along the Río Fuerte, *sahuiras* outnumber *pitahayas* (*aaqui* in Mayo; *Stenocereus thurberi*) and are an important source of nutritious fruits.

Güirocoba, Sonora, and north of El Fuerte cuttings were planted many decades ago as part of a living fence in combination with etchos and pitahayas. MEDICINE: Rosario Valenzuela of Baimena reports that each year he saves a supply of husks from the fruits, singes them, and keeps them as a remedy for hemorrhoids. They are applied directly to the anus. Alvaro Valenzuela reports that the fruits are especially effective for cleansing the insides and curing ulcers. ARTIFACTS: The arms of sahuira are used for making solid beds and tables.

***Stenocereus thurberi* (Engelm.) Buxb.**   aaqui (pitahaya; organpipe cactus)

A common columnar cactus in Sonora that changes life form from a multi-stemmed shrub branching from the ground in the Sonoran Desert to an 8–10 m tree in coastal thornscrub. It is probably the Mayos' most important plant, used for FOOD, MEDICINE, CONSTRUCTION, and INDUSTRY. See pages 82–90 for a more complete description.

CAPPARACEAE

*Capparis atamisquea* Kuntze         juvavena " 'how that smells!' bush"
[*Atamisquea emarginata* Miers]

This large spineless evergreen shrub, with an amphitropical distribution in both northwestern Mexico and Argentina, is common on the coastal plain. It is known to Yaquis as well as to Mayos. The foliage is dark green, the 2 cm oval leaflets leathery and smooth, resistant to desiccation, the margin entire. Juvavena tends to retain its leaves/leaflets throughout the spring drought. MEDICINE: For molar pain the tender branches are pounded to a pulp, then boiled. The resulting liquid is applied to the tooth. For mange, the same pulp is used and the affected skin bathed. While not as effective as nesco (see below), it is far more abundant and accessible on the coastal plain. Vicente Tajia pronounced it to be effective on dandruff as well. Leobardo López says that a tea brewed from the leaves provides relief from allergy symptoms.

*Capparis flexuosa* (L.) L.         tábelojeca "fallen shade tree"

A large shrub or small tree with oaklike leaves growing in the washes of extreme southern Sonora. The yellow flowers are bearded. The fruits grow into long pods, easily confused with leguminous bean pods. MEDICINE: Alejo Bacasegua of Camahuiroa identified the plant and noted that old people had used it for something medicinal, but he could not recall what. CONSTRUCTION: Felipe Yocupicio says the wood is excellent for fence posts and for use in housing, when straight poles can be found. LIVESTOCK: Felipe says it is good fodder for goats.

*Crataeva palmeri* Rose         capúsari

In Mayo country this strange tree is found only in tropical deciduous forest in Sinaloa and near El Chinal, Sonora. A single specimen was bulldozed from the roadside near Masiaca but has since resprouted and is growing once again. It is not known to most people in the area, however. Vicente Tajia, who lives 3 km away, knew of the tree/shrub, mentioning that he had not known what it was until he spoke with Mayos from Sinaloa who were familiar with it. The tree seldom exceeds 6 m in height, forming a spreading habit with a single main trunk and several smaller trunks. The branches may intertwine and grow horizontally as well, presenting a confusing small tree. FOOD: The lime-sized fruits that dangle from long peduncles ripen in June and July and are said to be sweet and satisfying. ARTIFACTS: The wood is ideally suited for carving. Natives fashion utensils and trays from the wood. A native of Los Capomos carves human figurines as well.

*Forchhammeria watsonii* Rose                                  jito (palo jito; lollipop tree)

A gnarled, strong-trunked, symmetrical tree up to 9 m tall, sometimes more, often resembling a lollipop in shape. It is endemic to southern Sonora and northern Sinaloa. Jitos give marvelous shade and edible fruit. See pages 105–6 for a more complete description.

## CARICACEAE

*Jarilla chocola* Standl.                                          chócola, tónaso "salty"

This remarkable plant occurs sporadically in tropical deciduous forest from southern Sonora and south to Jalisco. It is invisible much of the year, its root lying dormant. With summer rains it grows rapidly, up to 0.5–0.8 m, resembling a white-flowered herbaceous *Jatropha*. The strongly ridged, 6 cm long, oval green fruits turn pink after they fall to the ground. When the fruits drop, the herbage on the plant dies, and the plant disappears, leaving the fruits lying on the earth's surface as if cast there by the gods, undisturbed by the beasts, which avoid them. FOOD: The edible tuber grows to the size of a small egg. Gentry (1942:186) cited a laboratory analysis that touted the small tuber as having great nutritional potential. He noted that it is highly sought after among Guarijíos, who roast and eat it with enthusiasm. The same is true among those Mayos who live close enough to the tropical deciduous forest to be acquainted with chócolas. Mayos eat the fruits as well. Marcelino Valenzuela demonstrated for us, as did his uncle, Francisco Valenzuela, how the slightly sour fruits with whitish meat and black seeds are seasoned with lime juice and eaten with tortillas. They are known in Sinaloa, but we were unsuccessful in collecting any specimens there.

## CELASTRACEAE

*Maytenus phyllanthoides* Benth.                          pasio (mangle, mangle dulce)

This shrub, sometimes a small tree, with thick, succulent leaves, resembling a mangrove, is common on saline soils near the coast but does not grow in the water as mangroves do. It grows symmetrically to more than 5 m high and spreads out to an even greater width. LIVESTOCK: Pasio is viewed as good forage for goats. INDUSTRY: The reddish wood is easy to break off and makes good firewood. Near habitations or popular camping areas its populations have decreased due to overharvesting. This is a shame, for the dense foliage is an excellent refuge for wildlife, including the small Sinaloan whitetail deer (*ma'aso;*

*Odocoileus virginianus* subsp. *sinaloense*). MEDICINE: Alejo Bacasegua mixes the leaves with branches of tajimsi (*Krameria erecta*) or saituna, adds them to petroleum jelly, and applies the mixture to sores that will not heal, which he says is an effective treatment. Felipe Yocupicio agrees. Balbaneda López of Yavaros chews on leaves for stomach problems.

*Wimmeria mexicana* (DC.) Lundell. chi'ini "cotton" (algodoncillo)

A small tree in much of the forest, but under favorable conditions growing to 10 m tall, often with multiple trunks from a common root. The trunk is thick muscled and strong. With age, the bark peels away from the trunk in small, recurved sections, sometimes leaving a most agreeable multicolored mosaic pattern on the exposed bare trunk. When a piece of this peeling bark is carefully broken in two, tiny cottony fibers can be seen holding the pieces together, hence the common names. Tiny white flowers in September attract so many bees and flies that the trees can sometimes be heard before they are seen. CONSTRUCTION: The hard trunks are used for fence posts and for beams in homes. They must be replaced more frequently than other woods such as chopo (*Mimosa palmeri*), since they are said to be more susceptible to rot and infestation by termites.

## CELTIDACEAE

*Aphananthe monoica* (Hemsl.) Leroy (guasimilla)

A handsome spreading tree up to 25 m tall. In the Mayo region it is found only at Baimena, where it occurs in moist box canyons, sharing the habitat with other large evergreen trees such as *Drypetes gentryi* and *Sideroxylon persimile* (bebelama). It is found in similar habitats in extreme southern Sonora but is unknown to Mayos there, growing well outside Mayo country. The trunk becomes quite large, exceeding a meter in diameter and casting dense shade. CONSTRUCTION: At Baimena it is known to produce excellent lumber for housebuilding, but it is used only sparingly because its habitat is usually far from populated areas or inaccessible.

*Celtis iguanaea* (Jacq.) Sarg. cumbro (vainoro; tropical hackberry)

The tropical hackberry grows in great abundance, usually producing a growth form of huge, arching branches armed with stout, sharp, curved spines up to 12 cm in length. These branches often grow horizontally, finally curving over to reach the earth and form an impenetrable barrier. Sometimes cumbro combines with *garambullo* to form dense thickets along watercourses that imprison passersby within the floodplain, a condition that inspires impolite oaths from the traveler who carries no machete. Trees will assume a tangled, confused appearance in which other branches and vines are often intermingled, greatly confusing the onlooker. It seems to prosper in the aftermath of floods, rebounding with vigor and increasing its often considerable stranglehold on the margins of arroyos. Natives complain that it is more common now than previously, perhaps a testimony to the effects of channel erosion and widening due to degradation of watersheds. FOOD: The cranberry-sized light orange fruits are eaten with great relish by birds and humans, although the large seed occupies most of the fruit. A handful of the insipidly sweet fruits yields half a handful of husks and bony endocarps. Although the taste of the fruit is worth the energy expended collecting it, adults in the region appear to spurn the fruits. A Guarijío explained that eating them can make one dizzy. We have eaten many hundreds and have not experienced this phenomenon. At least, we think not; it could have been the heat. ARTIFACTS and CONSTRUCTION: The wood is used to make handles, and also for posts and roof cross-hatching. MEDICINE: Don Patricio Estrella maintains that freshly crushed leaves will heal the lesions of impetigo and other sores when applied directly to the affected area.

*Celtis pallida* Torr.        cumbro (cúmero; desert hackberry)

Mayos declared the name *cumbro* to be Mayo, probably the basis for *cúmero*, the Spanish name for hackberry, or perhaps vice-versa. Zavala (1989:160), however, attributes the term to the Cáhita, hence we include it. A spiny Sonoran Desert shrub or small tree, it is confined to the more arid parts of coastal thornscrub. Its distribution is the entire southwestern deserts of northern Mexico. FOOD: The fruits are relished by humans, and also by birds and foxes. MEDICINE: The roots are made into a tea used as a purgative. ARTIFACTS: The older Mayos used straighter branches for hunting bows. INDUSTRY: The wood can be used for firewood and other household uses as well.

*Celtis reticulata* Torr.        cumbro (cúmero; netleaf hackberry)

In the Mayo country, this normally rough-barked tree grows a straight trunk with smooth bark up to 10 m, assuming an unusual growth form unlike those in the more temperate areas to the north. We once spent the better part of a half hour puzzling over the identity of a specimen, confused by its smooth bole and its size. It is easily mistaken for guayparín. MEDICINE: The bark is boiled into a tea taken for tonsillitis. FOOD: The orange fruits are often eaten, and although sweet, they are too small and too few to fill one up.

## CHENOPODIACEAE

*Allenrolfea occidentalis* (S. Watson) Kuntze        bahue mo'oco "sea bush"

LIVESTOCK: Maximiano García of Agiabampo pronounced this common low perennial to be of no use but for goat feed.

*Atriplex barclayana* (Benth.) D. Dietr. bahuepo juatóroco
"sea-ash shrub" (saltbush)

A common subshrub or herbaceous perennial in sandy, salty soils along the coast. LIVESTOCK: Goats seem to be able to gain some nutrition from it.

*Chenopodium ambrosioides* L. lipazote (epazote; Mexican tea)

This green annual grows wild along arroyos and in disturbed places. It is also cultivated in gardens. FOOD: It is widely used for seasoning beans, soups, and stews. MEDICINE: It is also widely used throughout Mexico as a tea to alleviate stomach cramps and diarrhea and to purge intestinal parasites. Sara Valenzuela of El Rincón recommends it to alleviate menstrual cramps. Most families retain a good supply of the dried herb in their homes.

*Chenopodium neomexicanum* Standl. choal (quelite; fish goosefoot)

FOOD: After equipatas and las aguas these become rather common. They are gathered and eaten as a green when young and tender. They also provide an important green in the diet of LIVESTOCK, especially in spring, when few other annuals appear. As they grow older, the plants begin to exude a powerful odor of fish. Once the substance is on the hands, it requires several washings with soap to remove it. One of us had the unfortunate experience of being asked by associates to wash the stench from his hands before reentering the group vehicle.

*Suaeda moquinii* (Torr.) Greene bahue mo'oco (chamisa del mar)

A small shrub of salt flats near the sea. LIVESTOCK: It provides some sustenance for goats and burros.

### COMBRETACEAE

*Combretum fruticosum* (Loefl.) Stuntz (compio)

Compio is a common vine in the Arroyo de la Culebra near Baimena. The flowers resemble large, brilliant red toothbrushes covering the nearly leafless vines in May. They are a most conspicuous flower, blossoming in the brown gray of late spring, punctuating the riparian landscape with bright spots of red. In the land of the Mayos the plant has been collected only near Baimena, Sinaloa. To date it has not been collected in Sonora. ARTIFACTS: The tough vine is used to secure bundles to beasts of burden and hold ungainly loads together.

*Conocarpus erecta* L. (mangle; button mangrove)

A rare mangrove growing along with other mangroves at Huatabampito and at Naopatia. We have not found it to be locally distinguished from other mangroves. INDUSTRY: Mangle is valued as a firewood.

*Laguncularia racemosa* (L.) Gaertn. f.     moyet, pasio tosa "white mangrove nest" (mangle; white mangrove)

This small tree occurs in large numbers along estuaries in the southwest part of the Mayo region. FOOD: The leaves are brewed into a refreshing tea highly acclaimed for its flavor. The name "moyet" refers to the tree and a fish that lives among its underwater roots, according to Balbaneda "Nelo" López of Yavaros.

## COMMELINACEAE

*Commelina erecta* L.     osi (cuna de niño; dayflower)

A common pretty blue flower. No use has been ascribed to it, but several consultants have identified it and remarked on its beauty.

## CONVOLVULACEAE

*Ipomoea arborescens* (Humb. & Bonpl.) G. Don.     jútuguo (palo santo; tree morning glory)

A broad-boled tree up to 12 m tall, quickly tapering into narrowing branches that recurve gracefully and randomly from the gray white trunk. (At higher elevations the trunks of *I. arborescens* var. *pachylutea* Gentry are yellowish.) In portions of the Mayo region, especially in transitional areas between foothills thornscrub and tropical deciduous forest, it becomes the dominant tree or nearly so. The new growth at the tips of branches acts as a tendril and may grasp a nearby tree much as the morning glories will do. During the rainy season the large ovate leaves provide dense shade. During the dry season the tips of the bare branches burst into large, white flowers that stand out from afar. These flowers are an important early spring resource for northward-migrating hummingbirds. The trees grow quite fast and may reach a meter in diameter. When the trees die, the wood, which resembles wet cardboard, disintegrates rapidly, vanishing within a year. MEDICINE: The bark from branches is steeped into a tea that is said to be a dramatic remedy for insect bites, offering relief within an hour of drinking it. If a drop of the congealed sap is placed on an aching tooth, the pain is said to cease almost immediately. Care must be taken not to touch healthy teeth, however, for the affected tooth will slowly disintegrate over the following few weeks. Two of the consultants indicated bare gums where they had treated diseased molars with the gum of jútuguo. The fallen flowers are reportedly effective at controlling high blood pressure when brewed into a tea. At the request of Vicente Tajia of Teachive, we drove

## Convolvulaceae / 179

*Jútuguo* (*Ipomoea arborescens*) in leafless transitional tropical deciduous forest near Yocogigua, March. The understory has been destroyed by overgrazing.

him into the foothills so that he could gather the flowers and use them for that condition. CONSTRUCTION: Large strips of the bark are peeled from the tree and used a thatch for roofs. ARTIFACTS: The wood is carved into musical instruments in Sinaloa. The fallen flowers are eagerly consumed by LIVESTOCK and deer. The smooth bark is eaten by livestock, especially burros, which in years of drought (most years) causes heavy attrition in plant numbers. Some Mayos report jútuguo numbers to be dwindling due to overgrazing by burros and a general drying out of the climate. During the dry times when fodder is not available, large branches are cut and chopped into small pieces for cows to eat. Jútuguo constitutes a most important component of the Mayos' vegetation resources.

*Ipomoea bracteata* Cav. [*Exogonium bracteatum* (Cav.) Choisy ex G. Don]   tosa huira "white vine" (jícama)

A strong vine, it is most often detected by the bright pink purple bracts of the flowers contrasting with the gray brown of the dry forest; the flowers disappear in the rainy season. The blooms can be deceptive, leading the novice to believe that the host tree is a blooming amapa. It requires some patience to trace the vine to the ground. FOOD: The thickened roots are eaten, although all consultants agree they may not be worth the considerable effort needed to collect them. ARTIFACTS: The vine material is used as a rope. Francisco Valenzuela reported that it twists well and lasts for a long time. Many are the bundles he has brought home on the back of a burro, lashed together with the vine of tosa huira.

*Ipomoea pedicellaris* Benth.    jícure "twisted yarn" (trompillo; morning glory)

A fast-growing vine that explodes into action with the coming of las aguas, it will frequently cover shrubs and even some columnar cacti such as etchos. It produces a showy purple flower that enhances the summer and fall landscapes. LIVESTOCK: The herbage is valued as forage for cattle and goats.

*Ipomoea* sp.    jícure

MEDICINE: According to Patricio Estrella, "The black seeds are good for women who cannot conceive. I don't know what it is called." The plant material we collected was dry and much deteriorated, and we were unable to determine the species.

### CUCURBITACEAE

*Cucumis melo* L.    goy minole, hayá huíchibo (chichicoyota, melón de coyote; coyote melon)

A long, trailing vine that can remain prostrate or climb aggressively into trees. The melons, the size of a tennis ball, often remain suspended in a tree through the dry season, swaying in the wind. The seeds and flesh of the melon are foul smelling and taste like an improbably bitter cucumber. We tried to eat them but capitulated early. MEDICINE: Vicente Tajia and Braulio López report that the gourds are boiled, then crushed. The wash is rubbed in the hair to kill lice. Vicente says that he learned this practice from Yaquis.

*Cucurbita argyrosperma* Huber subsp.    tetaraca (chichicoyota; wild gourd)
*argyrosperma* var. *palmeri* (L.H. Bailey)
Merrick & Bates [*Cucurbita palmeri* L.H. Bailey]

Similar in appearance to *Cucumis melo,* these gourds remain green for many months after the vine has died. Mayo pascolas use the term *tetaraca* to refer to anything that is worthless. INDUSTRY: Vicente Tajia reports that in bygone times soap was made by chopping the gourd up and boiling it. The resulting decoction worked well for washing clothes, he maintains. The clothes turned out fresh and clean, but the treatment made them wear out more quickly than conventional soap.

*Güerequi* (*choya huani* in Mayo; *Ibervillea sonorae*) near Teachive in the dry season. Note the vines extending from the storage roots. *Güerequis* are widely marketed as a cure for cancer and rheumatism.

**Ibervillea sonorae** (S. Watson) Greene      choya huani "choya's brother"
[*Maximowiczia sonorae* S. Watson]      (güerequi)

This gourd—whose wire-stalked vine rises from a large, grotesque, semisubmerged storage root weighing as much as 5 kg—is surely one of the most interesting plants in the region. The texture of the tuber's skin is elephant-like. The fruits are bright orange or even reddish, the size of ping-pong balls, maturing well off the ground and appearing to be dangling from branches of the host shrub or tree. They stand out in the dying foliage of fall. The vine is sometimes so tendril-like as to escape notice, and it may be difficult to trace the vine from the highly visible fruits to the semihidden tuber. The fruits are devoured by birds but are of little interest to natives of the region. In spite of heavy harvesting for medicinal purposes, güerequis are quite common in the coastal scrub and foothills. They are less common in more moist habitats. They will prosper in desert landscaping in the Sonoran Desert of Arizona if protected from frost. MEDICINE: Vicente Tajia distinguishes between male (*o'o choya huani*) and female (*choya huani jamut*) by their shapes, pointing out that males are good for ailments in women and vice-versa. The root is widely used and highly valued for rheumatism, diabetes, and cancer. Don Vicente demonstrated how to cut off a small piece from the huge corm and stick it between the toes to alleviate the pains of rheumatism. Elsewhere the storage root is ground into a powder, poured into capsules, and marketed as a remedy for cancer, arthritis, and rheumatism. Natives hawk güerequis along major highways in coastal thornscrub, assuring miraculous healing properties.

### CUPRESSACEAE

*Taxodium distichum* (L.) J.M.C. Rich. var. *mexicanum* Gordon [*T. mucronatum* Ten.] — ahuehuete (sabino; Mexican bald cypress)

Widespread in Mexico (a specimen in the state of Oaxaca is reputed to be Mexico's largest tree). Ahuehuete is the Aztec term, and the Mayo name is the same, according to Wencelao Mendibles of La Cuesta. Sabino is a southwestern variety of the bald cypress of the swamps of the southeastern United States. Unlike its swamp-dwelling relative, sabinos in the Mayo region rarely exhibit the emergent knobby "knees" (roots). To date we have found this phenomenon only in Cañon Estrella on the Río Cedros near Tesopaco. Although the tree is well known in the region, we have been unable unequivocally to ascertain a Mayo name. Sabinos grow at the edge of permanent water, primarily the Río Cuchujaqui in the Alamos area, but along Arroyo El Tábelo and Arroyo Techobampo to the north as well. They also grow along permanent water well into the Río Yaqui drainage. The trees become truly massive, often in excess of 25 m tall, their roots an intricate, braided mass. In the land of the Sinaloan Mayos, notably along Arroyo Agua Caliente, they become gargantuan, in excess of 30 m tall and 2 m in diameter. The large specimens are undoubtedly very old, perhaps predating the Spanish conquest. In spring, the trees replace their old needles with bright green herbage, adding a touch of freshness to a browning, withering landscape. In the steep, narrow cajones (narrow canyons) of the Río Cuchujaqui, sabinos are the only riparian tree because their great roots mesh powerfully into bedrock, grounding them against flash flooding. During the massive floods in the Río Cuchujaqui in the wake of Hurricane Ismael in 1995, some huge sabinos were downed and swept away by the powerful currents that reached a depth of 8 m. With the gradual deforestation and conversion of forest lands to pastures taking place in Sonoran watersheds, floods and channel degrading are increasing and the future of the great trees seems each year to be in greater jeopardy.

The wood is very resistant to boring insects and rotting. A mycological survey of the tropical deciduous forest near Alamos carried out by Robert L.

Gilbertson demonstrated that wood-rotting fungi were very important agents in nutrient cycling of most forest trees. However, fungi were only found on fallen sabinos that had lain on the ground for many years. *Hiedra* (poison ivy; *Rhus radicans,* Anacardiaceae), normally a plant of oak woodland or pine-oak forest above 1200 m elevation, is inexplicably common at 240 m along the Río Cuchujaqui in the tropical deciduous forest zone, where it is a stout woody liana only found climbing sabinos. Indigo snakes frequent ledges where sabinos grow, and tiger herons often perch in the higher branches. ARTIFACTS: The aromatic lumber of the sabino is indisputably the best wood for making permanent tables and furniture. The trees are protected by the Mexican government, a measure that seems to be effective at protecting most trees from cutting, even though violations occur sporadically. In Mexico all watercourses are owned by the national government, so nearly all sabinos are national property. Natives living along the lower Río Cuchujaqui from time to time encounter driftwood logs of the tree in the meandering sands of the watercourse. With the help of beasts of burden they haul these to their homes and guard them as treasures. One consultant had stored a ponderous log for several years and was still contemplating using it to make an article of furniture.

## CYPERACEAE

*Cyperus compressus* L.        coni saquera "raven who knows how to toast corn" (coquillo; flatsedge)

A sedge growing along the Arroyo Masiaca into tropical deciduous forest. Vicente Tajia recalls that when he was younger, children would chew on the seeds.

*Cyperus perennis* (M.F. Jones) O'Neill        coni saca (coquillo; flatsedge)

Similar to *C. compressus.*

*Scirpus americanus* Pers.        (tule; bulrush)

A tule of the moist lowlands, it and *Typha domingensis* are widely used by Mayos in villages of the delta country. It grows in irrigation-fed wetlands and along canals. We are unsure of the origin of the name. CONSTRUCTION: Tule is often used in roofing for houses and in making walls. ARTIFACTS: It is also woven into mats.

## EBENACEAE

*Diospyros sonorae* Standl.        caguorara (guayparín; Sonoran persimmon)

A handsome straight-boled tree with a spreading habit, reaching 20 m tall. Uncommon; in the Mayo region it grows only along watercourses. The tree flowers sporadically through the year, producing cream-colored blossoms. Many bird species flock to the trees to consume the ripe fruits. Caguorara's origins are obscure, but it is never found far from human habitation, suggesting that it may have been introduced long ago; however, it is difficult to envision a regional endemic that is introduced. It is especially common in the Arroyo Camahuiroa, where its fruits are large and widely sought after. FOOD: It is valued for its plum-sized yellow fruits, which are eaten cooked or raw, and especially relished when eaten with milk. Locations of the trees are well known. The seeds are ground to produce atol.

### ERYTHROXYLACEAE

*Erythroxylon mexicanum* H.B.K.　　　　　momo ogua "beehive tree" (mamoa)

Normally a many-trunked shrub, in rich tropical deciduous forest it grows into a tree up to 8 m tall. The roundish leaves, which ordinarily do not drop in the dry season, harbor a hint of iridescence, helping distinguish the tree from *Esenbeckia hartmanii* and *Malpighia emarginata*, with which it is easily confused. Its bark is gray and slightly furrowed, which also contrasts with the latter two species, whose bark is smooth and mottled. CONSTRUCTION: The wood is said to be strong and durable, good for making fence posts and roof supports. Smaller trunks are used as cross pieces (latas) in roofs to support the layer of earth above. Although it is a close relative of coca (*Erythroxylon coca*), from which cocaine is produced, we have no reports of its being used as medicine or as a means of producing mind-altering experiences.

### EUPHORBIACEAE

*Adelia virgata* Brandegee　　　　　　　　ona jújugo "smells of salt"

A large shrub or small, compact tree up to 4 m tall, frequently with straight, brittle branches. It is rather common in the lower thornscrub of the region. Its Mayo name is a mystery to us. CONSTRUCTION: The wood is reddish and, though not especially strong, is often used in fencing. INDUSTRY: In an emergency, it is used for firewood. ARTIFACTS: Felipe Yocupicio of Camahuiroa notes that the trunks and branches are hard and straight, though skinny. This makes it ideal for use as a dibble, or planting stick, for poking holes in the ground for seed. It is also produces a good *choma*, the string heddle used in weaving.

*Bernardia viridis* Millsp.　　　　　　　　　　　　　　　　　juya jotoro

A medium-sized, nondescript shrub with tiny white flowers. Felipe Yocupicio of Camahuiroa was familiar with it, but viewed it as of little use except for kindling sticks.

*Euphorbiaceae* / 185

Stakes of *vara blanca* (*cuta tósari* in Mayo; *Croton fantzianus*) near Alamos awaiting shipment. Hundreds of men are employed cutting the small trees. (Photo by T. R. Van Devender)

***Croton alamosanus* Rose**  cuta chicuri (vara prieta)

See *C. flavescens*, below.

***Croton ciliatoglandulifer* C. Ortega**  tatio "it burns" (trucha, ortiga)

A common shrub to 2 m high with a curiously odd white inflorescence that resembles a small brush. Natives warn that it is toxic, cautioning against touching the eyes after coming into contact with the plant, for it can produce a most painful irritation and allergic reactions; in nonallergic individuals, however, it does not appear to damage intact skin. It is quite common on overgrazed bottomlands. MEDICINE: Vicente Tajia and the late Jesús Moroyoqui of Teachive demonstrated its use. The herbage is cut with a machete and then carefully pounded. The crushed herbage is applied directly to cysts and boils, which are immediately improved, they say. The irritation to the skin only helps the cure, they believe.

***Croton fantzianus* Seymour**  cuta tósari "white pole" (vara blanca)

A common small tree to 10 m tall. It was identified as *C. fragilis* H.B.K. (Gentry 1942) but has recently been referred by Grady Webster to *C. fantzianus*, whose type locality is in Nicaragua. The trunk of vara blanca is covered with lightly maculate bark. The leaves turn yellow, orange, and red in senescence, presenting an autumnal lilt to the landscape well into the winter months. CONSTRUCTION and INDUSTRY: The trees are cut into stakes and used in many aspects of house building, especially as latas (cross-hatching for roofs) before dirt is applied, but also for walls, ramadas, fences, and so on. Lengths slightly longer than 2 m are sold by the millions domestically and exported for fencing and as stakes for tomato and grape crops locally and in Baja California and Sinaloa. Cutting the vara provides many jobs in the tropical deciduous forests of Sonora and Sinaloa. The cutters are referred to locally as *vareros*. Vara blanca once occurred in thick, almost aspenlike stands, but populations of mature trees have been decimated everywhere in the region by excessive harvesting, resulting in government embargos and prohibitions on cutting, few of which are enforced. Indeed, the embargo appears merely to have fattened the pockets of the enforcement officials and to have shrunk the paltry bankrolls of the cutters. Huge truckloads still depart daily from the region. Some buyers and shippers are former enforcement officials. Natives claim that the stumps will regenerate into a harvestable trunk within two to three years. Field data do not support their claim (see Lindquist 2000). Although the harvesting does not kill the tree, the wholesale cutting has significantly altered large portions of the tropical deciduous forest of Sonora and Sinaloa.

*Croton flavescens* Greenm.　　　　　　　　　　　　　júsairo (vara prieta)

A shrub, sometimes a small tree up to 4 m tall, with white and gray bark. It is common in foothills thornscrub, less so in tropical deciduous forest, where it tends to be replaced by the similar *C. alamosanus*. Among other things the júsairo leaves turn red in the fall and spring, adding a rare touch of color to the landscape. MEDICINE: Júsairo is used for a variety of stomach ailments ranging from stomachache to diarrhea to *empacho*, a folk ailment of the digestive system perhaps synonymous with constipation. The bark and root are boiled into a rather bitter tea that is oddly refreshing. Cinnamon is often added to increase its palatability. CONSTRUCTION: Specimens that grow in excess of 2 m tall are often cut and sold to buyers as vara blanca, *C. fantzianus* (see above), with no apparent objection from the buyers. They are used in house construction in the Mayo region in the same manner as *C. fantzianus*.

*Croton sonorae* Torr.　　　　　　　　　　　　　　　júsairo (vara prieta)

This is a shrubby Sonoran Desert croton reaching its southern limits in drier parts of coastal thornscrub, where it is common. It is markedly similar to

*C. flavescens,* but with smaller leaves. It is said to be valuable as LIVESTOCK forage and as MEDICINE for the same ailments as *C. flavescens.*

### *Euphorbia abramsiana* L.C. Wheeler (annual)          cuépari (golondrina; spurge)

Numerous prostrate spurges, mostly annuals and herbaceous perennial with tiny white or yellow inflorescences, populate the Mayo region. They appear in great numbers following rains. Although ubiquitous in disturbed soils, the cuépari are collected and stored by many families in Teachive. LIVESTOCK: Goats eat the young plants quickly and appear unaffected by the sticky, milky sap. MEDICINE: The dried plants are brewed into a tea administered for ulcers and digestive problems. Lidia Zazueta of Teachive has successfully applied the tea to areas of skin that have lost pigmentation for whatever cause. Vicente Tajia assures us that he has used it for the same purpose and that it is highly effective. Santiago Valenzuela steeps leaves and branches of *E. abramsiana* into a tea that he applies to sores as a wash. Other cuépari in the Mayo region include *E. albomarginata* Torr. & A. Gray (herbaceous perennial), *E. capitellata* Engelm. (herbaceous perennial), *E. florida* Engelm. (annual), *E. gracillima* S. Watson (annual), *E. hirta* L. (annual), and *E. petrina* S. Watson (annual).

### *Euphorbia californica* Benth.          jímaro

A scraggly, succulent-branched shrub with small leaves borne on long, threadlike peduncles. It is widespread in Baja California, with scattered disjunct populations in Sonora. It is locally common on the southern portion of the coastal plain near Camahuiroa and Agiabampo. The sap is milky and flows with only a slight disturbance to the branch. Alejo Bacasegua warned that if the sap touches the skin, it will produce a blister. Similarly, he reported that if livestock brush up against the bush, the sap will cause their hair to fall out where it came into contact. Even with these dramatic properties, the plant is not to our knowledge used medicinally.

### *Euphorbia colletioides* Benth.          (candelilla)

This remarkable, usually scandent shrub (sometimes a small tree) growing up to 3 m tall has spindly, succulent gray green stems that exude volumes of milky sap when scratched or cut. The lush leaves of the rainy season quickly fall off during droughts. Although it is common in foothills thornscrub and well known in the region, we have been unable to obtain a Mayo name for the plant. MEDICINE: Mayos say that the sap is an excellent purgative. LIVESTOCK: It is also eaten by cattle and deer, without apparent disastrous results.

### Euphorbia gentryi V.W. Steinm. & T.F. Daniel. jímaro

This large shrub, growing in excess of 3 m tall, is closely related to *E. californica*. It is apparently endemic to the greater Mayo region, found only on the bouldery volcanic slopes that rise from the coastal plain as far south as near Los Mochis, Sinaloa. Vicente Tajia was interested in the population on Mesa Masiaca and suggested that its sap would burn the skin, as occurs with jímaro. It assumes a tree form and grows in nearly pure stands on the steep, rocky slopes of Cerro Sibiricahui north of Teachive.

### Euphorbia sp. cuépari (golondrina; spurge)

MEDICINE: This diminutive, prostrate annual is used to brew a tea applied to rashes and pimples. It is a strong and reliable remedy for these conditions, according to Alejo Bacasegua of Camahuiroa.

### Jatropha cardiophylla (Torr.) Müll. Arg. sa'apo (sangrengado azul; limber bush)

This is a Sonoran Desert shrub reaching its southern distribution limits in the more arid portions of the coastal plain. It grows no more than 70 cm tall, with several succulent, darkly barked stalks emerging from a common base. Its dark green, heart-shaped leaves appear in response to the summer rains. MEDICINE: Vicente Tajia reports that the juice was formerly said to be effective in curing leprosy. He also believes the sap to be more effective in curing cuts and sores than that of *J. cinerea*. The runny, staining sap of sa'apo is widely believed capable of removing cataracts, although the treatment is said to be somewhat painful.

### Jatropha cinerea (Ortega) Müell. Arg. sa'apo (sangrengado)

A large, succulent shrub up to 2 m tall with large, silvery, leathery leaves. It is typical of sandy habitats near the coast but is scattered throughout the coastal thornscrub. In some inland foothills it can be a small tree to 4 m tall. The blunt branches drip sap profusely when cut or broken. *J. cinerea* is one of only a few plants that livestock will not touch. As a result it has proliferated on the coastal plain and is extremely common, often sprouting in people's yards. It leafs out early and retains its leaves late, which is an aesthetic boon in a region where deciduation produces marked changes in the landscape. MEDICINE: The runny sap is placed directly on a bruise, cut, abrasion, fever blister, or sore and is believed to cure such things rapidly. It is used daily by people who work in the monte. Vicente Tajia demonstrated its application on a cut on his hand. His brother has used it to treat cold sores and the painful cracks some people are prone to get in the corners of their mouths. Vicente says that the sap burns and a layer of skin will eventually peel off, leaving the skin free of the sores. The plant is also used to treat diabetes; the root is cooked with cósahui to produce a liquid to be drunk daily. Vicente also claims that the root is mashed and steeped in water for several hours to produce a drink said to cleanse the bladder and the kidneys; in other words, the tea is another remedy for the urinary problems that abound in the torrid heat and dryness of the coastal plain. The sap is applied directly to the eyes to remove foreign objects, he and others report. Francisco Valenzuela claims that when he was young, some of the sap was mixed with mothers' milk and applied to cloudy eyes and seemed to help clear them. INDUSTRY: Julieta Nieblas Zazueta of Teachive, a *cobijera* (blanket maker), and others use liquid from the root as a mordant with brasil wood in preparing a bright red dye for woolen blankets. A mere drop stains clothing permanently. When the plant is cut or a leaf is snapped off, it bleeds profusely, hence its Spanish name (also said to be a contraction of *sangre de drago*, dragon's blood). It is so prone to bleed that those who wander through the coastal thornscrub can hardly avoid getting brown stains on their clothing.

*Jatropha cordata* (Ortega) Müll. Arg.     sato'oro (papelío, torote panalero; yellow torote)

A striking, irregularly branched, narrow tree reaching 8 m, with soft wood and a green to yellow to beige trunk from which peel large sheets of exfoliating tan bark. It is common in the region and used for a variety of purposes. See pages 100–103 for a more complete description.

*Jatropha malacophylla* Standl.     sa'apo (sangrengado)

This strange plant is a large shrub (or small tree), often reaching 5 m in height. It is found only in moist tropical deciduous forest. The stems are succulent, thick, and blunt. The large leaves are reminiscent of those of a sycamore (hence the old name *J. platanifolia*), turning yellow as they deciduate shortly after the end of the rainy season. MEDICINE: Don Francisco Valenzuela of El Rincón says that to clean the teeth and remove plaque people suck the sap from a branch, swish it around the mouth (*hacen buchis*), and then spit it out. Rosario "Chalo" Valenzuela of Baimena demonstrated another use: a few drops are added to water and then applied to the eyes to relieve redness. Because the sap stains clothing, we have been reluctant to test it, fearing it will stain mouths and eyes as well.

*Pedilanthus macrocarpus* Benth.     cantela oguo "candle tree" (candelilla)

A remarkable plant of the deep soils on the coastal plain, consisting of 10 to 40 upright, scapose branches reaching 0.5 m tall when the plant is solitary, up to 3 m tall when it grows in close association with other shrubs and trees. The branches resemble thin green tapers or candles. They produce bright red shrimplike flowers and fruits at the tips. When the skin of the branch is disturbed, it immediately exudes a milky sap. The Mayos warn that this latex is a powerful purgative even in minute traces and will remain effective until the hands are washed. Seferino Valencia Moroyoqui said that a mischievous lad from Las Bocas slipped a few drops into the wedding stew of a young couple. During the celebration numerous revelers had their intestinal tracts purged, he recalled with a laugh. MEDICINE: According to Alejo Bacasegua, treating a corn with the latex will soften and ultimately remove it. The treatment requires that a small piece of the plant be cut off and, with extreme caution, a tiny drop of the sap applied to the offending corn. Seferino Valencia extends this treatment to include calluses as well.

*Ricinus communis* L.  (higuerilla; castor bean)

The castor bean was introduced into Mexico many years ago and is now ubiquitous, often crowding out native shrubbery and growing in pure stands. It almost defies life form classification by growing as a perennial herb, shrub, or tree. MEDICINE: According to Marcelino Valenzuela of El Rincón, the leaves are cooked in pork fat and applied to blisters and sores. Vicente Tajia proclaims it to be a valuable remedy. Leaves rubbed on the head will alleviate *punzadas*, migraine or pulsing headaches.

*Sebastiania pavoniana* Müll. Arg.  túbucti (brincador, hierba la flecha; Mexican jumping bean)

A small tree seldom more than 8 m tall. Its smooth, shiny oval leaves of dark green turn red in the late fall and winter, making it easy to spot on the hillsides from ground or air. The leaves and stems give off milky sap when crushed. ARTIFACTS: The triangular seeds are used as toys. The movements of a moth larva (*Cydia deshaisiana*, Tortricidae) cause the seeds to tumble and turn of their own accord. Formerly found in curio shops on the U.S. border, they are harvested sporadically and sold to vendors.

*Tragia jonesii* Radcl.-Sm. & Govaerts.  nata'ari "burns the nose" (rama quemadora, ortiga; noseburn)

A climbing, herbaceous vine with tiny flowers. Insidious glandular hairs on the leaves and stems exude irritants that produce a most uncomfortable burn, hence the common names. Coastal plants of this genus have been referred to as *T. jonesii*. The more common inland plants are a new species awaiting formal description. MEDICINE: Marcelino Valenzuela reports that when leaves are boiled with cordoncillo leaves and cinnamon, the resultant tea is effective in reducing fever. He assured us that it would not burn the drinker. It is used along the coast for the same symptoms.

### FABACEAE

*Acacia cochliacantha* **Humb. & Bonpl.**   chírajo (güinolo, chírahui; boat-spine acacia)

An extremely common, spreading tree, seldom exceeding 8 m tall. In parts of the forest, it occurs in uniform stands as second growth, especially in abandoned milpas (cornfields), where it is easily discerned by the uniform light green color of the canopy. It plays an important ecological role in plant succession back to tropical deciduous forest and seldom survives for more than a couple of decades. The hollow red thorns turn gray with age, resembling silvery snail shells. As Gentry (1942:125) noted, in death the trees assume a roughly circular shape, the branches curling over upon each other on one side in a "death arch." When numbers of these skeletons are seen together, the landscape takes on an eerie tableau of death. It is one of the few plants able to compete successfully with buffelgrass in fields cleared of tropical deciduous forest. Once the plants attain sufficient size to shade buffelgrass, chírajo can win the contest. An aggressive invader of overgrazed pastures, its numbers are on the increase, and it is much condemned by ranchers. MEDICINE: The spines or thorns are brewed into a tea that is highly recommended for prostate and urinary problems. Francisco Valenzuela of El Rincón said it is more effective when combined with the spines of vinorama. Alejo Bacasegua of Camahuiroa, a specialist in herbal remedies for behavioral and physical disorders, recommends a tea of chírajo bark for children suffering from hypothermia, who tend to roll into a fetal position when they are excessively chilled. The tea helps heat them and relaxes their chilled muscles. INDUSTRY: It is used for firewood. LIVESTOCK: In times of scarcity burros strip long sections of bark from the tree and apparently consume it to their benefit. Most larger chírajo trees in the vicinity of homes bear heavy scars. ARTIFACTS: The legs of benches at San Antonio were fashioned from the tree. The sharp thorns are detached and used to prise splinters and foreign objects from the skin. A generation ago a Mayo cowman of Teachive was renowned for his wealth and stinginess. Although he was reported to have owned more than a thousand cows, he lived a slovenly and debauched life. He was said to be so cheap that he laced his shoes not with rawhide but with the bark of chírajo, which easily peels into long strips.

*Acacia coulteri* **Benth.**   baihuío "rushing water" (guayavillo)

In parts of the Mayo region this thornless tree resembles the more common mauto in habit and in its large leaves with tiny leaflets. It sometimes grows as tall as 10 m, leafing out a light green in late spring. It is found in a variety of habitats, from dry foothills thornscrub on Mesa Masiaca and the arid hills near Topolobampo on the Sinaloan coast, where it is often the most common leguminous tree, to moist tropical deciduous forest on the slopes of Cerro Tapustete near Alamos. It is often found in groves but is absent from large portions of the forest, a fact perhaps explained by its popularity as lumber. CONSTRUCTION: The wood of this exceptionally strong-trunked tree is ideal for beams and posts, among the best. INDUSTRY: It also produces fine firewood, a shame since it is such an excellent wood for construction.

*Acacia farnesiana* (L.) Willd.  cu'uca (vinorama; sweet acacia)

Normally a large, spreading shrub, it occasionally grows into a stately 10 m tall tree in the Río Mayo region. Its hardiness and drought tolerance work to its advantage, and it occurs in huge numbers in disturbed soils and overgrazed pastures. In late winter the tree produces a profusion of round yellow blossoms that attract humans by their perfume, sometimes in competition with bees. It tends to spread quickly in cleared pastures, its aggressiveness and its potent thorns making it much vilified by ranchers. Once it is established in a pasture it is difficult to eradicate. The village named Cucajaqui means "vinorama in the arroyo." INDUSTRY: It produces tolerable firewood. The wood is also used to produce a blue dye by prolonged soaking in water. MEDICINE: The root is cooked and the liquid drunk as a remedy for invertebrate stings and snakebites. The bark is wrapped in a rag and wrapped around the head for throbbing headaches.

*Acacia occidentalis* Rose  teso (Sonoran catclaw)

A large, solitary, spreading acacia reaching 10 m tall, flowering white in spring. At a distance it is distinguishable from mezquite and palo chino by its scraggly appearance. The branches seem to grow every which way, and errant limbs frequently escape from the general shape of the tree. At close range the dark catclaws are distinctive. It grows in pockets on low-lying flats, where it can be quite common or absent. The name of the town Tesopaco on the Río Cedros means "teso out there." Teso grows northward nearly to Arizona but only attains tree size in central Sonora and to the south. It is closely related to *Acacia greggii* A. Gray, the catclaw of the southwestern U.S. deserts. CONSTRUCTION: The hard wood is highly valued for ceiling beams (when straight lengths can be found) and posts. ARTIFACTS: It is also carved into *bateas* (wooden bowls). It is commonly used whenever a hard, solid wood is required.

*Acacia pennatula* (Cham. & Schltdl.) Benth.                    (algarrobo)

A medium-sized acacia with scrubby, thorny branches, it usually grows only on ash soils and at the upper margins of the tropical deciduous forest and higher oak woodland; hence, it is somewhat unfamiliar to most Mayos. In Argentina *algarrobo* is the name for several species of mezquite. INDUSTRY: Francisco Valenzuela said it makes acceptable firewood. Guarijíos use the bark to dye leather.

*Acacia* sp.                                        cochirepa "pig's earring"

An uncommon, squat, scrubby tree hardly exceeding 3 m in height, with a thick trunk tapering quickly at the apex. It superficially resembles *A. farnesiana,* but on closer scrutiny differs from it in that the pods are cylindrical rather than flattened, and the thorns are much shorter. Vicente Tajia says that it flowers in summer, but we have not collected the flowers. At first we knew it only from a few individuals located in areas where water pools after heavy rains around the southwestern base of Mesa Masiaca, but later we observed it in similar places near Yavaritos. We showed some of the pods we collected to Chico Leyba of Yavaritos, and he instantly identified it as cochirepa. He reported that it is not at all uncommon around that village. MEDICINE: Vicente, who pointed out the tree to us, notes that for teeth that are loose in the gums, the skin is removed from the pods, and the inner pith containing the seeds is chewed. When saliva accumulates, it is swished around the gums, strengthening both the teeth and gums.

*Albizia sinaloensis* Britt. & Rose                            joso (palo joso)

A graceful legume with a straight, yellow white bole rising to 20 m, where it branches into a canopy of lacy greenery. In southern Sonora it grows exclusively along watercourses, where it is valued for its beauty and shade. It also provides strategic perches for birds of prey and large waterfowl. Saplings rapidly reach maturity. A few individuals grow along the lower reaches of the Río Yaqui and its tributaries. It is common along some tributaries of the Río Mayo, and it is abundant in the Río Fuerte drainage. In Sinaloa, josos grow in upland locations where deep soils allow their roots to penetrate to groundwater. ARTIFACTS: The wood is used for making utensils such as spoons and bowls. At San Antonio the seat of a bench had been shaped from joso. INDUSTRY: The wood is sometimes used for roasting agaves. The flame it yields is not particularly hot. At Masiaca the bark is soaked and used to produce a white-colored leather that is marketed commercially. MEDICINE: In Sinaloa, the pulverized bark is thought to be a good remedy for snakebites.

*Brongniartia alamosana* Rydb.  cuta nahuila "worthless tree" (palo piojo [vara prieta, near San Bernardo])

A very common small tree in moist foothills thornscrub and in tropical deciduous forest of southern Sonora. The slender trunk rises irregularly to 8 m, the bark pocked with tiny lenticels that give rise to its Spanish name, translated as "the louse tree." It flowers dark blood red in June and July at the height of the dry season. Its drying fruits explosively dehisce following las aguas, their sudden popping sound often startling those nearby. *Nahuila* in Mayo means "good for nothing." It is also Mayo slang for a homosexual. INDUSTRY: Despite its name, it makes acceptable firewood.

*Caesalpinia cacalaco* Humb. & Bonpl.  huanaca (huisache)

A strong-trunked tree up to 10 m with spreading foliage and brilliant yellow flowers in December and January, it is widely grown as an ornamental. The only possibly wild specimen we know of grows in Arroyo Agiabampo roughly 2 km north of the village. While it may be a plant naturalized from seed spread by cultivated specimens, we believe it to be native to the wash, which harbors a diverse flora. Natives report it is more common in Sinaloa. AESTHETICS: It is a popular ornamental. MEDICINE: Felipe Zazueta of Agiabampo pronounced the bark to be an excellent remedy for bruises. It is steeped in hot water and the liquid is applied as a wash to the affected area. CONSTRUCTION: The wood is used for lumber in building houses. INDUSTRY: It is also used for firewood.

*Caesalpinia caladenia* Standl.  jícamuchi (palo piojo blanco)

Sometimes a 3 m shrub, other times a slender, elegant 10 m tall tree with yellow flowers, the straight-boled jícamuchi is absent from some areas and remarkably common in others. The thin, light gray bark is underlain by bright green that is revealed when the bark is scratched, a means of identification used by the Mayos. Flowers are pale yellow, beans black. CONSTRUCTION: The wood is said to make acceptable vigas if cut during the full moon. INDUSTRY: It makes acceptable firewood.

*Caesalpinia palmeri* S. Watson  jícamuchi (palo piojo)

This is a large shrub with bright yellow flowers. Its bark is dark red with white spots or lenticels resembling lice, hence its Spanish name. It grows abundantly in coastal thornscrub but is seldom seen in tropical deciduous forest. The Mayos do not distinguish *C. palmeri* from the tree *C. caladenia*, which is less common in much of the Mayo region. Only botanists can readily distinguish the two. Swarms of bees are attracted to the flowers. LIVESTOCK: Jícamuchi acts as a reserve fodder for goats, which resort to nibbling on the tough branches when more tender forage has been exhausted. MEDICINE: Reyes Baisegua of Sirebampo said that his family boils pieces of the branch with salt and chews on the cooked branches. The catalytic action produces a dental anesthetic, he says.

*Caesalpinia platyloba* S. Watson       mapáo, cuta síquiri (palo colorado)

Mapáo is a strong-trunked, dense-wooded legume crowned with large clusters of persistent broad pods that turn russet brown with age. It is common on steep rocky slopes in foothills thornscrub and in drier portions of tropical deciduous forest. The mottled bark often retains a reddish tint. The roundish leaves and pods are somewhat more persistent than those of other legumes, assuming a reddish color in senescence. Mapáo flowers in panicles of bright yellow in May, when the tree retains many old reddish pods but has few or no leaves. It is a most esteemed and valued tree in Mayo country, and despite a statewide Sonoran ban on harvesting, its numbers continue to be decimated owing to its excellence for home construction and fence posts. Vicente Tajia laments that ejidatarios (members of communally owned lands) will be punished if they are caught cutting a tree, while owners of large private estates clear thousands of acres with impunity. In Sinaloa the tree seems to be more abundant. In the late 1970s the Mexican government sponsored an attempt to raise the trees commercially near Masiaca. The project was a failure, even though the trees propagate prolifically. CONSTRUCTION: The wood is acclaimed as the best for fence posts, reputed to last a hundred years. It is also used for vigas, the beams used to support roofs, because it resists rot and insect infestations.

*Caesalpinia pulcherrima* (L.) Sw.       ta'áboaca "the sun consumed it"
                                         (tavachín; red bird-of-paradise)

This shrub is quite common in moister areas of the region. It is especially abundant near roads, but absent in areas remote from humans. It is native to the West Indies and Mexico. Showy plants such as this have been cultivated for so long in the New World tropics that their original distribution will never be known. The gaudy cultivar common in Tucson, Arizona, is Barbados Pride, presumably of Caribbean origin. AESTHETICS: Ta'áboaca produces showy red and yellow blooms up to 15 cm across, the largest and most colorful flowers in the region. It is often planted in yards and along well-used pathways. FOOD: The tender beans are eaten, often as a snack for those who pass along on trails. MEDICINE: For "twisted mouth" (*boca torcida, alferecía*—perhaps Bell's palsy), a poultice is made by mashing the root and rubbing it on the paralyzed part of the face. Vicente Tajia assured us that although the remedy and treatment sounded far-fetched, it does, in fact, work. Braulio López touted ta'áboaca for sores in the mouth and for aching or loose teeth. He grinds the root into a powder and then applies it directly to the tooth or sore. It can be dissolved in water as a healing mouthwash as well. Fausto López of Nahuibampo recommended the powdered root as effective on sores in general.

*Caesalpinia sclerocarpa* **Standl.**  ébano

A large, spreading tree to 15 m tall with a trunk up to 1 m in diameter, ébano has varicolored bark and very dense, hard wood. Its fruits are thick and tough, not particularly relished by LIVESTOCK unless they are fresh. It is a rare tree in the land of the Mayos, found only in the extreme southern part of the region and uncommon there. Its scarcity is probably due to a combination of its narrow ecological requirements and the persecution it suffers due to its excellent wood, said to be the best in the region for vigas. In Sonora it has been found only near Camahuiroa and Agiabampo. Four large trees are to be found on Arroyo Camahuiroa; a fifth was felled by the high winds of Hurricane Ismael in September 1995. It is somewhat more common in Sinaloa to the south. Large limbs have been sawed from one large Camahuiroa specimen, apparently with no harm to the tree. Although the seeds germinate readily in soils in the vicinity of arroyos, the smashing action of cattle hooves prevents the young plants from maturing. The name may be of Spanish origin. CONSTRUCTION: It is a pity that the tree is not more widespread, for it is much admired by natives, and beams from its wood will endure for decades. CULTURE: Alejo Bacasegua of Camahuiroa, known for his ability to remove curses and hexes, remarked that a small branch of ébano placed on the doorway or a windowsill of a house will keep rabid animals from entering. MEDICINE: He also mentioned that the bark ground up into a fine powder, or the ash from a fire, when cast onto a dog suffering from rabies, will cure the dog. Felipe Yocupicio has an alternative formula: he soaks ébano beans in water, dries them, and then grinds them into a powder. This he adds to water drunk by dogs or cows, and he believes it prevents them from contracting rabies or cures those who have it.

*Calliandra emarginata* **(Willd.) Benth.**  (guamuchilillo)
[*C. rupestris* **Brandegee**]

This is a rather uncommon thornless shrub that superficially resembles the far more common *Senna pallida* and young *Pithecellobium dulce*. It grows into a tall shrub or small tree in moist canyons and incised washes. LIVESTOCK: Its succulent leaves and tender stems are gobbled up quickly by cattle.

This great *ébano* (*Caesalpinia sclerocarpa*) provides shade for people and beasts along Arroyo Camahuiroa.

*Chloroleucon mangense* (Jacq.) Britt. & Rose
[*Pithecellobium undulatum* (Britt. & Rose) Gentry]

ooseo suctu "lion's claws"
(palo fierro, palo pinto)

A many-trunked, thick-muscled small tree with spreading branches that rise and descend umbrella-like to cover a wide area. Its height seldom reaches more than 6 m in the Mayo region. The convoluted, maculate trunk produces spectacular mosaic patterns of white, black, and gray splotches, sometimes layered over a greenish gray underbark. The boles are often hollow inside, providing shelter for a myriad of invertebrates, small mammals, *cachoras* (spiny lizards in the genus *Sceloporus* and iguanas). Spiny young plants in shady tropical deciduous forest can be confused with mezquites. INDUSTRY: The wood is valued for firewood. CONSTRUCTION: It is also used to make durable posts. The beans and leafage are eaten by LIVESTOCK. Its spreading habit also provides shade for livestock, though not much for humans since its descending thorny branches deter human access.

*Conzattia multiflora* (B.L. Rob.) Standl.　　　　　　　　joso (joso de la sierra)

A stately, uncommon tree of the higher foothills and mountain slopes, often reaching 20 m or more, with a smooth, silvery bole usually rising with great dignity to spread out above the canopy. It is the tallest tree in Sonora's tropical deciduous forest. It is leafless much of the year, its elegant form starkly accentuating the surrounding forest. It flowers a glorious yellow in late June, but the large compound leaves are present only during the summer rains. Shortly after the rains end, it sheds its leaves, the first large tree to do so. In September, its crown bursts into fruit, whose golden presence is seen from afar. In winter the persistent pods turn orange and add an unusual tint to the winter landscape. The Mayos (and Guarijíos) indicate that the tree, although handsome, has no use, perhaps because it is uncommon in the lower elevations of the tropical deciduous forest with which they are familiar. Its wood is considered inferior, soft and quickly decomposing. Bye (1995), however, states that it is one of two trees yielding commercially valuable lumber from the tropical deciduous forests in southern Mexico, where it is called *guayacán,* a name reserved for *Guaiacum coulteri* in the Mayo region and Sonoran Desert. We include it because it is known for its beauty, and hence has aesthetic value, but also because it is used elsewhere and may have been used by Mayos in the past.

*Coursetia glandulosa* A. Gray　　　　　　　samo, causamo "mountain samo"

A resilient, slender, thornless shrub (occasionally a tree) seldom more than 6 m tall, usually found with several to numerous trunks springing from a common root and quickly spreading and curving outwards. It flowers when leafless in the spring, the branches supporting a rather sparse harvest of white, yellow, and rose blossoms popular with bees. The leafage is also sparse, supplying hardly any shade. ARTIFACTS: The straight, springy wood is excellent for arrows. Baskets are still made from the branches, and chairs from the limbs and trunk. MEDICINE: A lac scale insect (*Tachardiella coursetiae,* Kerridae) occasionally attacks the branches, producing an orange, hardened sap commercially marketed under the name *goma de Sonora.* Mayos boil the gum with cinnamon and mistletoe gathered from mezquite. The resultant tea is administered to children to cure diarrhea reportedly contracted from their mother's milk. It is said to be a good remedy for stomach ailments. The Jesuit Ignaz Pfefferkorn (1949:67), writing after the expulsion of the Jesuits in 1767, observed: "When it is dissolved in water and swallowed, this gum is an excellent remedy for hemorrhages and bleeding. Even after the first swallow the patient is sometimes comforted, and his illness must be very stubborn if he finds it necessary to take this drink three or four times. . . . It is in truth to be regretted that this remedy, as well as many other very valuable ones with which Sonora is enriched by nature, are not made known more widely in the world."

***Dalea scandens*** (Mill.) R.T. Clausen  
var. ***occidentalis*** (Rydb.) Barneby

jíchiquia "broom"; yore cotéteca "sleepy" (dais, escoba)

This jíchiquia is a weak shrub up to 2 m high. ARTIFACTS: The straighter branches are cut when leafless, bound tightly together, and used as a broom. MEDICINE: At Los Capomos, Patricio Estrella reported that leaves placed under a pillow induce sleep.

***Desmanthus bicornutus*** S. Watson  
***D. covillei*** (Britt. & Rose) Wiggins ex. B.L. Turner

siteporo, jíchiquia "broom" (dais, escoba)

We lump these rather nondescript leguminous shrubs together. ARTIFACTS: Both are eagerly sought for making brooms, a function which seems rather pedestrian and uninteresting until one sees firsthand the Mayos' obsession with cleanliness and the need for inexpensive brooms. They are thus important shrubs. Vicente Tajia was pleased one day when in the field with us to come upon a large population of siteporo growing a couple of kilometers from his house. He cut enough branches to make a broom and vowed to return for more. The broom produced by his wife, daughters, and granddaughters was about 1.5 m long with the smooth parts of the branch bound tightly together and the shrubby ends spreading out. With such brooms, women often sweep twice a day, swinging the broom with a broad motion like a scythe, leaving the swept dirt looking clean and fresh. Siteporo is intimately known to the Mayos of the coastal plain, and a song about it is well known to those who participate in fiestas. It is sung by the deer dancer's musicians. It goes as follows:

SONG TO THE DEER

| | |
|---|---|
| Matchilia bétana | You come creeping to the fire |
| Segua choi, choi, huilama | You come slowly, nibbling flowers of siteporo |
| Siteporo so'jcu bétana | You come slowly through a grove of siteporo |
| Matchi choi, choi, huilama | You come slowly, nibbling flowers of siteporo |
| Segua yoreme | The flower of the Mayos |

***Desmodium scorpiurus*** (Sw.) Desv.

subuai muni "quail bean" (frijol de codorniz)

This low-lying perennial herb of moist canyon bottoms, where sabinos (*Taxodium distichum*) grow, produces pretty purple white flowers in the late spring. Although it has no human uses, Francisco Valenzuela pronounced it an important food source for *choli* (elegant quail; *Callipepla douglasii*, Phasianidae), which are eaten whenever they can be captured.

### *Diphysa occidentalis* Rose         huoquihuo "footprints"; huicobo (huiloche)

A many-boled large shrub or small tree growing on more arid or cut-over sites in foothills thornscrub and on the dry hillsides of the tropical deciduous forest. It flowers a delicate light yellow in early winter. It ranges from the desert-like hills around Topolobampo Bay in Sinaloa to the thick tropical deciduous forest on the upper Río Mayo and in innumerable locations between. While it cannot be said to be abundant, it is broad ranging. ARTIFACTS: The splendidly straight trunks, seldom more than 5 cm thick but springy, durable, and extremely dense, are cut to make canes and clubs of unyielding texture which resist aging of any sort. The trees would grow thicker but are so sought after that one seldom observes a plant without two or three stumps at its base. Gentry (1942) notes that the Spanish term *huiloche* also referred to a beating with a club of the same material. The trees are also cut in considerable numbers and split to make cross-braces for the attractive Mayo *taburetes* (stools), whose seats are fashioned from animal hides. These will endure decades of heavy use. Most rural men know a handy location to cut huiloche, and they harvest it regularly. CONSTRUCTION: Due to their strength and durability, the trunks are widely used as latas (cross-hatching) supporting dirt roofs.

### *Diphysa suberosa* S. Watson         (corcho)

A small tree with compact, spindly leafage and dark, deeply fissured corky bark reminiscent of some pine species. It occurs only sporadically throughout the Mayo region but is common in the localities where it grows. A specimen was not initially recognized by Francisco Valenzuela, who mistook it for algarrobo (*Acacia pennatula*). Upon examining the tree more closely, however, he affirmed that it was corcho, noting that it had no long spines and the bark was soft. ARTIFACTS: The bark is used to make corks (*corchos*) for bulis, the gourds used to by natives to transport water.

### *Erythrina flabelliformis* Kearney         jévero (chilicote, flor de mayo, peonía; coralbean)

A thick-boled tree with silvery gray, sometimes reddish orange or coppery bark similar in texture to the bark of palo barril, but with a noticeable vertical grain. The trees sometimes exceed 10 m in height. Usually, however, the trees are shorter and truncated. They frequently rise in several thick trunks from a common root, with large, broad, down-turned, harmless thorns protruding from the bark. The thick trunks taper quickly, thus resembling a candle. The coralbean tree was once common on hillsides throughout the region, but its many uses have led to overharvesting. Larger specimens are now found only in moister and more remote forests to the north and east. Unlike many tropical plants, the range of *E. flabelliformis* extends well into colder areas in the Sierra Madre Occidental to the east and to the mountains of southeastern Arizona, except that recurrent freezes prune the stately tree to a 1 m tall shrub on exposed south-facing slopes. In May and June the brilliant red spears of flowers can be seen from considerable distances in the dreary mojino of the parched forest. In fall and spring the persistent pods may linger, often dehiscing lazily to reveal a flash of bright red beans that fall to the ground and are left, thus suggesting that they are toxic when dried. Several Mayo consultants denied that the red beans are toxic, however, leaving us to speculate that the reputed toxic properties may be a North American superstition. Nevertheless, no one reports having eaten the red beans. MEDICINE: The young seeds are toasted and ground, and then a pinch is brewed into a tea for diarrhea and for the stings of scorpions. The bark is also directly applied to the stings. The thorns are boiled into a tea that is taken for urinary problems. Rosario "Chalo" Valenzuela of Baimena rubs the toasted beans directly on the forehead of patients to alleviate throbbing headaches. FOOD: Francisco Valenzuela pronounced the beans to be edible when green. We have not tried them. ARTIFACTS: The light, balsa-like wood is made into corks, masks, benches, and figurines. Don Serapio Gámez of San José de Masiaca still uses the wood for corks to plug holes in the *calabasas* (squashes) he makes into bulis (canteens). A bench built of the wood at El Rincón was light and quite comfortable. At Nahuibampo, rafts were formerly made of logs bound together to cross the Río Mayo in flood. In the Sierra de Alamos and near Güirocoba, straight branches cut from live trees are planted and grow to become a living fence, becoming most noticeable in May, when the brilliant red spearlike blossoms decorate the tree. INDUSTRY: The bark is boiled by Mayo weavers to produce a yellow dye.

*Eysenhardtia orthocarpa* (A. Gray) S. Watson    baijguo "breeze" (palo dulce; kidneywood)

Usually a shrub in thornscrub or Sonoran desertscrub, in protected canyons in the Mayo region baijguo becomes a tree up to 7 m tall with a myriad of uses. In the Sierra Madre it grows locally to 10 m. It flowers opportunistically, often remaining in flower from May through September. Honeybees frequent the dense spikes of small white flowers. ARTIFACTS: The wood is said to be ideal for tool handles. CONSTRUCTION: The durable wood, nowhere common, is said to be the hardest of local woods in the Sonoran tropical deciduous forest. It is used in walls, vigas, and cross-hatching, the straight sticks laced on vigas to support the earthen roof. INDUSTRY: Boiled alone, it produces a blue to pinkish dye, depending on the boiling time. Boiled with the root of tajimsi it produces a fine purple dye that is quite permanent when wool is steeped in it overnight. MEDICINE: Palo dulce is said to be capable of curing vomiting in hens when a small piece of the wood is placed in their drinking water. (How often or why hens vomit was not stated.) Francisco Valenzuela said that people make a tea from the branches to prevent recurring fever. FOOD: Domingo Ibarra attributes to the bark the power to cool water into which a few strips are placed. The water turns blue as well, he claims. LIVESTOCK relish the herbage, which is said to have curative powers for them. Cows are said to get fat on it, a possible explanation for the Spanish *dulce*, meaning "sweet." Such qualities would presumably render it scarce due to overharvesting by man and beast, but it is rather common in the monte, never growing in abundance, but well distributed. It is a candidate for cultivation as a forage plant and the sort of tree that many wish were more common.

### *Haematoxylum brasiletto* Karst. júchajco (brasil)

A common small tree seldom growing more than 6 m high, with a strongly fluted trunk of mottled gray and charcoal color, often an imposing work of natural artistry. (Brasil is in the same genus as the tree that gave the name to the South American country.) Its uses are many and varied, and the tree is so valuable that life in thornscrub and the tropical deciduous forest would be quite altered without it. Few homes exist that do not have some ongoing use for brasil. Unfortunately, it is heavily harvested and grows slowly. Populations are everywhere diminishing, and large trees have become scarce indeed. MEDICINE: The red part of the cambium is boiled and made into a tea for high blood pressure and for strengthening the heart. Some people drink it purely for its refreshing qualities. The liquid is bottled and a few drops used as a cure for tiricia (sadness or perhaps depression), especially in children. Don Francisco Valenzuela of El Rincón notes, perhaps with a trace of mirth, that three cooked lice from the head of a sick person can be placed in a tincture of brasil to diagnose tiricia sickness. If the solution turns bright yellow when applied to the palm, the donor of the louse is a victim of tiricia. Brasil's efficacy in curing tiricia is acclaimed throughout the Mayo region. INDUSTRY: The brownish liquid, becoming reddish with greater concentration (sometimes mixed with tajimsi root), is used as a potent red dye in making rugs and also in dyeing masks and basket materials. The more the wood is boiled, the greater the saturation of red. More than anything, though, the wood is cherished as firewood. It burns with a hot greenish flame, produces almost no smoke, and leaves little ash. Its desirability leads to excessive exploitation of the slow-growing trees. ARTIFACTS: Sharpened points of smoothed branches are used to shuck corn. These are also used as lance heads by participants in Lenten festivities. Mayo

Trunks of *brasil* (*Haematoxylum brasiletto*). (Photo by T. R. Van Devender)

blanket weavers consider brasil wood to be the best for carving into sasapayecas, the batten for tamping woven strands on their looms. Sections of brasil trunks also make corrals of a most agreeable appearance. A tinajera (tripod supporting a water jug) at San Antonio was constructed from the wood. CONSTRUCTION: In Sinaloa (and in inaccessible areas in Sonora) the tree grows large enough to be used for ceiling beams.

*Haematoxylum* sp.                                                      tebcho (brasil chino)

A large, multistemmed shrub without stem fluting, it is found in more xeric locations on the upper Río Mayo near San Bernardo, along the lower Río Cuchujaqui and adjacent uplands, and in the drainage of the Río Cedros. Although it is widely known, it has yet to be officially described by taxonomists. Its flowers are of a more orange hue and appear in the spring. Those of *H. brasiletto* are yellow and usually appear in the late summer. CONSTRUCTION: Straight pieces of the wood are used for latas (cross pieces) in roofs at Yocogigua. MEDICINE: Vicenta Valenzuela Bacasegua of Basiroa reported that a tea brewed from heartwood is effective when rubbed on varicose veins.

*Havardia mexicana* (Rose) Britt. & Rose chino (palo chino)
[*Pithecellobium mexicanum* Rose]

Similar to and often confused with mezquite, but readily differentiated by its tinier leaflets and more delicate aspect. In low-lying areas on the road between Alamos and San Bernardo, especially those areas with silty soil, it is the dominant tree, often exceeding 12 m, with a thick trunk and dark, fissured bark. The trunk often rises irregularly from the base. It is distinguishable from teso (*Acacia occidentalis*) by its more uniform, graceful, and open appearance and its paired, rather than single, catclaw thorns. Its fragrant white ball flowers attract hummingbirds and myriad bees and flies in March and April. The tree is rather slow-growing and subject to overharvesting. ARTIFACTS: The hard, reddish wood is widely used for furniture, ceremonial masks and rattles, wooden bowls, and planks. The late Benigno Acosta of Teachive carved the wood into dolls depicting Mayo festival dancers. His sons continue the tradition. Federico Badón of Rincón de Aliso, Sinaloa, used palo chino to construct the loom on which he weaves woolen blankets. CONSTRUCTION: The strong wood can be used for beams and posts, for which it is most attractive. Many Mayo homes retain small logs of chino for use in a variety of situations. INDUSTRY: The bark can be used for tanning hides but is generally considered inferior to mauto bark. Chino can be used as firewood but is not a preferred variety.

*Havardia sonorae* (S. Watson) Britt. & Rose jócona
[*Pithecellobium sonorae* S. Watson]

A small leguminous tree more or less endemic to the greater Mayo region, with vicious spines punctuating well-defined rings around the trunk and branches. Locally abundant, it grows in drier parts of the tropical deciduous forest and moister parts of thornscrub and desertscrub, reaching 8 m in height. It produces white globular flowers in June and July. The blossoms produce an intensely sweet perfume, attracting swarms of bees and hummingbirds and noticed by humans as well. It is viewed with disfavor by ranchers, who find that after clearing the trees to promote the growth of grasses, they quickly grow back in dense thickets with ever more vicious thorns, interfering with the pastures the ranchers wish to cultivate in the unruly thornscrub. Jócona serves an important ecological

A *palo chino* (*Havardia mexicana*) near San Bernardo in full flower in March.

function: its menacing thorns and bushy habit are a deterrent to livestock, permitting the survival under its sheltering branches of numerous plants and animals that would otherwise be nibbled or chewed away or have their habitats decimated. MEDICINE: Vicente Tajia said that the bark of the jócona is boiled with salt and poured in a nonplastic cup to be gargled and drunk to cure a sore throat. INDUSTRY: The wood makes satisfactory cooking fires. LIVESTOCK eagerly gobble the legumes, which are an important forage for goats.

### *Indigofera suffruticosa* Mill.   chiju (añil; indigo)

The indigo plant is a fast-growing 1–2 m tall perennial shrub found in deep soils, especially where summer rains are likely to create puddles. It is widespread from Sonora southward to South America and the Caribbean islands. Claims that it is introduced in Sonora are difficult to accept considering its distribution, its similarity to two other species in the Río Mayo region (*I.* cf. *densiflora* Mart. & Gal. and *I. sphaerocarpa* A. Gray), and the fact that the Mayos have a name for it. When the 2 cm long fruits ripen, they turn black like tiny overripe bananas and adhere to the plant in rows like stacked sickles on an assembly line. Although it is not rare, chiju tends to grow in isolated thickets, usually in disturbed soils, often on roadsides. Some Mayo weavers plant seeds in favorable localities to ensure a supply for their weaving. INDUSTRY: The leaves of chiju produce the deep blue found in Mayo cobijas (blankets). Nearly all weavers are familiar with it and its preparation. Doña María Teresa Moroyoqui of Teachive demonstrated how the dye is prepared. The fresh leaves are soaked in a large basin for an hour and then rubbed together briskly. After a while the dye begins to emerge. The resulting liquid is boiled for many hours to concentrate the dye. When it is sufficiently thick and concentrated, it is dried in the sun and wrapped in a cloth pouch. It is then submerged and stored in an olla (clay pot) filled with the urine of children, which acts as a mordant. (Apparently adult urine will not suffice. Vicente Tajia pointed out that the urine of children is "purer" than that of adults.) Buenaventura Mendoza warned that when handling the dye, it is important to thoroughly wash the hands with soap to remove oils from the skin that might make the dye run. An old woman who died in early 1996 was known to make the best chiju of all. Some inexperienced women find making the dye difficult. The more knowledgeable weavers believe that such problems stem from not gathering the leaves in the full moon, which they say "ripens" (*sasona*) the leaves. During the very dry years of 1995–97 summer rains were insufficient to ensure a good crop of chiju, making dye production difficult. Some weavers asked us to bring plants from other localities to ensure a good supply. MEDICINE: Francisco Valenzuela of El Rincón said that people used to boil chiju leaves to make a tea that they would drink on an empty stomach to cure diarrhea.

### *Leucaena lanceolata* S. Watson                              huique (guaje, palo bofo)

A slender tree with a branchless lower trunk and a broad crown of leguminous foliage. It produces many thin, persistent beans. It is rather rare, restricted to scattered specimens in well-developed tropical deciduous forest, and few individuals have a name for it. CONSTRUCTION: Although no specific uses for the tree were described in Sonora, due to its scarcity, Don Pancho Valenzuela mentioned that it would probably be useful for fence posts. INDUSTRY: Don Vicente Tajia claims it can be used for firewood. It is better known in Sinaloa, where it is used for both fences and firewood and grows to greater size, according to Patricio Estrella of Los Capomos.

### *Lonchocarpus hermannii* Marco                                 nesco (Venus tree)
### [*Willardia mexicana* (S. Watson) Rose]

A common tree of the forest, characterized by a somewhat twisted bole, the gray bark frequently invaginating in unusual and sensuous patterns. Usually a slender tree, it grows more than 10 m tall in mesic locations in the tropical deciduous forest or along moist washes in thornscrub. It is generally absent or shrubby in upland thornscrub. In late May and June it bursts into bloom, a spray of lavender to pinkish flowers visible from a distance. MEDICINE: The bark is brewed into a tea widely used as a bath for mange and ticks in animals and to get rid of lice in people (ticks are not known to infest humans in the region). Several consultants suggested an infusion as a remedy for dandruff. INDUSTRY: A concentrated brew is also said to have been used many years ago to stun fish.

*Lysiloma divaricatum* (Jacq.) Macbride                           mayo (mauto)

A stalwart of the tropical deciduous forest, possibly the most common tree in the uncut forest, found only marginally in thornscrub. It is also one of the most useful trees. Untold numbers rise to a height of 8–10 m, their anvil-shaped crowns meeting and forming a nearly uniform canopy. As the trees mature, small sections of stiff bark peel away into concave strips, presenting a characteristic rugged appearance. Natives distinguish between mauto blanco (white), mauto colorado (red), and mauto prieto (black), which, they assert, have entirely different woods; these varieties are not readily discernible to botanists. Individual trees attain heights in excess of 20 m in areas inaccessible to woodcutters. The tree grows rapidly, not as quickly as güinolo, but much faster than mezquite. The Río Mayo takes its name from the tree; thus the translation for "Río Mayo" is the Mauto River. Sobarzo (1991), an authority on Sonoran terminology, attributes the common name "mauto" to the Opatas, an extinct indigenous people of Sonora whose influence survives in many place names and terms for fauna and flora in the state. The name "mauto" is also used in Sinaloa and in Baja California. MEDICINE: A tea made from boiling the bark is said to be good for diarrhea. It is purported to cleanse the stomach and help expel phlegm. Many people chew on the bark to strengthen the gums and tighten loose teeth. INDUSTRY: The wood is greatly valued for firewood. The immense wealth in silver extracted from the mines near Alamos was smelted with fires that burned primarily mauto and brasil. The bark was formerly the basis of the southern Sonoran tanning industry and is still used in Masiaca, an important leather center in the heart of Mayo country, where large backyard areas of leather shops are given over to the drying of strips of mauto bark. The bark is shredded and pulverized, and the residue added to a soaking tank, where it is mixed with lime. There the hides are soaked. The barks of guamúchil and joso are also used, but that of mauto is preferred. According to those who gather the bark, it is harvested without obvious detriment to the tree, which they say recovers fully within a year. CONSTRUCTION: Only amapa and palo colorado are ranked superior for house construction. Vigas of mauto will last indefinitely. It is also used for fence posts. Someday a well-organized project to plant and raise mauto for firewood, bark, and poles will get underway.

***Lysiloma watsonii* Rose**     macha'aguo "man's thigh" (tepeguaje; feather tree)

Tepeguajes range in size from a small tree of 5 m to a large, prominent, graceful tree with dark, fissured bark and an angular, irregular trunk. In the Mayo region it is equally at home in tropical deciduous forest and oak woodland. Individuals may reach 20 m in height and spread equally wide. They are especially common in the drainage of the Río Fuerte, where they are one of the dominant species. The trees develop an individual appearance and apparently reach a great age. Tepeguaje superficially resembles mezquite with its dark trunk and leguminous appearance, but its longer, lacy leaf patterns, lack of thorns, and more open habit distinguish it from the latter. It leafs out in late April through June with light green, almost yellow foliage standing alone against the silver brown of the dead landscape. This contrasts with *L. divaricatum* and many other trees, which do not leaf out until the first rains in July. The leaves gradually turn darker green through the summer. It is an important source of shade in the powerful heat of late spring. MEDICINE: The bark is brewed into a tea used to treat gas pains, gastritis, and liver problems. As one Mayo explained, an indicator of the latter condition is the inability to eat and digest fats (which is often symptomatic of gall bladder problems). The tea is also believed to cure infertility in women who cannot have children because they are "cold" as a result of having eaten cold things (ice, a lemon, or an orange) during menarche. Hot tea brewed from the bark is believed to restore them to the "proper" temperature. This epidemiology is a manifestation of the folk diagnosis of disease as a function of the relative balance of hot and cold. The tea is also generally prescribed for "women's" problems. Only the inner bark is used. Many households maintain a supply of the bark for various medicinal purposes, especially since the larger individual trees are remote from most Mayo villages. CONSTRUCTION: Tepeguaje wood is widely used in building and in making posts (notably gate posts). It is a heavy and resilient wood, but not particularly hard. Along old *serrano* (mountain) trails one often encounters ancient gate posts of stout tepeguaje timbers, each with four or five holes 6–8 cm in diameter through which the horizontal cross poles fit. ARTIFACTS: The wood is used to make molds for producing *piloncillos* (brown sugar cones). Other artifacts made of tepeguaje wood—molds, drain spouts for clandestine stills, troughs, and so on, some of which are more than a century old—can be seen throughout the region.

*Mimosa asperata* L. [*M. pigra* L.]     yete ogua "sleepy bush" (rama dormilona)

This viciously armed shrub lines the banks of the old Río Mayo channel near the Sea of Cortés, making access to the puny current nearly impossible. This is just as well, for the runoff consists almost entirely of sewage, petrochemicals, pesticides, and the agricultural swill from chicken and pig farms. *M. asperata* is also found along streams and canals well to the south. The leaves are sensitive, folding rigidly when touched, as if in prayer. This shrub is a serious nuisance where it has been introduced in tropical areas of northern Australia. CULTURE: Several Mayos claim that placing the leafage under one's pillow will induce a profound sleep, hence the Mayo name. The remedy is said to be especially effective for children who have trouble sleeping. As Vicente Tajia cautioned, one should first remove all the thorns, for they are stout indeed and by puncturing one's head will make sleep impossible.

*Mimosa distachya* Cav. var. *laxiflora* (Benth.)     nésuquera (uña de gato;
Barneby [*M. laxiflora* Benth.]     wait-a-minute bush)

A large shrub or small tree seldom more than 3 m tall, with sharp, claw-shaped thorns. It flowers in late summer, producing fragrant pink to white blossoms. LIVESTOCK: In spite of its intimidating armor, goats eat the shrub's leaves and the beans. INDUSTRY: It also makes acceptable firewood. ARTIFACTS: Some women make brooms from it, which is quite a chore since the recurved thorns must be removed before it can be handled.

*Mimosa palmeri* Rose     cho'opo (chopo)

Chopo is a small tree with a straight bole and wispy leafage, well armed with potent hooked thorns. The tree has become scarce due to its renown for charcoal making and fence post durability. It is currently accorded protected status by government fiat, largely ineffectively, if piles of fresh-cut posts in various locations are any indication. Its curved thorns are an effective deterrent to close examination but fall readily to the harvester's machete. Elongate spikes of pink purple flowers adorn the tree from July to September, and sometimes in April and May. In November the myriad leaflets turn orange, giving the small trees a characteristic aspect of wispy rust color. CONSTRUCTION: It is renowned for making durable fence posts. INDUSTRY: It is also harvested to make charcoal. MEDICINE: Chopo bark is widely chewed to strengthen and clean teeth. Genaro Liso Jusaino of Nahuibampo said he uses it regularly. ARTIFACTS: The dense, hard wood is used to make handles for implements.

### *Olneya tesota* A. Gray                                         ejéa (palo fierro)

*Olneya tesota* is an important Sonoran Desert tree, reaching its southern limit in coastal thornscrub near Masiaca. There it grows to 8 m or more in open dry areas, but it becomes a dominant exceeding 10 m in height and equal width in coastal thornscrub to the west. Both because it is relatively scarce and difficult to cut and because many other varieties of wood are available, ejéa is less significant in the lives of Mayos living in foothills thornscrub and in tropical deciduous forest than in those of indigenous peoples in the drier habitats to the northwest (see, for example, Felger and Moser 1985). The tree is unknown in Sinaloa and in Agiabampo. Vicente Tajia observed that ejéas do not grow south of Teachive, which means their southern limit roughly corresponds with the southernmost locality for sahuaro cactus. Both plants are important dominants in the Sonoran Desert. Turner et al. (1995) locate the southernmost collection at Chinobampo, roughly 15 km north of specimens whose shade has cooled us at Teachive. We noted that tree one day when we were interviewing a Mayo consultant. We asked if there were any ironwood trees in the vicinity. He responded with a wry smile, "You're sitting under one." AESTHETICS: In the vicinity of the Mayo village of Sinahuisa some of the largest specimens are still alive because they grow in people's yards and are thus protected from woodcutters and livestock. The well-armed trees blossom in the late spring, adding a delicate pink lavender to the withering, deciduated landscape. Those individuals are highly valued for the shade they provide and grace many a yard in the inland coastal villages of Bacabachi, Buayums, Saneal, Sinahuisa, and Teachive. The massive dead stump of an ejéa stands protected as a monument on the plaza in Masiaca. In Teachive a large specimen has afforded shade to herds of cows for so long that Vicente Tajia refers to it as the Cow Hotel. ARTIFACTS: The extremely hard, dense wood is carved into items of personal adornment, charms for necklaces, medallions, and so on. Several residents of Teachive carve the wood into large spoons. INDUSTRY: The wood is occasionally used for firewood and produces the hottest flame of any wood in the region (too hot, some say). CONSTRUCTION: It is also used to make horcones (posts), and those fortunate enough to have a ramada or portal with posts of ironwood will never see them fail from rot or weakness. The dead wood used for posts (and also figurines) may have endured in a desiccated state as long as nine hundred years (Raymond M. Turner, pers. comm., 1994). LIVESTOCK: Goats relish the beans, which ripen and fall to the ground gradually, providing a time-released food source. The pods are tough and resistant to opening. Goats alone seem impervious to these defenses and casually munch away on them.

Fabaceae / 211

*Parkinsonia aculeata* L. guacaporo, bacaporo (retama; Mexican palo verde)

With age these normally green-barked trees with very long, narrow, almost needlelike leaves may develop a trunk with black bark belying the common name. Individuals 12 m tall and spreading over a similar diameter are not uncommon in the region. Trees in full flower in March are showy and attract bees and flies. The tree is commonly found in disturbed habitats along roadside arroyos. Although it is native to the New World tropics, it has spread widely into the southwestern United States. Its original distribution may never be known, but it is likely that the Mayos have known guacaporo for a long time. MEDICINE: The bark or leaves are brewed into a tea for cough and fever, and for mal de orín in humans and animals. Since the tree's thorns are nasty, other barks are preferred. LIVESTOCK consume the beans and leaves.

*Parkinsonia florida* (Benth. ex A. Gray) S. Watson caaro (palo verde;
[*Cercidium floridum* Benth.] blue palo verde)

A very large palo verde, nearly 10 m tall, it is common along Arroyo Masiaca. Its green bark is the source of its Spanish common name, which means "green pole." It often has a bluish cast in the northern portion of its range, in the Sonoran Desert in Arizona and California. Caaro gives comforting shade, even in the depths of the spring drought. An individual growing in an inhabited flat near Huasaguari is more than 13 m tall and helps cool the adjacent home. Other individuals are smaller, but many grow to great size throughout the drier portions of the Mayo region. It is nearly absent in the tropical deciduous forest, where *P. praecox* is occasionally seen. *P. florida* is a common riparian plant in the Sonoran Desert, apparently reaching its southern limit in the Mayo region. Caaro is an excellent resource for LIVESTOCK, who relish the beans and, in times of drought, the green bark. Its shade is also kind to overheated kine. ARTIFACTS: The wood is fashioned into saddle horns and forms the foundation for saddles. INDUSTRY: It is used for firewood when harder woods are not available.

*Parkinsonia praecox* (Ruiz & Pav.) Hawkins choy (brea)
[*Cercidium praecox* (Ruiz & Pav.) Harms]

A green-barked, bushy tree up to 8 m tall with numerous scraggly branches. It has a rather longer trunk and longer, more irregular branches than *P. florida* and is the leafiest of the palo verdes. The bark is a distinctive light green that is often speckled with a whitish crystalline exudate. Choy is a locally dominant plant in sections of coastal/foothills thornscrub, but less common in drier portions of the tropical deciduous forest. It also springs up in overgrazed pastures and where the soil has been disturbed. It flowers profusely in March and April, the petals a brilliant dark yellow, the stamens red. When the persistent pods of

*Brea* (*choy* in Mayo; *Parkinsonia praecox*) near Teachive.

the fruits ripen, they give the tree a reddish orange hue. MEDICINE: The sap (*chu'uca*) is used as a remedy for arthritis, diabetes, bronchitis, and asthma, and is eaten as a sweet as well. Vicente Tajia produced several ounces of the sap and offered it to one of us who suffers from asthma, assuring him that the chu'uca would provide relief. A tea brewed from the bark is said to be good for bruises, sprains, and strains but supposedly turns urine dark. INDUSTRY: Doña Buenaventura Mendoza, born in 1905, reports that long ago older women would burn the root of *P. praecox,* mix the ashes in a tub with a pig carcass (she rather giggled at this) and water, and cook the godawful mess for a couple of days. Out of this would come a liquid that was poured into clay molds to make a tolerable soap. The branches are used as firewood. LIVESTOCK: Domestic animals eat the beans and the bark as well, especially goats, who gobble up the fallen beans as though they were popcorn and stand on their hind legs straining to reach the lower branches. CULTURE: On Good Friday, groups of Mayo youths dressed in fantastic fariseo (Pharisee) costumes run through all the villages, knocking over the three-foot-tall house crosses that the villagers planted in the ground the previous night. The children in Vicente Tajia's home in Teachive gathered handfuls of the yellow, red-stamened blossoms and sprinkled them on the fallen cross, completely covering it. Thus adorned with a golden blanket, it lay throughout the day.

*Piscidia mollis* **Rose**                                                              **joopo (palo blanco)**

A ragged-trunked leguminous tree resembling a grayish oak, sometimes growing in excess of 10 m tall with large oval leaflets. Trunks in excess of 50 cm in diameter are not uncommon. The large, roundish, crinkly gray oval leaflets and the large four-winged pods distinguish it from most other trees of the region. Joopos often retain their leaves well into the spring drought, all the while appearing craggy and laden with age, reminiscent of old oaks. The tree's spreading habit provides welcome shade in hot weather. A village near Masiaca is called Jopopaco, "Joopo out there." CONSTRUCTION: The strong wood is used as a base for porch posts, for posts themselves, and for beams. MEDICINE: The white flowers with a pale magenta petal appear in June and July and are brewed into a tea for epilepsy. INDUSTRY: A tea from the bark was formerly used to stun fish (and is reportedly in use in Sinaloa), hence the generic name (the Latin *piscis*—fish, plus *caedere*—to kill). The leaves are spread over a roasting agave to protect the steaming head and leaves from dirt and to help keep steam from escaping. The leaves are not bitter and not believed to be poisonous to humans, and the heating does not impart any flavor to the succulent agave.

*Pithecellobium dulce* **Benth.**                                         **maco'otchini (guamúchil)**

A venerable leguminous tree prized for its fruits and shade, the guamúchil is one of the best-known trees in the Mayo region and throughout tropical Mexico. It often grows to great heights, exceeding 20 m along streams and in arroyos, where it is fast growing. It sometimes succumbs to invasion by strangler figs. Its seeds readily germinate, and sprouting plants and small trees are common throughout the region. A row of trees planted on a riverbank near Chinatopo on the upper Río Mayo attained a height in excess of 20 m in thirty years, according to a local rancher. Rings of small thorns or thorn scars on the gray bole help identify this excellent tree. It is common in and near villages and is greatly valued by persons in whose terrain it grows. It is an important food source for birds. In May, Sinaloa crows (*Corvus sinaloae*, Corvidae) and white-fronted parrots flock to the high branches and feast on the sweet fruits. With age it often becomes scraggly and sickly, and in times of drought it may shed large branches, creating a hazard underneath. A prominent woman of Los Capomos—tired of the constant shedding of twigs, leaves, branches, flowers, and legumes—had an old tree in her yard cut down, eliminating her shade source and rendering her yard less attractive.

Considerable speculation surrounds the tree's geographical origin since it is found only in the general vicinity of human habitation. The currently favored theory places its origin in the New World tropics, from which it was carried to the Philippines by Spanish galleons, and then to the coast of India, where it was discovered and described by botanists. Apparently the Spanish were more interested in shade trees than botanical novelties. The fact that the indigenous Mayo and Guarijío names are similar and wholly different from the Spanish suggest it has been known to the native peoples for a long time. FOOD: Guamúchil yields edible pods (*buaruchia*), which most natives eat eagerly, many toasting them first in a comal (concave griddle). They have a sweet taste, reminiscent of a rather dry plum, but can pucker the mouth when harvested prematurely. Children gather the corkscrew-shaped pods, which turn pinkish red when ripe, and chew on the pulp for a snack. Some natives object to the pods, complaining that they produce halitosis and intestinal gas. CONSTRUCTION: The reddish wood is sometimes used for vigas, albeit not high-quality ones. INDUSTRY: The bark is used in curing leather, though not widely, since it is smells rather bad, somewhat reminiscent of vomit. MEDICINE: The bark is made into a tea for diarrhea.

***Pithecellobium unguis-cati*** **(L.) Mart.**     jodimaco'otchini (guamuchilillo)

This large shrub or small tree is a tropical species reaching its northern limits along coastal washes in extreme southern Sonora and northern Sinaloa. Its flowers, leaves, and fruits bear strong resemblances to those of *P. dulce*, but it remains shrubby or begrudgingly arborescent in the region. Unfortunately for the plant, its habitat is fast disappearing as agricultural development proceeds rapidly. In 1997 the abundant and rich gallery growth along the incised Arroyo Agiabampo was bulldozed clean before a detailed study of the plants along the wash was completed. FOOD: The legumes are said to be edible, similar to those of *P. dulce*, but Reyes Baisegua of Sirebampo warned that if you eat them you can easily become lost (!). Vicente Tajia observed that the plant makes excellent fodder for LIVESTOCK.

***Platymiscium trifoliolatum*** **Benth.**     tampicerán (from "ta'ampisa" [numbs your teeth])

A straight-boled tree growing to 15 m, producing bright yellow flowers in March when it sheds its yellowed leaves. It is found in moist riparian areas above 400 m. When its large leaves fall, the ground becomes covered with a leafy yellow carpet. ARTIFACTS and CONSTRUCTION: While uncommon in Mayo country, specimens at Rancho El Guayabo on the Río Cuchujaqui were known to Francisco Valenzuela. The wood is valued for making furniture and buildings. At

Jurinabo on the Río Taymuco above San Bernardo, it was the preferred wood for making molds for piloncillos (brown sugar cones). Near Tepopa, in the Sierra Sahuaribo, it was used to make gate posts. Don Francisco was familiar with these uses, although the producers were non-Mayos. Natives of Baimena report numbers of large trees on the Arroyo de la Culebra. They use the wood for construction, dishes, utensils, and carving. We decided not to seek vouchers to verify the report after being warned that we could be shot at by "culebras" (snakes), a euphemism for guardians of poppy fields. This consideration dimmed our botanical ardor.

*Prosopis articulata* S. Watson sanéa, juupa (mezquite)

*P. articulata* is common in seasonally flooded areas on the coastal plain. Around Saneal, for which it appears to have been named, it is more common than the larger-leafed *P. glandulosa*. It shares uses with *P. glandulosa* but does not achieve the latter's great size and thus is less useful to the Mayos. The honey of sanéa is believed to be sweeter than that of juupa. CONSTRUCTION: The wood is said to make fence posts superior to those made of *P. glandulosa*.

*Prosopis glandulosa* Torr. var. *torreyana* juupa (mezquite; honey mesquite)
(Benson) I.M. Johnst.

This versatile tree grows to great proportions in the tropical deciduous forest where human clearings have permitted seedlings sufficient sunlight to survive. Individuals may reach 12 m in height (occasionally more), and the branches encompass an equal width, with trunks a meter in diameter. A windrow of large mezquites intermingled with chinos grows just south of San Vicente, near Basiroa. It is worth the trip simply to walk among the giants and hear and see the many birds that take advantage of the thick foliage along the Río Cuchujaqui. This variety has huge distribution from western Texas to southeastern California and south to Sinaloa. Its life form ranges from sprawling multistemmed shrubs on coppice dunes in overgrazed desert grasslands in the Chihuahuan Desert to the giants in the Mayo region. Mezquite is not found in intact tropical deciduous forest, where it apparently cannot compete successfully with other trees for sunlight. Where forest has been cleared for homes or pastures, however, it takes root, whether planted deliberately or because the seed is transported through animal manures, and soon achieves great size. In such forests the presence of mezquites is always an indicator of human habitation. It is probably far more common now than in the past, benefiting from seed dispersal by burros, mules, cows, and goats, and is able to withstand the onslaught of livestock better than other trees. Large specimens are mercilessly harvested, however, and are becoming increasingly rare.

It is hard to imagine a more useful plant. It is irreplaceable as a shade tree and has a multitude of uses. CONSTRUCTION: The wood is used in all phases of building, and in the construction of corrals and fence posts. A home built in Teachive in the 1930s was constructed around a dead mezquite tree whose massive trunk is now the principal support for the roof. INDUSTRY: Juupa is esteemed as firewood and consequently much persecuted. The root and bark are used to produce a brown dye for blanket weaving. ARTIFACTS: Although its virtues as a shade tree are widely acknowledged, the desire to exploit the wood often outweighs the need for shade and the tree is sacrificed. The wood is carved into utensils, bowls, figurines, and any required wooden implement, as well as fine furniture. The beans are much sought after by LIVESTOCK. During long periods of no rain when all grass has been eaten by cattle, the pods, called *péchitas,* fall to the ground in a steady sprinkle and are eagerly gobbled up by goats, burros, horses, and cows. Chickens scratch for remnants of the beans in the other animals' leavings. During the prolonged drought of 1995–97 the sustained release of péchitas meant the difference between life and starvation for thousands of cows, goats, and burros in the Mayo region, while on cleared pastures sowed with buffelgrass, cattle starved. FOOD: The white gum (called *chúcata* in Mayo) is kneaded into a ball, mixed with sugar, then roasted and eaten as candy. Wounds to the tree often produce a dense brown gum that is not eaten due to its bitterness. The péchitas were formerly ground into flour and eaten, but the beans of *P. glandulosa* are said to be less sweet than those of the velvet mezquite (*P. velutina*), which is common in the Sonoran Desert to the north, but absent from the Mayo region. MEDICINE: The sap of the root is used to treat sore, itching eyes and as a cure for conjunctivitis. Several Mayos from Teachive report that their mothers chopped up mezquite leaves in the month of May, sprinkled them into sweetened water, and administered the liquid to children to cleanse the stomach and prevent fever in the coming months. Women also mashed the whitish inner bark and boiled it with salt. This was given to children as a mild purgative to cleanse the system. The leaves are boiled to produce a liquid used to wash a face afflicted with acne, and to rub into the scalp for dandruff. CULTURE: On Good Friday the Mayos of Masiaca weave branches of mezquite into diminutive arches over their uprooted house crosses after the fariseos have passed and knocked them all out of the ground.

*Rhynchosia precatoria* (Humb. & Bonpl.) DC.  chana pusi "blackbird's eye" (chanate pusi; rosary bean)

Chanate pusi is a widespread tropical liana that reaches southernmost Arizona. It is well recognized throughout the region. It grows quickly with the rains and engulfs low shrubs and the lower part of trees, sometimes climbing up the sides of houses. Light yellow flowers appear in May and in the fall after the rains. The distinctive flexible woody stems are flattened, with lateral ribs. The leaves are large and pealike. CULTURE: The legumes mature and dry, producing an attractive shiny, half-black, half-red bean somewhat larger than a BB. These are collected by natives and carried as good luck charms or occasionally made into necklaces. Chana pusi is familiar to most natives of the region, who regard the plant with delight, as do we. We find the seeds irresistibly charming. MEDICINE: According to Marcelino Valenzuela of El Rincón, the beans are pulverized in a mortar, mixed with lard, and applied under the eyelid to heal conjunctivitis. We have not tested this remedy. Elsewhere the root is cooked and made into a tea for asthma and bronchitis. It is also recommended to alleviate arthritis. Sara Valenzuela of El Rincón said that the seeds are ground with the seed of an ornamental flax (*Linum* sp.) and mixed with kidney fat to produce a mixture placed on the forehead to treat *punzadas* (severe or migraine headaches).

*Senna atomaria* (L.) Irwin & Barneby         jupachumi "skunk's ass" (palo zorillo)
[*Cassia emarginata* L.]

A common tropical deciduous forest plant, rising to 10 m tall. Jupachumi grows into a symmetrical, spreading tree with a decided crown and large, dark green, roundish leaves, a lighter green on the underside. The tree is often hung with lengthy, skinny, and persistent beans up to 30 cm long that dangle for many months like so many dead snakes hung from a line. The dark gray to black trunk is often blotched with lighter gray or white, and in older individuals is dented with slots. It flowers a dense yellow on leafless branches in April and May. The Mayo name is possibly based on the color of the trunk, but more probably on a musky odor from the root, for the crushed leaves smell only faintly aromatic. The tree is sometimes left standing when pastures are cleared. The shade it gives and the flowers make it too nice to fell. CONSTRUCTION: The wood is used for vigas and horcones in house building, and also for fence posts. INDUSTRY: It is also used for firewood. LIVESTOCK relish the beans as they fall to the ground, an important forage food in the dry spring months. MEDICINE: The beans and leaves are boiled to produce a tea that people drink to remedy diabetes.

*Senna obtusifolia* (L.) Irwin & Barneby         (ejotillo cafecillo)
*S. occidentalis* (L.) Link.

These fast-growing plants (*S. obtusifolia* is annual; *S. occidentalis* is annual or weak perennial) shoot up by the thousands in disturbed lowlands following the summer rains, reaching more than 1 m in height and often growing in such dense thickets that passage through them is nearly impossible. Livestock avoid them, suggesting that they have toxic properties. FOOD: The beans are dried, ground, and roasted to make a coffee substitute, according to Francisco Valenzuela.

*Senna pallida* (Vahl) Irwin & Barneby  huícori bísaro "iguana
var. *shreveana* Irwin & Barneby [*Cassia biflora* L.]  flower" (flor de iguana)

This common woody shrub is found on disturbed soils throughout the region. It is easily identifiable by the bright yellow double flower. Spiny-tailed iguanas are said to be especially fond of the flowers, hence the Mayo name. LIVESTOCK: Its only reported use was as a fodder for goats. Cows appear to avoid it. We find it odd that such a well-known plant has no more uses.

*Sesbania herbacea* (Mill.) McVaugh  baiquillo (ejotillo)

At maturation, this annual reaches from 20 cm to 2 m, depending on the abundance of las aguas. One may see a low-lying field devoid of vegetation and return a month later to find it a jungle of baiquillo. The flowers are light yellow with dark lines and spots. The name is apparently the Spanish diminutive of a Mayo name. CULTURE: Peasants look upon the proliferation of the weed with favor, for they plow it under and find that it enriches the soil, or so the late Jesús Nieblas informed us.

### FAGACEAE

*Quercus albocincta* Trel.  cusi (encino)

Only a small number of these spreading oaks grow in Mayo country, where they are found on red hydrothermally altered soils of Cerros Colorados near El Tábelo. The name "cusi" is used by Guarijíos for this species, and by Mountain Pimas for various oaks. Although elsewhere in Sonora cusi acorns (*bellotas*) are harvested and eaten among Guarijíos and mestizos, we have no records of consumption by Mayos. INDUSTRY: We took Francisco Valenzuela to see these oaks near El Tábelo, and he said he was familiar with their use as firewood but had no further acquaintance with them.

*Quercus chihuahuensis* Trel.  (encino; Chihuahua oak)

A spreading oak tree up to 10 m tall but commonly smaller. Normally a tree of the higher slopes, several of them grew at about 340 m elevation in El Cajón del Sabino, where they were identified by Francisco Valenzuela. Because oaks seldom grow in the tropical deciduous forest, Don Pancho was unfamiliar with their uses except for firewood and the acorns as food for *guíjolos* (wild turkeys; *Meleagris gallopavo,* Meleagrididae). In that area, however, pigs eagerly gobble up the acorns. We were unable to ascertain a Mayo name, a further testimony to the uncommon appearance of oaks in Mayo lands.

## FOUQUIERIACEAE

*Fouquieria diguetii* (Tiegh.) I.M. Johnst.  murue (ocotillo, palo adán, palo pitillo)

*F. diguetii* occurs primarily in Baja California, where it is called *palo adán,* with a few disjunct populations along the Sonoran coast from around Guaymas, near Yavaros, and near Naopatia, perhaps into coastal Sinaloa. Our field studies suggest that it is not uncommon in coastal thornscrub north of Agiabampo and should be expected near Topolobampo. *F. diguetii* tends to branch from the base and remains shrubby, lacking the typical arborescent form of *F. macdougalii.* The inflorescences are shorter and stouter than in *F. macdougalii.* Our consultants did not distinguish the two species and ascribed to *F. diguetii* the same uses as for *F. macdougalii.*

*Fouquieria macdougalii* Nash  murue (jaboncillo, ocotillo macho, palo pitillo; tree ocotillo)

Shrubby in the desert, the plant becomes a tree in the Mayo region, sometimes rising to 10 m in a straight trunk of luminescent light green, covered with exfoliations, the bark and former attachment points of branches forming a dark mosaic. Large specimens in thick forest may be confused with *Bursera fagaroides* or *Jatropha cordata,* but the spine-covered, greenish, exfoliating bark, the scraggly, curving branches, and the bright red flowers are identifying characteristics. The thorns can be hazardous: one of our consultants lost an eye when it was pierced by a murue thorn. Tree ocotillos are an important food resource for northward-migrating hummingbirds in late February and March. The fast-growing trees are planted to make living fences for LIVESTOCK. Although they are not a strong tree, planted thus they give off numerous well-armed, intertwined branches, an effective deterrent to larger animals. Some individual plants in fences have lived for more than two decades. INDUSTRY: The bark is widely used as a source of soap and shampoo. It is said to be effective on dandruff. Vicente Tajia recalls that if sitavaro is added to the soap mixture, it will disinfect as well as clean. MEDICINE: Small sections of the bark applied to a cut will stanch the bleeding and promote healing, according to Francisco Valenzuela. One seldom sees large specimens of this plant that do not show signs of bark harvesting in one way or another. Felícita Valenzuela of San Antonio claimed to have eliminated cataracts using the sap of murue. She suffered the condition for several years and then began washing her eyes with a liquid derived from steeping murue bark in water overnight. The application stung mightily, but the burning ceased when she rinsed the solution from her eyes with clear water. She said that she continued this treatment for two weeks and her cataracts disappeared.

## HYDROPHYLLACEAE

*Nama hispidum* A. Gray                    goy tabaco (tabaco de coyote; sand bells)

MEDICINE: A low-lying annual, goy tabaco (its name is a mystery to us) is not smoked, but the leaves are scorched and applied to stiff joints. According to Domingo Ibarra, the toasted leaves are also steeped for a few days in alcohol, and the resulting liquid is massaged onto aching joints. He maintains that it is highly effective.

## KRAMERIACEAE

*Krameria erecta* Willd. ex Schultes        tajimsi "sun beard" (tajuí; range ratany)
[*K. parvifolia* Benth.]

Range ratany is another widespread desert species ranging from Texas to Arizona and south to southern Sonora and Sinaloa, where it is called tajimsi. Its flowers are pink. It appears to be confined to the flattest portions of the coastal plain in the Mayo region. MEDICINE: Bark from the roots of this low, nondescript, often leafless shrub is used to help heal sores that refuse to heal naturally. A tea from the root is often recommended for diabetes. Vicente Tajia, when visiting another community on the coastal plain, dug up a specimen as a demonstration, gave us half, and took the rest home. Reyes Baisegua of Sirebampo assured us that tajimsi is an excellent remedy. He keeps a supply of the dried root in his home, even though it is abundant in the nearby monte. INDUSTRY: Tajimsi is also a source of dye when combined with leaves of palo dulce or brasil or the root of sangrengado. It apparently enhances the color of the plant with which it is brewed.

*Krameria sonorae* Britt.                    cósahui (tajuí; Sonoran ratany)

Sonoran ratany is a weak shrub up to 2 m tall with irregular, thin branches usually bare of leaves or flowers. *K. grayi* Rose & Painter is widespread in the southwestern United States from Texas to southern Nevada and Arizona and south into Sonora, but is replaced by *K. sonorae* near Guaymas. *K. grayi* differs from the latter in having densely hairy stems rather than green smooth ones, and rose purple flowers rather than white flowers turning pink. The shrub shows up in odd locations in tropical deciduous forest as well as thornscrub and desertscrub. In villages on the coastal plain the plant is readily identified by most older children as well as adults. MEDICINE: The branches are collected and widely used to make a tea taken for kidney and back pains, apparently acting as a diuretic. Cósahui is regularly used in many households, where the branches are stored in a shed or tucked into the rafters for easy access. Teresa Moroyoqui of Teachive found it helpful when she suffered from an aching back while weaving cobijas.

### LAMIACEAE

*Hyptis albida* Kunth [*Hyptis emoryi* Torr.]     bíbino (salvia; desert lavender)

This is an uncommon shrub to 2 m tall with purple flowers and fragrant foliage. Previously plants in thornscrub and tropical deciduous forest with densely lanate pubescence and elongate inflorescences were thought to be a different species from the widespread desert lavender (*H. emoryi*). Recognizing the clinal gradation in these characteristics from dry desert environments to moist tropical ones, we consider all of the plants to be *H. albida*. MEDICINE: All three members of the *Hyptis* genus found in the region have medicinal uses such as cleaning foreign objects from the eye, curing earaches, and curing *pasmo*, a neurasthenic disorder in which the victim appears disoriented and irrationally fearful. Francisco Valenzuela recommended a bath into which a tea from bíbino is added for soothing the nerves. The water must be cold or the cure will be neutralized. Don Vicente Tajia collects bíbino for regular use in his home. On one occasion he snared some branches from a hillside, saying it would make a good tea to help his wife sleep. AESTHETICS: Branches are often gathered and burned inside a home, where the smoke acts as a pleasant fumigant.

*Hyptis suaveolens* (L.) Poit.     comba'ari (chani)

Chani is an upright annual growing as much as 1 m in height with white to pale lavender flowers. It is confined to tropical deciduous forest and its fringes. MEDICINE: The fruits of chani are often used in the field to remove foreign objects from the eye. A seed is put directly into the eye. It quickly absorbs moisture and apparently sucks the foreign body into it, becoming large enough to be easily removed with the foreign body attached. The hygroscopic fruits are also drunk in an atol to cure empacho (a digestive system that has ground to a halt). It acts as a mucilage agent similar to psylium seed and relieves the problem, usually considered to be something lodged in the digestive tract, or simply constipation. FOOD: Cowboys near Arroyo El Mentidero, a side drainage of the Río Cuchujaqui, reported that they mix the seeds with water to make a pleasant drink.

## LOASACEAE

***Eucnide hypomalaca* Standl.**  ma'aso arócosi "deer gourd" (rock nettle)

This is an attractive plant with large, showy creamy white flowers growing on vertical cliff faces and roadway cuts. It blossoms in the winter and spring and exhibits bright green, extremely brittle leaves. Its Mayo name is a mystery, for it produces no gourd. FOOD: Vicente Tajia said that the plant can be eaten, but he knows of no one who now eats it.

## LORANTHACEAE

***Struthanthus palmeri* Kuijt**  chíchel "bird that goes 'chi chi chi' in the night" [perhaps a kingbird] (toji; mistletoe)

This mistletoe has long, narrow, flat leaves and grows into huge bundles of plant material parasitizing a variety of trees, mostly legumes. The branches often grow to more than a meter in length. *S. palmeri* is the northernmost species of the Loranthaceae, which includes typical mistletoes with showy flowers. *Psittacanthus calyculatus* (DC.) G. Don is an orange-flowered species that is common in the álamos along the Río Mayo. *S. palmeri* is a truly catholic parasite, found on many trees, including ahuehuete, Bonpland willow, guásima, mezquite, palo blanco, and others, as well as on cusi oaks at low elevations. Another mistletoe, *Phoradendron californicum*, is described under the family Viscaceae, page 269. MEDICINE: Chíchel is said to be good for many things, including a tea for diabetes and an infusion for healing sores. Marcelino Valenzuela of El Rincón reports that the leaves of chíchel growing on guásima are effective at healing sores in the mouth or on the tongue. They need only be chewed. In the case of a bite or sting of unknown origin, he recommends a tea from the same chíchel. Santiago Valenzuela ("El Teco" [squirrel]) of Yocogigua believes that for diarrhea only a tea made from the chíchel from a mezquite host will be effective. Chíchel is also an important fodder for LIVESTOCK in the dry season. Cowboys will climb trees and lop off clumps of the parasite for the waiting cows below.

## MALPIGHIACEAE

***Bunchosia sonorensis* Rose**  goy carabanzo, tóroco (garbanzo de coyote)

This is an attractive large shrub or small tree, rather uncommon in the Sonoran tropical deciduous forest, but common near Los Capomos, where it is conspicuous for its waxy orange red fruits. FOOD: The fruits are well known and universally pronounced edible. We tried them and agreed. The yellow flowers are highly aromatic and attract numerous bees.

*Callaeum macropterum* (DC.) D.M. Johnson      sanarogua "ribs joined"
[*Mascagnia macroptera* (Moç. & Sessé) Nied.]                    (mataneni)

Mataneni is an aggressive vine, often posing as a shrub. With opportunity it climbs well up into trees. It blooms in spring, its bright yellow flowers often covering fences and other plants. The flowering may continue well into the starkest days of the spring drought. The large, conspicuous winged fruits are light green. It is very common throughout the region, spreading rapidly into waste places and disturbed soils. MEDICINE: Livestock with sores that refuse to heal are treated with a mixture of the root and rattlesnake fat, applied directly to the infected area. Vicente Tajia said the sores then heal quickly, and that this treatment is effective on humans as well. Although rattlesnake fat on its own is judged to have excellent powers of healing and invigorating (it is widely collected and rendered, and sold in vials), combined with herbs like mataneni, its powers and those of the herb are believed to be enhanced. Leaves of mataneni are applied directly from the plant to ant bites. Bone breaks and sprains are bound with a rag containing many leaves. Braulio López of Yocogigua described a similar treatment for sprains, bruises, and arthritis. In some areas the leaves and branches are boiled first and then applied to breaks, sprains, bruises, or arthritic joints. For extreme bloating, Francisco Valenzuela recommended heating some leaves and packing them on the stomach and abdomen. Relief will soon follow, he assured us, a wry smile on his face.

*Heteropteris palmeri* Rose      (compio)

This is a tropical liana spurred into rapid leaf production by summer rains. By September it often engulfs trees. Its flowers are pink, maturing into bright red winged fruits that adhere to the vine in large clusters. The strings and clusters of brilliant red light up the green forest and are seen from far away fluttering like helicopters when they fall to the ground. The leaves and vine also turn red in maturity. ARTIFACTS: At Los Capomos, it is collected and woven into the palm matrix of *guaris* (baskets) to impart a decorative reddish color. MEDICINE: According to Don Patricio Estrella, the fruits are boiled into a tea taken for aches and pains, especially *punzadas* (sharp headaches or migraines). He said it is more effective if the crushed fruits are mixed with a lard or petroleum base and rubbed on the forehead.

*Malpighia emarginata* DC.      sire (granadilla, morita)

Sire is a large shrub or small tree seldom more than 4 m tall, often branching densely near the base, with arching and irregular branches. It has gnarled trunks and maculate, flaking bark. The tree's peculiar growth pattern bestows it with the appearance of great age. It grows in the drier parts of the Mayo region, from foothills thornscrub through tropical deciduous forest. FOOD: The bright red fruits are eaten, becoming available during las aguas (the summer rains), but earlier in some locations. They are usually rather dry and taste better stewed. For several years we sampled them when they ripened in June and July but were not inclined to add them to our regular diet, finding them dry, tasteless, and of a waxy texture. In September 1997, however, Reyes Baisegua of Sirebampo reported that the sire in the arroyo immediately west of the village were especially sweet. We verified his report, finding the trees growing in the moist soil of a densely vegetated swale. The fruits, reminiscent of small cherries, hung in clusters from the branches, the red and green contrasting handsomely. The fruits had a rather insipid but satisfying taste. Our Mayo friends joined us in harvesting a couple of kilos of the fruits, most of which we devoured on the spot. Don Reyes noted that the berries only become succulent when rains are abundant during their maturation. This would explain why over much of their range they are barely palatable. They appear to grow especially large near Sirebampo, hence the appropriateness of that village's name ("sire in the water"). Their red color attracts many birds, which rapidly devour each year's crop, often before humans can reach the berries.

*Malpighia glabra* L. sire

This is a rare 3 m diameter shrub in coastal thornscrub. It produces bright red edible berries even larger (1–1.5 cm) than those of *M. emarginata*.

### MALVACEAE

*Abutilon abutiloides* (Jacq.) Garcke ex Britt. & Wilson

jíchiquia jéroca; ta'ri sorogua "broom" (pintapán; Indian mallow)

This tall (more than 2 m with adequate rains) shrub is common in disturbed areas. Its yellow orange flowers can be showy, but the cat-urine odor of the foliage makes it less attractive. MEDICINE: For children with bloody stools, the root is boiled into a tea that they continue to drink until the condition is cured.

*Abutilon incanum* (Link.) Sweet    jíchiquia to'oro cojuya "ash weed" (escoba, malva; Indian mallow)

This common shrub grows to more than 2 m tall following good rains on the coastal plain. ARTIFACTS: This malva is harvested in great quantities for making brooms. At La Cuesta, on the lower Río Cuchujaqui, several women had gathered hundreds of the straight stems and carried them homeward on their heads. Harvesting the plant for brooms must be done at the right time. If the

Balbina Nieblas sweeping with an *escoba* of *jíchiquia* (*Abutilon incanum*), Teachive. (Photo by T. R. Van Devender)

plant is green, the stems will be weak and the herbage ineffective. If they are left too long in the ground, they will become brittle, and the herbage will fracture. LIVESTOCK eat considerable quantities of the plant as well.

**Abutilon mucronatum J. Fryxell**
**A. trisulcatum (Jacq.) Urban**

These are closely related species in tropical deciduous forest. They are lumped together with *A. incanum* by natives.

**Bastardiastrum cinctum (Brandegee) D.M. Bates**     (malva, escoba)

This low subshrub with white to pale blue flowers blooms throughout the year. ARTIFACTS: According to Santiago Valenzuela of Yocogigua, it is used there and on the Río Cuchujaqui for making brooms.

*Malvastrum bicuspidatum* (S. Watson) Rose       to'oro jíchiquia (malva)

A common subshrub to 1 m, it is said to be excellent forage for LIVESTOCK. Vicente Tajia said that goats will grow fat on it. ARTIFACTS: Vicente uses a leafy branch to beat harvested tunas to remove aguates (glochids), and it appears to be highly effective for this admirable purpose.

*Malvella lepidota* (A. Gray) Fryxell       tori naca "rat's ear" (oreja de ratón)

This is a rather uncommon low plant of disturbed clay soils on the coastal plain. MEDICINE: Jesús Muñoz Valdés of Los Buayums recommends a tea made from a root as a remedy for bad stomach and intestinal distress. His wife noted that the remedy is rendered more effective if it is combined with root of chíchibo (*Ambrosia confertiflora*).

*Sida aggregata* Presl.       to'oro jíchiquia "broom bull"
       (rama lisita, sayla de pintapán)

ARTIFACTS: This shrub to 2 m with yellow flowers is also used for making brooms. At El Rincón it was the most commonly used malva, reflecting the use of the locally available species for this most basic household utensil.

## MARTYNIACEAE

*Martynia annua* L.       tan cócochi "cat's claw" (aguaro)

A fast-growing annual with large, sticky leaves, it reaches over 1 m high, producing delicate and showy lavender to pinkish to white flowers. It is common in moist, disturbed places in tropical deciduous forest and along moist arroyos in thornscrub. The hard black fruits, up to 3 cm long, resemble large beetles with grasping claws. These are irresistible curiosities to children and adults alike. We have spent hours playing a game seeing who can hook together the greatest number of the fruits with one held in the hand without dropping any. ARTIFACTS: The claws of the fruits are used for sewing *costales* (sacks) used for transport on burros and mules. FOOD: The seeds are eaten by people, according to Francisco Valenzuela of El Rincón. MEDICINE: The fruits are also brewed into a tea taken as a diuretic and for stomach pain. According to Paulino Buichílame of Las Bocas, the tea is also applied to an infant's infected umbilical chord. Others recommend a tea of the boiled fruits for diabetes, others as a remedy for an enlarged prostate.

*Proboscidea altheaefolia* (Benth.) Decne.       aguaro (desert unicorn plant)

This prostrate herbaceous perennial, its leaves covered with sand adhering to the sticky surfaces, is common in coastal thornscrub and in 1999 grew in abundance near the sea. It bears attractive yellow flowers. Its uses are the same as those of the annual *P. parviflora* (below).

*Proboscidea parviflora* (Woot.)                aguaro (cuernito; devil's claw)
**Woot. & Standl. subsp.** *parviflora*

This sensational, sprawling, upright annual produces fruit up to 15 cm long with a pair of long, curving, springy claws that will grab passersby, causing momentary terror. The large delicate flowers are pink to lavender. The leaves and stem are covered with a mucilaginous substance that is hard to remove. FOOD: The black, chewy seeds are casually eaten, though not without some concentration. CULTURE: María Soledad Moroyoqui Mendoza of Teachive described how the clawed fruits are gathered with great care and preserved for festivals. At that time, they are soaked in hot water until the long claws are soft and workable. The ends are carefully worked into the cut-open end of poultry or hawk feathers, which are then dyed and attached to a stick or rod to form the wand used by matachines, the ritual dancers who sport gaudily colored headdresses and legwear. Vicente Tajia said that a softened claw can also be used as a needle to pull thread to repair a fractured buli (canteen) or jícara (gourd bowl).

## MELIACEAE

*Cedrella odorata* L.                                                                   (cedro)
[*C. mexicanum* M. Roemer]

This fine tree with dense, dark green leafage grows to more than 25 m tall in some areas. In the Mayo region it is found only in extreme southern Sonora and in the Río Fuerte drainage in Sinaloa. The Sierra Cedro east of Güirocoba is reputedly so named because of a population of the fine trees in a canyon therein. CONSTRUCTION and ARTIFACTS: Here at its northern limits the local distribution of cedro is probably influenced by excessive harvesting, for the wood is widely viewed as one of the finest for construction and for making furniture. In a side canyon of Arroyo de la Culebra near Baimena, Sinaloa, a small forest of the trees grows in which individual trees reach more than 25 m in height and shade area. Recruitment appears to be successful, for small individuals appear to be prospering. Our consultants reported that the trees are watched to make sure that they are not stolen. In economic hard times individuals from the community will request and receive permission to harvest a tree and sell the wood. Even at a harvest rate of only one or two a year, however, it is doubtful that the population could be sustained.

*Trichilia americana* (Sessé & Moç.) T.D. Penn.  síquiri tájcara "red tortilla"
*T. hirta* L.  (bola colorada, piocha)

These tropical chinaberries are handsome, spreading small trees seldom reaching more than 5 m in height. They are restricted to shady canyons in tropical deciduous forest. The large compound leaves and the young green fruits resemble those of the Arizona walnut (*Juglans major* [Torr.] Heller, Juglandaceae). Luis Valenzuela of Las Rastras identified the trees and believed they had uses, but could not recall them.

## MORACEAE

*Chlorophora tinctoria* (L.) Gaudich  (mora)

This handsome tree with yellowish bark, dark green mulberry-type leaves, and an attractive spreading habit is found in the Mayo region only in extreme southern Sonora near El Chinal and in the Río Fuerte drainage, where it is well known. Near Los Capomos it grows to more than 15 m high on granite outcrops and in granitic soils. The tree provides excellent shade and is most valued by natives. It has a greater variety of uses than most fig trees, which it resembles, but it is unknown to Mayos of the Río Mayo. It appears to be reproducing successfully, if the numbers of small specimens growing in the uplands are a good indicator, and it is well represented in the tropical deciduous forests of Sinaloa. We have not come across a Mayo name for the tree. FOOD: The figlike fruits are eaten, although we have not tried them. CONSTRUCTION and ARTIFACTS: The wood is said to be excellent for beams and for making into dishes. INDUSTRY: A yellow dye is made from the bark. MEDICINE: The bark is brewed into a tea taken to soothe ulcers. The liquid is also said to be effective in expelling amoebas.

*Dorstenia drakena* L.  na'atoria "root spread out" (baiburilla)

This large-leafed perennial herb grows in moist, shady places throughout the tropical deciduous forest and in shady arroyos in thornscrub. Baiburilla no longer grows in the vicinity of Teachive owing to the drying out of the climate, according to Vicente Tajia. It still can be found on moist banks and in shady places upstream from Teachive near San Antonio. It is common near El Rincón, Los Capomos, and Baimena, and its virtues as a healing plant are widely regarded. MEDICINE: The radish-sized root is dug up (sometimes with considerable effort) and then mashed into a poultice that is rubbed on the face to heal boca torcida or alferecía (twisted mouth), probably a reference to Bell's palsy. The malady is supposed to be caused by a contrary wind, according to local healers. For a cure, the patient must lie still in a place where there is no wind. The na'atoria is twisted until juice runs out. The mashed starch is then massaged onto the paralyzed face. This procedure must be followed daily until healing is complete. One of Vicente Tajia's grandchildren used it with satisfying results, even though the treatment took more than a month. CULTURE: Consultants recall that as youngsters they would play a game for two with the fruiting stems. The players hooked the mushroomlike fruiting heads over each other and then gently tugged on the end of the long peduncle until one head popped off. The player whose stem retained its head was declared the winner, much as in playing with a wishbone.

*Ficus cotinifolia* H.B.K.    nacapul "ear lobe" (nacapuli; strangler fig)

This great fig often grows to immense size, more than 25 m tall, with massive entanglings of roots and buttresses. It may spread as wide as it is tall. This is the most tropical fig in Sonora, mostly restricted to moist, shady canyons in tropical deciduous forest. The tree is greatly valued for shade, a blessing in the stifling heat of late spring and summer. A solitary 20 m tall individual in a basaltic boulder field at 200 m elevation on Mesa Masiaca is anomalous. The use of the tree for lumber would destroy entire ecosystems, so great are the trees and so vital to wildlife. Although not rare, they are nowhere common, and the loss of just a few individuals would depauperate large areas and destroy the habitat of a myriad of creatures from cholugos to tiny wasps. FOOD: The fruits are relished not only by humans but also by beasts, birds, and insects. ARTIFACTS: The wood was formerly used for furniture, but this is now uncommon.

*Ficus insipida* Willd. [*F. radulina* S. Watson]    tchuna (chalate, higuera)

The leaves of *F. insipida* are rather large, oval, and pointed. As do other figs, it grows to great size, more than 25 m tall, especially along watercourses in the foothills and mountains. In late spring a constant light rain falls as evaporative cooling from the high branches as the tree transpires tons of water (or, as it is rumored, untold millions of leafhoppers excrete simultaneously). At certain times of day beams of sunlight highlight the tiny drops, resembling a mist. Many species of birds, but especially orioles and parrotlets, feast on the fruits. FOOD: The strangler fig produces the best fruits of all native fig trees, according to Mayos. Some natives will travel considerable distances to gather the succulent fruits. Although tasty, none are as sweet or as large as the fruits of *F. carica*, the domestic fig. ARTIFACTS: The wood may be used for making household implements, but only when a branch breaks off spontaneously, for the Mayos value the great trees for shade and food, and they are never felled.

## Moraceae

Tescalama (rock fig, *báisaguo* in Mayo; *Ficus petiolaris*) near Chorijoa.

***Ficus pertusa* L. f.**     nacapul "ear lobe," tchuna (nacapuli)

This large fig is decidedly less common than others. It has smaller, pointed leaves but reaches great proportions, up to 30 m tall and equally wide. Its common name is a mystery since its leaves do not resemble human ears, as do those of *F. cotinifolia*. A canyon near Guaymas bears its name for the large specimens found there in riparian thornscrub (Felger 1999). The fruits are popular with wild creatures. Fructivorous bats consume them in great numbers and then deposit voluminous quantities of dung over most unprotected surfaces as they dart and flutter about. AESTHETICS: It is highly valued for its shade. FOOD: The fruits are eaten.

***Ficus petiolaris* H.B.K.**     báisaguo "three saguaros" (tescalama; rock fig)

The rock fig is found on more-arid slopes and watercourses in desertscrub, thornscrub, and tropical deciduous forest. Its bright yellow white bark and large, roundish dark green leaves are distinctive. No other fig grows brazenly from cliffs and rock outcroppings. Its pale roots assume grotesque shapes and may cover great areas, cascading down rocks like a frozen yellow waterfall. The trees are notable for their sheer beauty and cool shade. Though it is a smaller fig, individuals may attain heights in excess of 15 m. The origin of the Mayo name is a mystery. The trees propagate easily and in the most improbable places. In Arroyo Gochico near San Bernardo, one sprung up on the sheer wall of a box canyon, perhaps 75 m above the canyon floor. It sent a ropelike root straight down the canyon wall into the wet sediments to provide water for its growth. Even on the very arid north-central coast of Sonora, trees will take root on rock masses and survive in the wilting desert heat. MEDICINE: The sap is viewed as particularly efficacious in curing bruises and hernias. The sap is rubbed on a rag or the bark is wrapped in a rag to bind hernias. The same method is used to correct dislocations and sprains and to heal bruises. Vicente Tajia collected a bagful in Arroyo Las Rastras, some 20 km north of Teachive, and took it home to replenish his pharmacopoeia. Don Vicente demonstrated how the bark of the roots are carefully scraped and the gum collected to be rubbed onto a rag and wrapped around an injured area. He keeps some of the stuff on hand. The popularity of the remedy may explain the tree's greatly diminished range. Whereas it once was found at Teachive, one must now travel upstream nearly to San Antonio to find the great trees. FOOD: Vicente notes that one can eat the figs in a pinch, but they are best left to the animals.

*Ficus trigonata* L. [*F. goldmanii* Standl.]   (higuera, chalate)

A huge fig tree, with some specimens in excess of 30 m tall, producing immense numbers of small fruits to the delight of numerous birds, insects, and mammals. Humans value it more for its beauty and shade than for its dry and tasteless small fruits, which are seldom harvested. As with *F. pertusa,* bats consume the fruits in enormous numbers and shortly thereafter rain down their dung on rich and poor alike. We have been unsuccessful in ascertaining a Mayo name for this species, which is rather less common than the other wild figs. ARTIFACTS: The wood is sometimes used for carving utensils.

## MYRTACEAE

*Psidium sartorianum* (Berg.) Nied.   (arrellán)

An uncommon small, smooth-barked tree growing in moist and protected places in arroyos or in tropical deciduous forest. Seldom more than 6 m tall, it produces an edible, though piquant, fruit well known in the region. Its distribution, usually associated with present or former human habitations, suggests that it may have been introduced and has since become naturalized. It is closely related to guayabo (*P. guajava* L.), the domestic guava. Although none of our consultants visited a tree with us, we have located them in the Sierra de Alamos and elsewhere in the region. MEDICINE: According to Doña Amelia Verdugo Rojo of Los Muertos, the fruits are a good remedy for curing kidney stones. She said people have brought her fruits "from the mountains."

NYCTAGINACEAE

*Boerhavia coccinea* Mill.           juana huipili, mochis (red spiderling)
[*B. diffusa* L.]

This mochis is a perennial with red flowers; all the others are annuals. Its uses are the same as those of *B. spicata* (below).

*Boerhavia spicata* Choisy                              mochis (spiderling)

A low-lying, spreading summer annual abundant on disturbed soils, where it grows in great numbers. It has numerous glandular hairs that make it adhere to clothing, skin, and animal hair. Were it not for the goats and cows that eat and trample it, it could be a nuisance. Its flowers are white. Other annual species in the Mayo region with pink or white flowers that are called *mochis* are *B. erecta* L., *B. intermedia* M. E. Jones, *B. maculata* Standl., *B. rosei* Standl., and *B. xantii* S. Watson. MEDICINE: The leaves are boiled into a tea drunk to treat measles.

*Bougainvillea spectabilis* Willd.                              (bugambilia)

This spectacular flowering, woody vine was introduced from Brazil. The large flower bracts add color to many local houses. It has no Mayo name. MEDICINE: The flowers and bracts—the typical purple magenta and varying shades of white, orange, and pink—are brewed into a tea taken for coughs. We include it because, although not native, it has become ubiquitously cultivated in the region and widely used herbally.

*Commicarpus scandens* (L.) Standl.                              miona

Miona is an open, weak shrub that climbs in and on other shrubs to a height of 2 m. It is commonly found in disturbed soils and open places throughout the region. It has greenish white flowers that can be found throughout the year. MEDICINE: The leaves, branches, and roots are boiled into a tea taken for sore kidneys. The tea's effect is said to be enhanced when it is brewed with cósahui and a touch of bugambilia blossom as well. Teachivans use the tea frequently to alleviate kidney pain and mal de orín.

*Pisonia capitata* (S. Watson) Standl.         baijuo "dew" (garambullo, vainoro)

An unpredictable plant braced heavily with enormous thorns, it often grows into vast, forbidding thickets impenetrable even to cattle. Less often it grows as a straight-trunked tree. The intimidating spines, resembling ice picks with heavy hooked thorns on them, are said by natives to be poisonous. Baijuo is a most unfriendly plant. Still, with pruning the arching tree can be shaped to provide dense shade, as was done by Alejandro Ramos of Rancho Guadalupe near San Bernardo. The Mayo name is a complete mystery and may be a linguistic coincidence. LIVESTOCK: It is planted to make a living fence that excludes all but the smallest creatures, or such culebras as the brown vine snake (*Oxybelis aeneus*), indigo snake, or babatuco, parrot snake (*Leptophis diplotropis*), and *chicotera*, or Sonoran red racer (*Masticophis flagellum* subsp. *cingulum*), all Colubridae. ARTIFACTS: The wood is carved into masks and musical instruments. In times of little forage, the leaves can be eaten by goats.

*Salpianthus arenarius* Humb. & Bonpl.             cuey oguo (guayavilla)
[*S. macrodontus* Standl.]

A common shrub up to 2 m tall or more, common on disturbed soils and along roadways throughout the region. When the flowers are crushed, they exude an aroma reminiscent of guavas, hence the Spanish common name. MEDICINE: A woman from Mexiquillo on the Río Mayo reportedly brews the flowers and leaves into a tea effective for curing hepatitis. Branches are boiled into a tea that is drunk as an antidote for scorpion stings. LIVESTOCK: Goats and desperate burros will eat it, while cows seem to avoid it. FOOD: Gentry (1942) reported that the roots were an important source of starch for the Guarijíos prior to the advent of corn. This use, however, appears to be unknown among Mayos.

## NYMPHACEAE

*Nymphaea elegans* Hooker           capo segua (flor de capomo; water lily)

This water lily with elegant white to bluish purple flowers grows only in quiet, permanent to seasonal ponds and pools. Its roots remain dormant when the water dries up. FOOD: Mayos from Teachive dig up the roots in the dry season and eat them. Vicente Tajia and the late Jesús José Nieblas demonstrated the harvesting for us. It is not an easy chore, for the mud is usually rich in clays and adheres in great clumps to footwear while resisting the bite of a shovel. If the soil is drier, it becomes nearly rock hard, resisting excavation. CULTURE: Bañuelos (1999) cites an older Mayo who recalled that children used to make necklaces and adorn themselves in general with the lilac-colored flowers. With the drying of the climate in recent decades and the pumping of groundwater, many wetlands have disappeared and, with them, the water lilies.

## OLACACEAE

*Schoepfia schreberi* J.F. Gmel.  cuta béjori "lizard tree" (palo cachora)

This is a symmetrical tree, often small, but in the Sierra de Alamos growing to nearly 10 m tall. Its leaves are small and crinkly, superficially similar to oak leaves. Its tiny flowers may be either yellow or red on different trees. "Cachora" refers to large lizards such as Clark's spiny lizard (*Sceloporus clarkii*) and the spiny-tailed iguana, which often live inside the hollow trunks. CONSTRUCTION: Possibly because it is scarce in the land of the Mayos, our consultants appear hesitant to ascribe uses to this tree, although they acknowledge its possible value for fence posts. The late Luis Valenzuela of Las Rastras believed that it used to be used regularly for that purpose.

*Schoepfia shreveana* Wiggins  cuta béjori "lizard tree" (palo cachora)

A coastal inhabitant, this small, symmetrical tree, up to 7 m tall, superficially resembles an olive. Early in the study we collected it but once and considered it uncommon. Much later we were shown a larger population near Bachomojaqui on the coast in the Masiaca comunidad. Consultants suggested that it is fairly common along the coast in that region. *S. shreveana* has relatively linear leaves, quite different from the oval leaves of *S. schreberi*. We consider it remarkable that natives label both species cuta béjori based on the similarity of the small flowers, red in *S. shreveana*, yellow and red in *S. schreberi*, while the trees are quite dissimilar. INDUSTRY: Seferino Valencia Moroyoqui of Las Bocas reports that its trunk and branches make excellent firewood. It is used in ovens for baking bricks and other ovens that bake clam shells for producing lime. The poor tree's hard woods may be its undoing, for its range is uncertain, and it may be limited to the coasts of extreme southern Sonora. A boom in construction along the sea in the Mayo communities has created a large demand for *ladrillos* (fired bricks). Manufacturing them requires enormous piles of firewood. MEDICINE: It was also used in a medicinal remedy, but Seferino could not recall precisely what.

## OPILIACEAE

*Agonandra racemosa* (DC.) Standl.  úsim yuera "kills little boys" (palo verde, palo huaco)

A slender, solitary tree with a wide distribution, it is usually found along watercourses in Mayo lands, sometimes growing into a tall, thin-trunked tree more than 10 m tall. In tropical deciduous forest, it grows on canyon sides; in thornscrub, along arroyo bottoms. It bears shiny, dark to bright green oval leaves that tend to droop, an aid in identification. The leaves appear to remain affixed to the tree even into the heat of June. Its Mayo name may derive from its brittle branches, for, although it appears to invite climbing, it can scarcely support the weight of many lads. The wood is used in all phases of house CONSTRUCTION. MEDICINE: The leaf is cooked as a cure for snakebite. The bark is boiled into a tea said to alleviate bronchitis.

## PAPAVERACEAE

*Argemone ochroleuca* Sweet subsp. *ochroleuca*  táchino (cardo; prickly poppy)

This spiny annual with a creamy white or occasionally sulfur-yellow flower is common in disturbed, overgrazed soils and waste places, where it can survive droughts that wither lesser plants. MEDICINE: Benigno Buitimea of Teachive drank a tea from the leaves of this prickly poppy to alleviate the symptoms of diabetes. It is also rumored to be good for prostatitis. At Nahuibampo táchino was believed helpful in curing pinkeye. A drop of the canary yellow sap was prescribed for each eye, a common use in the region. The plant is abundant in waste places and is spurned by LIVESTOCK until there is nothing else left to eat.

## PASSIFLORACEAE

*Passiflora arida* (Mast. & Rose) Killip  ma'aso alócossim "deer gourd" (talayote; passion flower)

This is a coarse, perennial aboveground vine with showy white flowers, relatively common throughout the region. It is ephemeral and an inconsistent producer. FOOD: Vicente Tajia reports the fruits are edible and are eagerly gobbled up by deer. We tried them and found them edible.

## PEDALIACEAE

*Sesamum orientale* L.  (ajonjolí; sesame)

Many Mayos of the tropical deciduous forest plant ajonjolí as a cash crop. Thousands of acres of sesame are planted in the Mayo region and have been for many decades. Almost all of the seeds are sold on the international market, for the Mayos find them little to their liking. They have learned to interplant beans, squash, and, above all, watermelon. The seeds often escape, and the plants are found growing adventively here and there throughout the region.

## PHYTOLACCACEAE

*Petiveria alliacea* L.  jupachumi "skunk's ass" (cola de zorillo)

An uncommon perennial herb growing in moist places in foothills thornscrub and tropical deciduous forest, it sends out spikes of white flowers up to 30 cm long. MEDICINE: The root of this herb is peeled, chopped, and placed in a small flask of alcohol. After several weeks the preparation is ready. The vapors are inhaled to cure colds, sniffles, and asthma. A tea is also brewed to cure empacho, or bad digestion. Don Vicente Tajia of Teachive took us to a location on the Arroyo Masiaca where it grows, and he demonstrated how to collect and prepare it. He was quite enthusiastic about its powers and urged us to use it. He believes that it will prove beneficial for any asthma sufferer.

*Rivina humilis* L.      bacot mútica "snake's pillow"
(coralillo cimarrón; pigeon berry)

A slender perennial herb, it usually grows among dense annuals or shrubs in shady, moist soils near watercourses. Spikes of white flowers and red berries glow through the greenery of late summer. LIVESTOCK: It is touted as good goat fodder by Don Chico Leyba of Yavaritos.

*Stegnosperma halimifolium* Benth.      jamyo olouama "old woman's back"
(hierba de la víbora)

This shrub, with oval, succulent leaves, white flowers, and maroon berries, grows on occasion to more than 2 m tall and spreads out over several square meters. Its leaves tend to stand upright, perpendicular to the ground, reminiscent of the bony plates on the back of *Stegnosaurus* dinosaurs. It is common in coastal and foothills thornscrub as well. MEDICINE: According to Santiago Valenzuela of Yocogigua, the ground-up leaves applied directly to a snakebite will reduce and sometimes prevent swelling. He had not used it himself but said he knew people who had. His neighbor Miguel Moroyoqui of Yocogigua agreed. The Spanish name is widespread in the region, so the plant's reputation must accompany it.

### PLUMBAGINACEAE

*Plumbago scandens* L.      toto'ora "old rooster" (cresta de gallo; leadwort)

This low herb with sticky white flowers is common in foothills thornscrub and lower tropical deciduous forest. MEDICINE: The root is crushed and applied to the skin to heal *jiotes* (spots without pigment on the skin).

### POACEAE

*Aristida ternipes* Cav. var. *ternipes*      ba'aso (zacate; spider grass)

Spider grass is a widespread rounded bunch grass in the Sonoran Desert. In the summer rainy season in tropical deciduous forest, plants grow rapidly into 2 m tall erect grasses. INDUSTRY: This common grass is added to adobe to strengthen it. The grass is also woven into dense mats in walls and roofs. Properly constructed, such structures will last fifteen years. They constitute a distinct fire hazard, however, especially given the ubiquity of cooking fires.

*Arundo donax* L.  ba'aca nagua "carrizo root" (carrizo; giant cane)

This giant cane is widely believed to be introduced from the Old World and has since apparently replaced the native *Phragmites australis* as the region's basic carrizo. It is found in dense colonies along still or standing freshwater in the lower Río Mayo and along washes. It constitutes the dominant cane along the Sea of Cortés, and its apparent ability to crowd out the presumably native *Phragmites* is poorly understood. At the time of Contact, Mayos along the lower Río Mayo were living in homes constructed primarily of cane, presumably *Phragmites*. Upstream and in Sinaloa, *Arundo* is found in freshwater habitats where it will not be washed away by flooding, but *Phragmites* appears to be rare. If *Arundo* managed to supplant *Phragmites*, it was a remarkable regional replacement of a species by one distinguished from it only with considerable difficulty by botanists. ARTIFACTS: The dried reeds are split and woven into covered and uncovered baskets of many sizes, some hamper-sized and capable of storing sizeable objects, others perfect for tortillas. Natives of Mezquital de Buiyacusi, a village north of Navojoa, are specialists in the production of woven containers. Smaller baskets are used for all manner of household necessities. These handsome artifacts are marketed in and around Navojoa. Women of El Júpare also produce baskets from the cane, though on a lesser scale. The split green reeds are also woven into the *petates* (sleeping mats) that are used in nearly every Mayo home in and near the delta region. The strong dried stems are laid together side by side to make a tapeste (bed) said to be most comfortable. Weavers also use the dried hollow reed as the na'abuia, or shed stick, when weaving blankets. The end is hollow and an extension, usually made of sina wood, can easily be fitted into it to accommodate a wider textile. CULTURE: The hollow canes are carved into flutes used during fiestas. CONSTRUCTION: Ba'aca nagua is widely used for constructing walls and roofs. For walls, the long canes are interwoven while still green and then daubed with mud. When dried, they are also split and woven into partitions and temporary walls inside houses. For roofs they are used as latas, laid directly on top of beams and then covered with dirt. MEDICINE: The grassy root is brewed into a tea drunk to relieve ulcers.

*Bouteloua aristidoides* (H.B.K.) Griseb.  yemsa ba'aso (saitilla; six-weeks needle grama)

This short annual grass grows rapidly following summer rains and often covers thousands of acres of flatlands that only a few weeks before were devoid of any hint of grass. The needlelike florets dry out within a couple of weeks of the last rain and lie in waiting to attach themselves to clothing and shoes of passersby. They become a great nuisance, for they are irritating and can even penetrate the skin. INDUSTRY: Ants gather the fruits and carry them into their colonies, where they consume the seeds but eject the husks. Worker ants lug the husks from inside the earth, leaving them piled as a tan collar around the anthill. This detritus is gathered by pottery makers and worked into the clay to strengthen the pots made by, among others, Leonarda Estrella of Los Capomos.

*Bouteloua barbata* Lag. var. *barbata*         ma'as ba'aso "deer grass"
(six-weeks grama)

*B. barbata* var. *sonorae* (Griffiths) Gould         (Mayo grama)

*B. barbata* var. *sonorae* (Mayo grama) is an endemic perennial variety of the widespread annual *B. barbata* var. *barbata* (six-weeks grama). Common on the coastal flats and uplands following summer rains, both taxa appear as though out of nowhere, for all sign of grasses disappears from the monte in late spring. Most other perennial grasses have long since been extinguished by catastrophic overgrazing. The short-lived grama grasses (*B. aristidoides, B. barbata,* and the perennial *zacate borreguero* (*Cathestecum brevifolium* Sw.) form dense stands but do not accumulate enough fine fuel to carry fires. Their root masses are comparatively shallow and largely ineffective at retaining runoff from rains. Ma'as ba'aso is an excellent forage grass for LIVESTOCK, according to their owners.

*Cenchrus brownii* Roemer & Schultes         na'a chú'uqui (guachapori; sandbur)
*C. echinatus* L.
*C. palmeri* Vasey

These common, rather low annual grasses all have small and bothersome, but not damaging, burs (*guachaporis*). The grasses are valued as forage for LIVESTOCK and are very common. *C. echinatus* is widespread and especially common in estuarine environments and on the sandy margins of the sea.

*Cynodon dactylon* (L.) Pers.         (zacate de lana; Bermuda grass)

Bermuda grass has no Mayo name, perhaps a reflection of its introduction to Sonora from Africa, the English common name aside. MEDICINE: Vicente Tajia reported hearing several older people mention that the roots brewed into a tea provide relief from kidney pain.

*Digitaria bicornis* (Lam.) Roem. & Schult.         huilanchi "very narrow, the root"

A large field near the volcanic hill Huichabiricahui, lying fallow after a crop of ajonjolí, had been taken over by this common grass. LIVESTOCK: Ramón Morales of Los Muertos pronounced it to be decent, but not the best, grass for cattle feed.

*Gouinia virgata* (Presl.) Scribn.     ba'aso síquiri "red grass"

This grass grew thickly and nearly a meter high among rocks on the mountain called Huichabiricahui west of Basiroa. LIVESTOCK: Luciano Valenzuela of Los Muertos pronounced it to be excellent for cattle feed.

*Lasiacis ruscifolia* (H.B.K.) A. Hitchc.     baca cupojome jicto "little thrasher grass" (carrizito de huitlacoche, negrito)

This bamboolike perennial grass with broad leaves and shiny black seeds was once used for making mats and walls but is no longer used, according to Francisco Valenzuela. It grows up to 4 m tall and 3 cm thick along the banks of the Arroyo Huasaguari, persisting only where LIVESTOCK have difficulty trampling it and gnawing it down to the roots. ARTIFACTS: Domingo Ibarra demonstrated how boys make flutes from the canes, lopping off an end at an angle and adding holes to produce varying tones. The instrument he produced was hardly mellifluous, but in the hands of a practiced musician could be incorporated into Pan's orchestra.

*Monanthochloë littoralis* Engelm.     bahue huitcha "sea thorns"

A low, spreading grass ubiquitous in mud and wet sand along the coast. It prospers in salty environments, exuding precipitated salt on its leaves. LIVESTOCK: It is eaten by goats and burros. MEDICINE: Maximiano García of Agiabampo recalled that old people used to brew it into a tea for colds or fever, but he does not know anyone who still uses it.

*Otatea acuminata* (Munro)     ba'aca (otate; bamboo)
C.E. Calderón & Soderstr.

A native bamboo found in higher places in tropical deciduous forest throughout the region, especially in moist box canyons, but in upland locations as well. ARTIFACTS: Where otate is abundant, the poles are used to make tapestes, formed by laying poles across benches or platforms made of etcho or pitahaya wood, and then sewing them into place. These are remarkably comfortable beds and are valued for use in warm weather. During the colder months, however, they provide little protection against cold from below. Although we did not see the poles used otherwise, they are widely known as acceptable material for fishing poles and the building of walls, roofs, and anything else for which long, light poles are required. In upland places, *O. acuminata* is used in the same way that *Arundo* is used (and *Phragmites* formerly was used) in the lowlands. Benjamín Valenzuela reports that it was used at Las Rastras by his ancestors.

***Pennisetum ciliare*** (L.) Link [***Cenchrus ciliare*** L.]   (zacate buffel; buffelgrass)

This tough, aggressive African grass has been introduced into pastures covering hundreds of thousands of acres in Sonora below 800 m and where rainfall exceeds 250 mm, as part of a government-sponsored program to increase beef production. A savannah grass, it is relatively woody, leafing out and flowering rapidly in response to minimal rains, and recovering rapidly after fire. It has become ubiquitous and invasive, found on disturbed soils everywhere and invading coarse soils lacking vegetative cover. Mayos, along with most other folk, gather great loads of the grass, which they cut along highway medians and shoulders. This becomes an important source of fodder for LIVESTOCK (horses, mules, burros, goats, sheep, and cows) during much of the year. Mayos also harvest the seed with long combs. The seed is winnowed and bagged in (among other places) Masiaca, and sold in other parts of northwestern and northeastern Mexico. Collecting seed often constitutes an important income in some villages near the coastal plain. The grass is highly aggressive, moving into disturbed and coarse soils when rainfall is adequate, displacing native vegetation in coastal thornscrub and Sonoran desertscrub, and inviting fires, which cause it to grow even more vigorously. Intact foothills thornscrub is not as dramatically affected, for the canopy during las aguas can deprive buffelgrass of needed sunlight. Tropical deciduous forest seems immune from infestation, which is why ranchers have taken to total *desmonte* (clearing) to ensure growth of the seed.

***Setaria liebmannii*** E. Fourn.   hayas guasia (cola de la zorra)

A common and attractive annual bristlegrass so named because it resembles the upright tail of a running female fox. The size of the grass is highly correlated with rainfall, ranging from 10 cm tall grasslets in dry areas to 1 m tall cola de la zorra in tropical deciduous forest. LIVESTOCK: It is considered decent forage for cattle.

### POLEMONIACEAE

***Bonplandia geminiflora*** Cav.   (ciática)

Ciática is a sprawling, low perennial herb, often densely leaved, with pink or purple flowers, growing in shady, moist areas of tropical deciduous forest. MEDICINE: As its name suggests, ciática is valued as an analgesic for arthritic and other joint pains. The leaves and stems are boiled, and the liquid is applied to the affected region. Francisco Valenzuela, who lives in tropical deciduous forest, professes to have used it himself. Vicente Tajia asked us to find him some since he had heard of its powers but none grows in the vicinity of

Teachive on the coastal plain. When he requested some of a Mayo from Chorijoa, he was presented with a bunch of chuparosa (*Justicia candicans*).

## POLYGONACEAE

*Antigonon leptopus* Hook. & Arn.  masasari "the root has a fruit" (sanmiguelito; queen's wreath)

This vine, common in thornscrub and tropical deciduous forest alike, is mostly bare in the spring drought but comes back to life with the first rain and grows over all available surfaces within a few weeks. The bright pink to rose red flowers illuminate many landscapes. It is cultivated throughout the warmer parts of Sonora and into Arizona. By September much of the Mayo region is turned pinkish red by the multitude of flowering sanmiguelito. In spite of the beauty of the flowers, Vicente Tajia warned us, the vine is a host for *baiburines* (chiggers), which lurk in untold millions for passersby and render them miserable with their insufferable bites. FOOD: Vicente Tajia and Santiago Valenzuela of Yocogigua demonstrated how the small tuber is excavated and peeled for eating. We found it mildly tasty and somewhat refreshing.

## PORTULACACEAE

*Portulaca oleracea* L.  huaro (verdolaga; purslane)
*P. umbraticola* H.B.K.

FOOD: The green, succulent leaves and reddish stems of these fast-growing annuals are gathered and eaten as a green in summer. When the rains come, many thousands of the plants appear, especially on disturbed and overgrazed soils. Virtually all Mayos enjoy huaro and consider it a natural part of summer. In most of Sonora the verdolaga sold in markets is *P. oleracea*, which is more common than *P. umbraticola*.

*Talinum paniculatum* (Jacq.) Gaertn.  cochizayam "pig saya (*Amoreuxia*)" (pink baby's breath)

A common perennial herb with pink flowers and succulent leaves appearing in the shade of shrubs after summer rains in coastal thornscrub and tropical deciduous forest inland. LIVESTOCK: Pigs and goats eat it with relish, according to Alejo Baisegua.

## RHAMNACEAE

*Colubrina triflora* Brongn.  cuta béjori "lizard tree"

A large shrub or small, spindly tree growing up to 10 m high in moist tropical deciduous forest, cuta béjori has angular branches and sparse, weak, often serrate leaves. It is nowhere common, yet it is found throughout the tropical deciduous forest down into foothills thornscrub, where it assumes a shrubby form. It is abundant on Mesa Masiaca and noticeable after heavy summer rains. Few Mayos climb the slopes of the mesa, so few are familiar with the plants that grow on it. Because *C. triflora* is sparsely represented in the forest, only a few consultants were able to identify it. CONSTRUCTION: The straight trunks are used for cross-hatching in houses and as fence posts.

*Colubrina viridis* M.E. Jones             ono jújugo "smells of salt"

A large shrub or small, bushy tree growing up to 8 m tall in coastal thornscrub. Seferino Valencia Moroyoqui of Las Bocas believed it had medicinal uses but could not recall them. The pea-sized fruits are pink to red, ripening in December. The common name is the same as that given to *Adelia virgata*, which it superficially resembles. INDUSTRY: The wood makes satisfactory firewood.

*Condalia globosa* I.M. Johnst.             juupa quecara "flaky mesquite"
(mezquite hormiguillo; bitter *Condalia*)

A dense shrub of the desert, seldom more than 2 m tall in the Mayo region, its distribution is confined to coastal thornscrub. It is well armed with strong, thorny branches. Vicente Tajia warned that the thorns are wicked and venomous. CONSTRUCTION: The longer trunks are used for fence posts, with the spines left on to increase their efficacy. When the plant is young, the tender branches are nibbled off by LIVESTOCK, and goats eat the leaves. INDUSTRY: It makes decent firewood as well. ARTIFACTS: The branches, though seldom more than 5 cm wide, are used to make chomas (shuttle sticks) and sasapayecas (tamping sticks) for looms. In Sinaloa one fellow carves figurines from the wood. CULTURE: Alejo Bacasegua claims that the bark decocted into a tea is an effective antidote to evil spells.

*Karwinskia humboldtiana*             aroyoguo "crippled tree"
(Roem. & Sch.) Zucc.             (cacachila, tullidora)

Along the northern Río Mayo in Sonora, cacachila is normally a leafy shrub. In the Mayo region it grows into a tree as tall as 10 m, especially in well-developed tropical deciduous forest. The leaves are a most agreeable smooth, gray green color, typically with black dashes along the veins on the underside and bright red berries. The chemistry of cacachila was studied extensively in the United States in the 1960s and 1970s (Mark A. Dimmit, pers. comm., 2000). One of its toxins is especially virulent, attacking the myelin sheaths of nerves and bones. Effects of cacachila poison are not noticed for several weeks after ingestion but are progressively severe, hence the Spanish name *tullidora* (crippler), widely used outside the Mayo region. If the patient survives, recovery can be very slow. The toxin likely is localized in the seed, as Guarijío children eat the fruit pulp without harm, and other parts of the plant are used medicinally. MEDICINE: A tea is made from the leaves for liver problems. According to Vicente Tajia, when the eyes become yellow and urine very yellow, you have hepatitis and should use the tea. Victims of the latter are also bathed with the infusion. For colds, the leaves are steeped in alcohol during the rainy season. Thereafter, the liquid is stored in a vial and inhaled. Patricio Estrella mentioned that a strain of ayoroguo that produces extremely large fruits grows on a hill near Los Capomos. He said a tea from these is effective in curing rabies. CONSTRUCTION: The wood is used for posts in houses and fences. CULTURE: Alejo Bacasegua uses the root as a medicine to overcome malpuestos, or hexes. He uses aroyoguo (and other herbal remedies) to undo the bad influences and spells placed on individuals by hechiceros, who, he lamented, are all too frequently found in the region.

*Ziziphus amole* (Sessé & Moç.) M.C. Johnst.

báis cápora "three corporals" (saituna)

A spiny tree with large, dark green leaves whose many branches form intricate arches. It is common in coastal thornscrub, where it stands out among the columnar cacti, reaching a height under ideal conditions of more than 13 m. Near the coast the branches often become laden with epiphytes (huírivis cu'u; *Tillandsia exserta, T. recurvata*), which are harvested as livestock feed. Saituna disappears in tropical deciduous forest, apparently unable to compete for sunlight outside of the coastal thornscrub. It appears to be replaced in that habitat by cumbro (*Celtis reticulata*), which assumes a similar habit and produces similar fruits. White-fronted parrots flock to the trees in December through February to feast on the ripening fruits. FOOD: The fruits are eaten by people and by many birds. MEDICINE: The bark is made into a tea taken to expel amoebas and said to be quite effective. The same tea is said to alleviate stomach ulcers and intestinal gas. LIVESTOCK: The leaves are a principal source of fodder for goats. They persist well into the dry season, making the tree even more prominent than its mere size would suggest. It is capable of sustaining considerable abuse from livestock without apparent harm. It is a very important plant in its habitat in terms of human uses.

*Ziziphus obtusifolia* (Hook. ex Torr. & A. Gray) A. Gray var. *canescens* (A. Gray) M.C. Johnst.

jó'otoro, jutuqui (ciruela del monte, huichilame; graythorn)

This large shrub is widespread in the deserts and desert grasslands of the southwestern United States and northern Mexico. It is infrequent in coastal thornscrub in the Mayo region near the southern edge of its range. For much of the year it is entirely leafless. Its thin, green gray branches have numerous sharp, greenish spines up to 5 cm long. When rains come, the branches become covered with leaves that largely conceal the spines, thus giving the plant an entirely different aspect. FOOD: The blue-black fruits are eaten by humans and by LIVESTOCK. MEDICINE: The plant seems to exude an aura of healing. Alejo Bacasegua informed us in a most confidential tone that a tea brewed from the branches is a treatment for chancres of syphilis or for gonorrhea. Others claim a tea made from the stems is effective for cancer, for stomach infections, and for helping alleviate the effects of stroke.

## RHIZOPHORACEAE

*Rhizophora mangle* L.         cánari, pasio síquiri (mangle rojo; red mangrove)

In southernmost Sonora and northern Sinaloa extensive mangrove communities (manglares) dominate coastal estuaries. *R. mangle* is an important component of these dense thickets. MEDICINE: According to Maximiano García of Agiabampo, the leaves of the reddish-barked tree are brewed into a tea taken for diabetes. Balbaneda López García of Yavaros noted that the roots are brewed into a tea for treating kidney ailments. Leobardo López of Chomajabiri recommends a tea for cancer victims to drink and use for bathing. The wood can be used for firewood, but so many superior woods are available from nearby thornscrub that only fishermen or beach dwellers resort to it. ARTIFACTS: The wood is strong enough that oars and poles are made from it.

## RUBIACEAE

*Cephalanthus occidentalis* L. var. *salicifolius*         (mimbro; button bush)
(Humb. & Bonpl.) A. Gray

Easily mistaken for a willow, this shrubby tree with showy white ball flowers is found in moist riparian settings throughout most of temperate North America. The variety *salicifolius* (Latin for the willowlike leaves) is only occasional in the Mayo region, often at the edge of streams or rivers. The variety *californicus* Benth. in Arizona is not palatable to livestock and is reputed to be poisonous, containing glucosides such as cephalanthine (Kearney and Peebles 1969). MEDICINE: Francisco Valenzuela recognized specimens growing in the Río Cuchujaqui and noted that the flowers are made into a tea for stomachache. It also grows in the Arroyo de la Culebra near Baimena, where it is used for the same purpose.

*Chiococca alba* (L.) Hitchc.         tori naca (oreja de ratón)

Although we have collected this rare shrub along Arroyo Agiabampo, we have not obtained a name or uses for it. Bañuelos (1999) reports that a tea is steeped from the root and administered to alleviate diarrhea.

## *Hintonia latiflora* ( Sessé & Moç.) Bullock            tapichogua (copalquín)

A highly esteemed thin tree up to 10 m tall, with rough, brownish bark (often one side of the tree has smooth bark as well) and a dense crown of leaves occurring in whorls toward the tips of the branches. It blooms from late May through September. The pale green club-shaped buds with six sharp ribs and ivory-colored, trumpet-shaped flowers hang pendulously from the tips of branches, presenting a most odd appearance. The showy white flowers open at night with an overpowering jasmine scent. When the tree is leafless in early spring, the walnut-sized fruits persist, assisting in identification. CONSTRUCTION: The durable wood is used for vigas in houses and, despite the tree's medicinal value, for fence posts. MEDICINE: A homeopathic taste of the bitter bark easily identifies the tree. The bark is made into a tea for numerous ailments, including malaria and stomach problems. José and Fausto López of Nahuibampo reported that it had cured people of that village of malaria. They also drink the tea or chew the bark directly from the tree to "thicken the blood" to alleviate or prevent anemia. It is widely hailed as a tonic. The bark is widely believed to have the power to kill microbes in the blood and is considered an excellent remedy for coughs. It is also thought to help control diabetes. An individual in Teachive keeps a bundle of it in his house and drinks the bitter tea daily to help combat his diabetes, routinely mixing it with other curative barks. The bark is also ground into a powder and applied to chigger bites and other sores. Some trees are thought to be more potent than others. These show scarring from repeated harvesting of bark. Trees seem to tolerate considerable bark harvesting without appearing much the worse for wear. Ground tapichogua is sold commercially as the remedy copalquín in markets throughout Mexico. The trees are heavily harvested farther south in Mexico. The abundance of unscarred trees in the Mayo region is viewed with wonder by southern herbal healers. Tapichogua is often the first plant mentioned when one asks Mayos what sorts of medicinal plants they know. The tea is usually drunk with a pinch of cinnamon to combat the bitterness.

## *Randia echinocarpa* Sessé & Moç.                                 jósoina (papache)

Found primarily in tropical deciduous forest or well-developed foothills thornscrub, papache is a strange large shrub or small tree with four stout spines at the end of each branch. The branches seem to grow in arches rather than straight into the air or horizontally. The plant yields even stranger tennis-ball-sized fruits with giant, dull, recurved spines protruding randomly. Papache fruits are popular with animals, including a variety of birds, skunks, raccoons, coatimundis, and coyotes. Insects wait for larger animals to penetrate the tough rind and then feast on the interior pulp. FOOD: When yellow and ripe, the fruits are eagerly harvested and enjoyed, especially by children, who suck in mouthfuls of the black pulp and spit out the many seeds. Many adults also claim they are good to eat. In much of the region they are not consumed because most Mayo villages are a bit low and arid for the papache bush or tree. Vicente Tajia requested that we take him to the forest northeast of Teachive to harvest some. He brought home a couple of dozen, and his grandchildren gobbled them up. We found the fruits to have contradictory flavor, both bitter and sickly sweet, and could eat but one, and preferred to make it an annual affair. Still, the fruits are popular with most natives, especially with young boys. To determine if the fruits are ripe, natives shake them to see if the contents move about inside the skin. MEDICINE: The pulp is said also to be good for many ailments, including stomach distress and diabetes, for which it is highly recommended by many Mayos. María Soledad Moroyoqui, a Teachive weaver, claims that when eaten on an empty stomach, the fruits kill amoebas and other parasites. Santiago Valenzuela of Yocogigua, born in 1914, attributes to papache pulp the power to lower blood pressure. Francisco Valenzuela told us that the fruit is peeled and burned and the ashes are then ground and mixed with *chorogui* (cholugo, coatimundi) fat and applied it to a cut or bruise to help it heal more rapidly. He told us this with a candid expression. FOOD: Vicente Tajia claimed that if you make a hole in the ripe fruit and let the juice drip out, the resulting liquid is refreshing. We let him enjoy it without any comment from us.

*Randia laevigata* Standl.　　　　　　　　　　　　　　　　　　　　sapuchi

An irregularly formed, weird little tree up to 5 m tall with blunt, unarmed angling branches and relatively sparse foliage consisting of very long leaves. It is usually found in upper tropical deciduous forest, or on hydrothermally altered soils or indurated ash at lower elevations. We found it growing on the red soils of Cerros Colorados, near El Tábelo north of Alamos, at 200 m elevation in association with *tornillo* (*Helicteres baruensis* Jacq., Sterculiaceae), cusi, and negrito. At higher elevations (up to 1,000 m) it is associated with *Quercus tuberculata*. It produces a yellow orange flower in late spring and a green avocado-shaped fruit in summer. The leaves are lanceolate, up to 20 cm long. FOOD: Francisco Valenzuela identified the tree as sapuchi and ate some of the fruits, which we found had an interesting flavor.

*Randia obcordata* S. Watson　　　　　　　　　　　　pisi (papache borracho)
*R. thurberi* S. Watson

Papaches borrachos are large, slim, straight-branched shrubs sometimes to 3 m tall, rarely taller, with abundant paired spines on the ends of the branches. *R.*

Fruits of *papache borracho* (*pisi* in Mayo; *Randia obcordata*), Huasaguari.

*obcordata* is less common in coastal thornscrub in the Mayo region than *R. thurberi* and has smaller, more numerous fruits. In most years the fruits cluster like green grapes along the branches in late winter. White-fronted parrots and woodpeckers devour the fruits, leaving few seeds. The fruits of both papaches borrachos are said to induce dizziness. *R. thurberi* seems to thrive in xeric habitats in foothills thornscrub, where it is rather common. FOOD: The cherry-sized fruits are edible, and nearly all the natives in the region have tried them. ARTIFACTS: A piece of branch is shaved and sharpened and attached to the end of an agave pole to make a bacote for harvesting pitahaya fruits. Another artifact is produced in part by larvae that infest many of the fruits. Woodpeckers come searching for the larvae, making pea-sized holes while extracting their prey. The fruit then dries out in the arid winds, persisting on the branches for months in this dried state. Boys pick these desiccated fruits and blow forcefully into the hole, producing a whistle that can be heard over long distances.

## RUTACEAE

*Amyris balsamifer* L.                          úsim yuera "kills little boys" (palo limón)

This small tropical tree with shiny, aromatic compound leaves and tiny white flowers reaches its northern limits in coastal thornscrub in southern Sonora. Its limited distribution and lack of recognized virtues render it somewhat obscure. The Mayo name, the same given to *Agonandra racemosa*, is of doubtful application here. AESTHETICS: Alejo Bacasegua of Camahuiroa notes that it is sometimes burned as firewood because the smoke is laden with a refreshing lemony smell, hence its Spanish name, which means "lemon tree."

*Casimiroa edulis* La Llave ex Lex.               ja'apa ahuim (chapote, zapote)

This dioecious, majestic, spreading tree, often growing to 15 m in height, is valued for shade. It is often seen near human habitation, suggesting that its origin in Sonora may be connected with human migrations. It has been heavily persecuted by agricultural officials, who accuse it (and all members of the family) of being an intermediate host to fruit fly pests and bacterial pathogens lethal to citrus. A local campaign of extermination has been a failure, thanks in no small part to the Mayo practice many decades ago of planting the trees in remote areas. FOOD: The candy-sweet fruits are relished. Gentry (1942:204–5) noted that youths of San Bernardo would walk many kilometers to collect them. Judging from the hordes of honeybees visiting the female flowers, it is an important source of honey. CONSTRUCTION: The wood is sometimes used for vigas and horcones in houses. MEDICINE: The leaves make a tea effective in lowering blood pressure, according to Patricio Estrella. Luciano "Chano" Valenzuela of Baimena mashes the bark and boils it in water for several hours. He uses the resulting liquid to treat domestic animals for lice and mange.

*Esenbeckia hartmanii* B.L. Rob.              jójona "ripe"; momoguo (palo amarillo)

A small tree or large shrub, often many-trunked, with peeling, maculate bark, growing inconspicuously in the forest. When the branches are broken, the wood appears yellowish, hence the Spanish name. CONSTRUCTION: The wood is sometimes used for posts and building. INDUSTRY: Seferino Valencia Moroyoqui of Las Bocas collects the dried fruits for firewood. They closely resemble miniature Osage oranges (*Maclura pomifera* [Raf.] Schneid., Moraceae). He said that they burn hot and smokeless, and are second only to brasil wood for cooking. MEDICINE: Seferino also noted that for those suffering from muscle cramps, any part of a branch boiled into a tea offers rapid relief, but he warned that pregnant women should not touch any part of the plant since it may lead to miscarriage.

*Zanthoxylum fagara* (L.) Sarg.        o'ouse suctu "lion's claws" (limoncillo)

Typically a well-armed shrub, under moist and sheltered conditions it grows into a small tree up to 8 m tall. Although widely distributed in tropical deciduous forest and along moist arroyos, it was known to only a few consultants. MEDICINE: Its bark is said to be chewed for toothache. CULTURE: Alejo Bacasegua said it is one of the plants he uses to undo malpuesto, an ability for which he is well known. CONSTRUCTION: The wood is quite hard and durable. At Baimena specimens grow sufficiently large to provide posts for fences and small vigas for houses.

*Zanthoxylum mazatlanum* Sandwith        yocutsuctu (jaguar's-claws)

A large shrub or small tree growing in tropical deciduous forest. It is rare in Sonora, but not uncommon in Sinaloa. Our specimens were collected by Ignacio Baesmo Cota near La Misión, Sinaloa. MEDICINE: The bark is shredded, dampened, and applied directly to aching teeth. It is also chewed on to strengthen teeth. INDUSTRY: The trunk is said to provide acceptable firewood.

*Zanthoxylum* sp.        (palo limón)

This large shrub or small tree was found growing in Arroyo Camahuiroa. Alejo Bacasegua thought it was once used for something but could not recall what it was. When the leaves are crushed, they give off a citruslike smell, hence the common name.

## SALICACEAE

*Populus mexicana* subsp. *dimorpha*        aba'aso (álamo; cottonwood)
(T.S. Brandegee) J.E. Eckenwalder

The cottonwood is a tall, majestic spreading tree growing to 25 m or more where groundwater is readily available. This species has remarkably dimorphic leaves, hence the varietal name, with linear willowlike leaves on juvenile growth and typical heart-shaped cottonwood leaves on older growth. Huge specimens still grow along the floodplains of major rivers, but recruitment has virtually ceased on major watercourses because the construction of upstream dams has eliminated annual floods. Seedlings of plains cottonwood (*Populus sargentii*) in North Dakota establish on river sandbars deposited during floods (Everitt 1968); the recruitment ecology of álamo is likely similar. Many enormous old trees have become heavily infested with mistletoe (*Psittacanthus calyculatus*) and tent caterpillars, indicators of high stress. Natives point out that there are fewer cottonwoods now throughout the region than formerly. Several large specimens growing on the Río Cuchujaqui south of Los Muertos have been washed away by flooding. This suggests that removal of vegetation from the watersheds is causing faster runoff and increased downstream flooding. While the flooding may enhance recruitment of seedlings, it means death for the old trees. The seedlings are usually eaten by LIVESTOCK or trampled and do not survive. MEDICINE: For blows and bruises the bark is cooked with salt and the wound washed with the liquid. ARTIFACTS: The wood of the extensive root is used to carve figurines, masks, and household implements. Federico Badón of Rincón de Aliso uses it exclusively to carve pascola masks. It is also sawed into planks for tables. CONSTRUCTION: Wencelao "Mono" Valenzuela of La Cuesta has become an expert in the use of cottonwood for lumber. CULTURE: Near the rivers, natives use the newly leafed-out branches in Easter ceremonials. The bright green leaves bring an atmosphere of newness to the festivities. The tree plays an important role in Mayo mythology as well. A large álamo growing on a bank near the Sea of Cortés is said to have marked the place where the cross from the Júpare church was hidden during the church-burning excesses of the Calles presidency (1928–34).

*Salix bonplandiana* H.B.K.             huata (sauce; Bonpland willow)
*S. gooddingii* Ball                     huata (sauce; Goodding willow)

The Bonpland willow is a large willow. Don Pancho Valenzuela observed one growing in the Arroyo Cuchujaqui. He believed it had uses but was unfamiliar with them. Goodding willow is the sauce of the lower drainages, associated with the álamos. Huatabampo means "willow in the water." We have not encountered medicinal uses for the bark. Willows are used in CONSTRUCTION in various places along watercourses, but are everywhere considered inferior to other materials. ARTIFACTS: A Mayo family in the upper Río Mayo hamlet of Chorijoa had fashioned a house cross from the small branches. In villages along the lower Río Mayo the springy sauce branches are woven into wicker baskets. In Tesia and thereabouts rustic hats are woven from the new shoots. They are agreeable to look at but of doubtful virtue in screening out harmful rays of the sun, nor can they be said to be very comfortable.

### SAPINDACEAE

*Cupania dentata* DC.                         (diente de culebra)

This tropical tree resembles a vine in some aspects, but quickly grows into a tree. In the land of the Mayos it has been collected only in Arroyo de la Culebra in a box canyon with large semideciduous trees. Rosario "Chalo" Valenzuela mentioned that his father knew of a use for the plant but he could not recall it.

*Sapindus saponaria* L. tupchi (abolillo, amolillo; soapberry)

A common tree becoming quite large when growing in or near watercourses, reaching 12 m in height and nearly as wide. Its leaves often remain green when most other trees' leaves have withered and dropped. CONSTRUCTION: Logs are sometimes used in home construction but are inferior to others such as amapa and palo colorado. The wood is made into crucifixes and beads for rosaries and necklaces. ARTIFACTS: The black seeds of the fruit are edible, but more commonly they are strung into attractive necklaces worn with pride by Mayo fiesteros as a mark of their office, along with a fox skin, which is draped over a shoulder. The fruits are still made into soap in a few places, hence the tree's Spanish common name, amolillo.

*Serjania mexicana* (L.) Willd. caamugia "it lacks a [weaving] shuttle"
*S. palmeri* S. Watson (güirote de culebra)

These aggressive vines are ubiquitous in tropical deciduous forest and common in moist areas in thornscrub from the coast to the mountains. During las aguas they leaf out rapidly, the new growth revealing quickly how they penetrate to the tops of the tallest trees and engulf them, sometimes obscuring the host leaves with their own. In spring both produce a show of pretty white flowers. The stem of *S. mexicana,* more common in tropical deciduous forest, is heavily striated, often square or rectangular. It is covered with sharp barbs quite capable of tearing the skin. When first walking through dense forest, one is tempted to use the overhanging vines of caamugia as a handhold. One experience is usually sufficient to encourage one to do otherwise, as a bloody hand will prove. MEDICINE: The branches of caamugia, stripped of barbs, are chewed to strengthen the gums. Vicente Tajia of Teachive stated, apparently seriously, that a tea made from the branches of the unarmed *S. palmeri,* which is more common in thornscrub, is administered to cows to relieve constipation.

### SAPOTACEAE

*Sideroxylon occidentale* (Hemsl.) T.D. Penn. júchica
[*Bumelia occidentalis* Hemsl.]

At Huasaguari this tree grows near Mayo homes, rising to 12 m, with checkered bark and *Lycium*-like leaves. The rather scraggly-looking tree is common in a few areas around Teachive (and in canyons farther north in Sonora) but has not been found in other locations in Mayo country. ARTIFACTS: The wood is carved into dishes and utensils.

*Sideroxylon persimile* (Hemsl.) T.D. Penn. subsp.                 **bebelama**
*subsessiflorum* (Hemsl.) T.D. Penn.

A very large tree growing more than 25 m tall in canyon bottoms, with long dark spines and large, dark green, oval leaves. It affords excellent and profound shade. Huge specimens are found, appropriately enough, in the Arroyo Las Bebelamas north of Alamos. These trees are said to be very old and have been identified for many generations. Water flow in the arroyos where the trees grow may be ephemeral, but the bottom soils remain moist throughout the year from the water table that is near the surface. FOOD: The fruits, similar to those of ca'ja (*S. tepicense*), ripen in June. They must be cooked so that they will not burn the mouth. They are said to be inferior to the fruits of ca'ja.

*Sideroxylon tepicense* (Standl.) T.D. Penn.                          **ca'ja (tempisque)**

A spreading tree up to 20 m tall, with a star-shaped, palmate leaf pattern. The trees are highly valued for their fruits. The location of trees is maintained as family traditional knowledge. They often grow near homes, suggesting that they may be cultivated or, alternatively, that home sites may have been selected near a tree; however, they also occur commonly on remote uplands in moister parts of the tropical deciduous forest and thus should be considered native to the region. Their habitat is more wide ranging than that of bebelama, which is confined to the bottom of moist canyons. FOOD: The orange fruits are relished, though some say they should be cooked, for they tend to burn the lips and mouth when eaten raw. MEDICINE: The bark is boiled and the resultant tea drunk for fever. INDUSTRY: The bark can be used in place of rennet for coagulating cheese.

*Sideroxylon* sp.                                                    **júchica, bebelama**

We found but one specimen of this tree growing wild in the monte near Bachomojaqui on the coast of the Masiaca comunidad. Seferino Valencia Moroyoqui stated that it was well known in the area and pronounced the wood to be useful for firewood. The fruits, however, are not eaten. Other residents of the region agreed with his comments. The tree is similar to *S. occidentalis* in aspect but appears to represent a species new to science. We found it extraordinary that both consultants identified the tree's genus correctly. One consultant pronounced it similar to *S. occidentale*, while the other associated it with *S. persimile*. The leaves of the tree resemble neither.

## SELAGINELLACEAE

*Selaginella novoleonensis* Hieron.  teta segua "rock flower"
(flor de piedra; resurrection plant)

Occasionally found on shady rock faces in arroyos in moist tropical deciduous forest, teta segua is abundant on cliffs at 1,220–1,460 m in oak woodland and pine-oak forests north of the Mayo region. Those we examined were gathered at 100–200 m from Cerro los Algodones near San Antonio. These herbaceous perennials, up to 20–25 cm in diameter, close tightly when dry, opening into bright green rosettes with the arrival of rain. MEDICINE: Vicente Tajia keeps several stored in his home. The entire plant is soaked in water and the water then drunk to relieve kidney problems, kidney stones, and gall stones. Vicente is convinced it works, explaining that since the roots can dissolve rock, the plant can transfer that power to humans. He has treated himself with it, combining it with a tea of cósahui. Kidney problems appear to be common among natives of the region. The intense heat of the coastal region, which promotes rapid loss of body water through sweating, may be related to the common pathology.

*Selaginella pallescens* (C. Presl) Spring  teta segua (flor de piedra; resurrection plant)

This species is similar to *S. novoleonensis* in resurrecting into green herbs after rain, but it is a more delicate mat former, with elongate scaly stems rather than a rosette. It is widespread on shady banks throughout tropical deciduous forest up into oak woodland and pine-oak forest. It grows in large numbers on protected cliff faces in the Arroyo de la Culebra near Baimena. MEDICINE: Sinaloan Mayos, including Patricio Estrella, stated that the plant is brewed into a tea taken for diabetes.

## SIMAROUBACEAE

*Alvaradoa amorphoides* Liebm.  guaji, paenepa (palo torsal)

A small, uncommon tree to 10 m high, with spreading branches and lengthy (10 cm) spikes of small white flowers, blooming from December to March. Both the flowers and the persistent reddish fruits make this a handsome tree. The flowers and the compound leaves are reminiscent of legumes such as palo dulce. The tree appears to flower only sporadically, and years may pass in which the inflorescences are not noticed. A 10 m specimen near Tepahui bloomed and fruited spectacularly in 1994. This is a tropical tree in a very tropical family. It reaches its northern distributional limits just north of the Río Mayo region. In Sonora it is mostly found in moister tropical deciduous forest canyons. Francisco Valenzuela concurred with its common name and said it has uses, but he could not recall what they were. The tree is more common in the Sinaloan Mayo country. CONSTRUCTION: At Baimena natives use the wood for posts and firewood. MEDICINE: Rosario "Chalo" Valenzuela said the bark is boiled and used as a rinse to kill lice and nits (*piojos*).

## SIMMONDSIACEAE

*Simmondsia chinensis* (Link.) Schneid.  (jojoba)

A symmetrical, dioecious shrub characteristic of the dry Sonoran Desert north of Guaymas. In the 1970s thousands of plants were introduced in two large plots, one near Huebampo and one near Jopopaco. The Mexican government hoped to spur the local economy of the Masiaca comunidad and envisioned harvesting the oily jojoba nuts for sale on the lucrative international market. According to the *comuneros* (members of the indigenous community), however, the plants have done poorly. They say the female plants bear the fruits but that 85 percent of the plants provided by the government were male, hence production was limited. Also, the climate is too humid and the rainfall too great for ideal growth. Although most of the plants are still alive, they have yielded little in the way of fruits, and prospects for improved production are poor. The Mayos of the region view them with curiosity, recognizing the value of the oil, which is of a high quality, but have little idea how production might work. Still, they harbor hopes of future harvests.

## SOLANACEAE

*Capsicum annuum* L. var. *aviculare* (Dierbach) D'Arcy & Eshbaugh  có'cori (chiltepín; red bird pepper)

The fiery little chile is found in every rural household in the Mayo region. The plants grow wild along arroyos and in more mesic sites in uplands, from coastal thornscrub into the oak woodland. The weak shrubs may exceed 2 m in height if protected and supported by other shrubs, and produce dozens, sometimes hundreds, of fruits each. FOOD: No meal is complete without có'cori. In most households a small dish of the dried pea-sized red fruits is placed on the table. The fruits are crushed between the thumb and forefinger and sprinkled on food, especially meat. One must not rub ones eyes or other sensitive body parts for many hours thereafter. Yetman recalls the following experience: "On the Arroyo Masiaca a few kilometers upstream from San Antonio, Vicente Tajia, Tom Van Devender, and I were snooping among the plants along the stream. I happened upon a robust bush of ripened chiltepín fruits resembling tiny persimmons. I picked and ate one rather complacently and was instantly overcome by a molten eruption in my mouth. My companions gave me little sympathy

*Chiltepines (có'cori* in Mayo; *Capsicum annuum*) being dried near Nahuibampo.

(everyone knows chiltepines are hot and are not for sissies). When the burning did not subside after fifteen minutes, Don Vicente investigated the plant more closely. After examining it, he looked at me in wonder and exclaimed, 'Ay, David, no wonder. This is a *chiltepín de la sierra*! They are very, very hot! [¡Son muy, muy bravos!]'" And thus many Mayos discern differences in the degree of heat of chiltepines among different plants.

The Jesuit missionary Ignaz Pfefferkorn, stationed in Sonora in the mid-eighteenth century, remarked of the chiltepín:

> A kind of wild pepper which the inhabitants call *chiltepín* is found on many hills. It grows on a dense bush about an ell in height and is similar in shape and size to the thick juniper berry, except that when ripe it is not black, but all red, like the Spanish pepper. It is more bitingly sharp than the latter, yet it is manna to the American palate and is used with every dish with which it harmonizes. It is placed unpulverized on the table in a salt cellar and each fancier takes as much of it as he believes he can eat. He pulverizes it with his fingers and mixes it with his food. The chiltepín is the best spice for soup, boiled peas, lentils, beans, and the like, where Spanish peppers cannot be used. The Americans swear that it is exceedingly healthful and very good as an aid to the digestion, a conten-

tion more easily believed than the claim that the Spanish pepper is cool and refreshing. (Pfefferkorn 1949:48)

The Spanish verb *enchilar,* meaning "to spice up with chile" and from which the name of the dish *enchilada* is derived, is appropriately applied to the action of chiltepines on any food to which they are added. In late summer and early fall the chiles are ready to harvest, and at times entire families will venture into the monte to collect the cherished fruits. No special care is taken to collect only the fruit. Sometimes entire branches are yanked off the shrub, a practice that led Gentry (1942) to voice concern that such mistreatment would cause extinction of the plant. In some areas they have been overharvested and abused and have thus become scarce. Elsewhere, however, they seem to flourish despite the crude gathering practices, and large quantities are gathered and marketed each year. Most herbaceous plants are well adapted to recover from short-term physical damage from drought, frost, grazing, or harvesting. At Yocogigua an entire family—men, women, and children—sat at a table piled high with chiltepines and associated herbage to cull the leaves and stems. Some of the fruits were kept and eaten fresh, while others were dried in the sun and stored. Any extras would be sold. They will last several years if stored in an airtight jar. At La Labor, a mestizo ejido near Alamos, a government program has provided chiltepín seedlings to women in the hopes of establishing a cottage industry. At La Aduana, in the vicinity of Alamos, and throughout the region, children and adults alike sell unripened chiltepines along the highways, offering plastic bags full of the small green currant-sized fruits. In this manner tons of the fruits are harvested and sold each year.

The fruits are an important source of vitamins A and C (Nabhan 1985:126), a potentially important food in a region where fresh vegetables are often absent from the normal diet. A researcher from the Universidad de Sonora suggested to us that a domesticated form of chiltepín could be an important source of additional vitamins in the Mexican diet, which is traditionally deficient in fresh vegetables.

Many horticulturalists have been frustrated by the difficulty in getting chiltepín seeds to germinate. Considering that only reptiles, birds, and primates can see colors, the bright red fruits are most likely dispersed by birds and humans. In nature they must first pass through a digestive system. This dispersal method has worked to great advantage for human gatherers, for the plants are common along trails in tropical deciduous forest where generations of humans have relieved themselves. Birds eagerly devour the ripe fruits and excrete the seeds whole, with only the seed coat softened. Although domestication of chiltepín has eluded enthusiastic promoters of the plant, nature seems to be blessing the plant's natural tendency to proliferate. As long as some forest is left intact, the delectable, fiery chile seems destined to prosper.

At the Arizona-Sonora Desert Museum crushed chiltepines are being tested as a biorepellent to control house mice colonies in aviaries.

*Datura discolor* Bernh.   tebue "ground squirrel" (toloache; desert thorn apple)

Apparently all species of *Datura*, variously known as jimsonweed, thorn apple, moonflower, sacred datura, or toloache, have renowned medicinal and hallucinogenic properties. *D. discolor* is a widespread annual in the Sonoran Desert from southern Arizona and southeastern California south into Sonora. In southern Sonora it is common in disturbed soils in thornscrub and tropical deciduous forest. Like many desert annuals, it can vary greatly in size depending on rainfall. It may be only 4–5 cm tall in the desert, but as much as 80–90 cm tall in tropical areas. The large tubular flowers, varying in color from white to light purple, open at dusk to attract nocturnal moths. Livestock assiduously avoid the plant, so it is free to proliferate in the rich soils around corrals and watering places, free from competition with more edible forbs. The fruits are similar to ping-pong balls covered with heavy thorns. All parts of the plant are said to be toxic, even though it is also said to have powerful hallucinogenic properties. MEDICINE: According to Fausto and José López of Nahuibampo, the leaves are heated and applied directly to an injury or blow to alleviate swelling and bruising. CULTURE: Santiago Valenzuela of Yocogigua warned that the plant can be used to work evil, but he did not specify the evil. All consultants agreed that the plant has considerable power and commands their respect.

*Datura lanosa* Barclay ex Bye   tebue "ground squirrel" (toloache; sacred datura)

*Datura lanosa* is a large perennial herb from a woody rootstock. It is a hairy southern relative of *D. wrightii* Hort. ex Regel, which is widespread in the western United States. It has pure white flowers larger than those of *D. discolor*, up to 20 cm in length. It appears to have the same toxicity, for neither humans nor livestock will eat it. MEDICINE: Vicente Tajia told us in all seriousness that its leaves are used in an excellent treatment for hemorrhoids. The leaves are scalded and then rubbed on the anus. Better yet, he says, the leaves are rubbed or ground, pressing out the oil, which is then applied to the anus.

*Lycium andersonii* A. Gray   sigropo "pile of red" (wolfberry)

This shrub, thick with spinescent branches, is widespread across the southwestern United States from Utah and New Mexico to California and south into northwestern Mexico. It is occasional throughout the Mayo region, except in coastal thornscrub, where it is common. The flowers are elongate white tubes that can appear any time the plant is well hydrated. It produces red berries the size of a pea. These are eaten elsewhere, but not especially relished by Mayos. ARTIFACTS: The thin branches are cut into small sections and threaded into necklaces, especially rosaries, for they are nearly hollow inside, and a needle can easily pass through. Local artisans use the sigropo beads to weave imitation ténaborim (strings of cocoon rattles) on dolls resembling fariseos.

### *Lycium* aff. *berlandieri* Dunal     jejiri, jotónari (wolfberry)

This shrub is similar to *L. andersonii* in overall distribution and general growth form, differing in its short lavender flowers and often shiny maroon bark on new growth. It is common in the vicinity of Las Rastras. MEDICINE: Benjamín Valenzuela of Las Rastras, who is familiar with the shrub, reported that the tiny leaves are mashed and soaked in water, and the liquid is applied to snakebites. Benjamín may have confused this *Lycium* with *Stegnosperma halimifolium*.

### *Lycium brevipes* Benth.     bahue mo'oco "sea sagebrush" (chamisa del mar; wolfberry)

A wolfberry with thick succulent leaves, it is confined to the coastline in Sonora, northern Sinaloa, and Baja California. Its flowers are short lavender to blue tubes. FOOD: The bright red berries, which often ripen in great numbers in late winter, are rather tasty, although, as Paulino Buichílame of Las Bocas pointed out, one expends considerable energy collecting such a tiny fruit. LIVESTOCK: The shrub is also viewed as good goat fodder, although the leaves are somewhat salty.

### *Lycium* aff. *fremontii* A. Gray     to'oro bichu "bull's testicles"

This *Lycium* is so similar to *L. brevipes* that they are virtually indiscernible except when in flower. It occurs from southern Arizona and southeastern California south to southern Sonora. FOOD: The edible fruits are brilliant orange and tend to be elongated rather than spherical as in *L. brevipes*. The leaves are succulent and slightly salty, as are those of *L. brevipes*.

### *Nicotiana glauca* Graham     (cornetón, marihuana)

Cornetón is an introduced species, a native of South America. Domingo Ibarra of Huasaguari recalls his father saying that it was a good remedy for something, but he could not recall what. The local term for marihuana is *mota*. Only a few people use the term "marihuana" for this plant.

*Nicotiana obtusifolia* Mart. & Gal.          goy biba (tabaco de coyote; wild tobacco)

This wild tobacco is commonly seen in moist waste places, on disturbed soils, and in overgrazed pastures. A perennial herb, it seldom exceeds 50 cm in height but can flower soon after germination. Its greenish yellow flowers bloom opportunistically. We have been unable to find anyone smoking goy biba or recommending it for smoking or chewing. A native tobacco, *Nicotiana tabacum* L., is still cultivated to the north by Guarijíos, who use it ceremonially. Older Mayos recall using this *tabaco rústico* but believe it is no longer grown or used in the region. MEDICINE: According to Vicente Tajia, the leaves are boiled into a tea applied to stiff joints and drunk for rheumatism.

*Physalis maxima* Mill.                                            tombrisi (tomatillo)

A rather delicate annual with equally delicate white flowers with a yellow center. The fruits develop into a grape-sized green mini-tomato enclosed within a loose Japanese lantern–like husk. Although it does not resemble a tomato from the outside, when the husk is removed and the fruit cut, it reveals tomatolike characteristics. FOOD: Although the fruits are widely gathered and eaten (made into salsa and other sauces) in the region, Mayos appear to be uninterested and view them as something mestizos eat. The Mayo name may be of non-Mayo origin. Other members of the genus *Physalis* in the region that are commonly called tomatillo include *P. acutifolia* (Miers) Sandwith, *P. crassifolia* Benth. var. *versicolor* (Rydb.), *P. leptophylla* B. L. Rob. & Greenm., *P. philadelphica* Lamarck (the cultivated tomatillo), *P. pubescens* L., and *P. subulata* Rydb.

*Solanum americanum* Mill. [*S. nodiflorum* Jacq.]        mambia (chichiquelite;
*S. nigrescens* Mar. & Gal. [*S. douglasii* Dunal]                        nightshade)

These fast-growing herbs with delicate white flowers appear after summer rains, often in enormous numbers, covering in a few days' time the bottoms of recently flooded washes and fine-soiled flats. After especially heavy rains, vast numbers of them will appear to grow as one, reaching a height of more than half a meter in a fortnight. FOOD: Mambias are relished as greens when harvested early, before they become tough and dry. They are especially popular prepared with sautéed onions, garlic, tomatoes, lime, and salt. (With that combination, almost anything would taste good.) Following a very heavy rain Vicente Tajia took us to an arroyo in which the rapid-growing plant abounded. We gathered large bags and presented them to his wife, Doña María Teresa Moroyoqui, who accepted them with delight. She prepared the greens by stewing them and by sautéing them as above. Unfortunately, mambias appear in abundance for a very few weeks each year at most, and then only if there are heavy summer or fall rains. MEDICINE: In addition to their value as a food, María Gonzalez, born in 1905, reported that they are an effective medicine for kidneys and lungs. She treated her children with mambia for infirmities of these organs. Others maintain that a tea made from the leaves can be administered to victims of sunstroke, dehydration, or high fever and will gradually return them to normal if they drink it regularly for several hours. *S. americanum* is less common than *S. nigrescens*; they are similar species, differing only in the size of the flower. Filemón Navarete, a deer singer, sang for us the following song.

MAMBIAM

| | |
|---|---|
| Jita juya síalita | There is a green plant |
| Amani yota pa'tcuni bétana | Growing where no other plants grow |
| Amani huécali | There is that plant |
| Sialeya jatijeca | It waves in the wind |
| Jita jueja taca? | What plant can it be? |
| Sialeya jatijeca | It waves there in the wind |
| Sehaylo ta'ta yohue bétana | There, where the sun comes in, at sunrise |
| Síaleya jatijeca | There it waves, in the wind. |

***Solanum seaforthianum* Andrews** (bellísima)

This aggressive vine with showy blue lavender or white clusters of flowers and brilliant red fruit has been widely planted in the Mayo region as an ornamental and thus locally has escaped into the countryside. It is native to the Caribbean islands but is cultivated as an ornamental and widely naturalized in tropical America. AESTHETICS: It is an ornamental in many Mayo yards but has no Mayo name.

***Solanum tridynamum* Dunal**      paros pusi "jackrabbit eye" (sacamanteca)
**[*S. amazonium* Ker-Gawd]**

***S. azureum* Fern.**

*Solanum tridynamum* is a prolific, prickly subshrub that grows to 1.5 m tall on disturbed soils throughout the region. For much of the year it features attractive dark purple flowers, about 3 cm across, with a yellow center. Around Alamos, occasional individuals with albino white flowers are encountered. At times it is one of the few plants in flower. Both the purple- and white-flowered forms are being introduced as ornamentals in Tucson, Arizona. *S. azureum* is a very similar plant with more lobed leaves that is endemic to coastal thornscrub in southern Sonora and Sinaloa. MEDICINE: The leaves are made into a concentrated tea and poured into the ears of people who suffer from deafness. It is said to open up the canal. Vicente Tajia and Alejo Bacasegua both have used it and assured us that it could work on people who have been deaf for a long time.

### STERCULIACEAE

*Byttneria aculeata* Jacq.                  bacot mútica "snake's pillow"

This is a wiry perennial vine, often common in canyon bottoms in tropical deciduous forest, with stout recurved thorns. It forms dense thickets, quite suitable to be a "snake's pillow," as the common name translates. It is distinguished by a macelike fruit armed with spines. This tropical species does not extend north of the Mayo region. None of our consultants identified a use, noting only its virtue as snake habitat.

*Guazuma ulmifolia* Lam.                        agia (guásima)

A common, spreading, elmlike tree in valleys and most riparian habitats in tropical deciduous forest, growing as high as 12 m. In thornscrub it grows as far north as central Sonora in arroyos that have permanent flow. It usually retains its leaves through the heat of spring, even though they may curl up and be practically dead. The spinescent fruits, the size of large cherries, cling to the branches for months after ripening. Then they slowly fall to the ground. Agia is a most important tree in the Mayos' lives. ARTIFACTS: Guásima—strong and durable but not excessively hard—is the wood of choice for constructing moveable furniture such as chairs, stools, tables, cribs, and cradles. The color varies from a handsome white to a delicate pink. Also, the branches are easily cut into strips up to 6 cm wide that can be worked into odd shapes such as the cylindrical base for a drum. The time of appropriate harvest is considered critical and should coincide with the full moon. September is said to be the best month for cutting the wood. Furniture of guásima wood harvested in a timely fashion will last more than forty years. Native carpenters say that if harvested at other times, the wood will crack, will be less strong, will be subject to boring insects, and will fail within a single generation. Natives claim (somewhat optimistically, we fear) that branches or boles harvested for lumber will regenerate in two years. MEDICINE: The inner bark is boiled and made into a tea given for liver sickness. The fruits are similarly boiled into a tea taken for kidney problems. FOOD: Strips of the bark are chewed on for the refreshingly sweet flavor and as a tonic. The dried fruits are ground and brewed for a coffee substitute. Domingo Ibarra and Vicente Tajia both claimed that the fruits are replete with vitamins. They recalled a fellow from San José de Masiaca many

Chair of *guásima* (*agia* in Mayo; *Guazuma ulmifolia*) made by Marcelina Valenzuela, El Rincón. (Photo by T. R. Van Devender)

years ago who made his livelihood from agia fruits. He would mash them, add water, and boil the liquid into a syrup that he marketed as a tonic, and he enjoyed much esteem for his labors.

*Melochia speciosa* S. Watson (té de lila)

*M. tomentosa* L.

*Melochia speciosa* is a common shrub on the southern coastal plain, while *M. tomentosa* occurs higher, in disturbed areas in tropical deciduous forest, north as far as Sonora's Seri country, where it is well known. The flowers range from pink to pink lavender. MEDICINE: According to Alejo Bacasegua, it is a good plant for brewing a tea to reduce fever. Felipe Yocupicio recalled that it was

The sharply pointed leaves and fruits of *sanjuanico* (*tásiro* in Mayo; *Jacquinia macrocarpa*). In the past the fruits were pulverized to make soap.

formerly believed that younger women who touch the leaves would become promiscuous, a belief shared by Seris.

### THEOPHRASTACEAE

*Jacquinia macrocarpa* Cav. subsp.                      tásiro (sanjuanico)
*pungens* (A. Gray) Stahl.

A small tree, often symmetrical in outline, with a typically thick trunk bearing smooth gray bark. The prolific small (around 4 cm long), persistent, leathery green leaves are ovate with a needle-sharp tip capable with gentle probing of puncturing the skin with minimal pain. On the other hand, if one approaches the herbage too quickly, numerous skin punctures can ensue with disagreeable results. The flowers are a bright, almost gaudy orange, though they are but a centimeter in diameter. Harvesting flowers and fruits is hard on the hands, for it is impossible not to receive a multitude of tiny stabs from the needle-sharp leaves. ARTIFACTS: Natives of the desert to the north string the dried flowers into necklaces. When soaked, the flowers return to their original shape.

Some women in the region also make flower necklaces for little girls. INDUSTRY: The mustard-colored fruits, the size of large grapes or oak galls, used to be mashed in water and allowed to stand. The resulting liquid was used as a shampoo that left sheared wool and human hair squeaky clean and nice smelling. Now detergents are simpler to obtain. Teresa Moroyoqui and other women boil the dried flowers to make a yellow dye of considerable beauty for their weaving. The flowers, they say, must be harvested before las aguas or the dye will run. MEDICINE: Doña Buenaventura Mendoza was out in the monte one day gathering the flowers when we went to visit her, and she had collected quite a pile. She said a tea from the flowers would strengthen her heart. The flowers must be harvested before the rains come, she said, or they will be no good for heart medicine. She was born in 1905, so we paid close attention. Felipe Yocupicio, who is well known in the region for his use of herbal medicines, recommends a tea made from the bark as a cure for whooping cough. He said he has used it regularly and has recommended it to others who have found it effective. One consultant stated with a straight face (and others corroborated) that the leaves are mashed into a ball and inserted into the anus of a constipated cow for immediate relief to the cow.

### TILIACEAE

***Heliocarpus attenuatus* S. Watson**　　　　　　　　　　　　sa'amo (samo baboso)

A slender, springy tree up to 7 m tall with small, white terminal flowers and rather large, but thin and weak, heart-shaped leaves. The disk-shaped fruits, smaller than a penny (about 0.8 cm in diameter), persist, turning brown and resembling a tiny image of the sun, hence the generic name. *Baboso* means "drooling" in Spanish. Mayos explain that the name is derived from the fact that bark mixed with water produces foam. INDUSTRY: Pieces of the bark are soaked in water and mixed with lime to make whitewash adhere to plaster. According to many Mayos, if the sa'amo is not added, the whitewash will slowly chip away, producing a blotched color. Long strips of the bark are used for lashing poles together. They are reported to shrink upon drying, making a secure binding that must be cut to be removed. Francisco Valenzuela reports that pieces of bark are mixed with feed to relieve constipation in cattle and other LIVESTOCK, which sometimes occurs in the dry season when green forage is not available and they eat the péchitas (pods) shed by mezquites.

***Triumfetta semitriloba* Jacq.**　　　　　　　　　　　　　　　　　(guachaporito)

A small shrub with tiny yellow flowers growing in tropical deciduous forest. The Spanish name refers to the spiny, sticky fruits. Francisco Valenzuela said the plant is excellent forage for LIVESTOCK.

## TURNERACEAE

*Turnera diffusa* Willd. (damiana)

A small subshrub with delicate, small white flowers and small leaves. On hydrothermally altered soils at Cerros Colorados, the plants are low and spreading. MEDICINE: The herb is brewed into a tea said to abet fertility in women. It also helps purify the blood, according to Francisco Valenzuela. It is widely marketed in Mexico as an aphrodisiac and is said to be especially helpful to men with impotence. It apparently is uncommon for men to acknowledge that they have taken it, however.

## TYPHACEAE

*Typha domingensis* Pers. (tule; cattail)

This robust aquatic perennial is found in standing or slow-moving water throughout the Mayo Delta. ARTIFACTS: Its leaves are used to weave mats, utilitarian baskets, and even walls.

## URTICACEAE

*Pouzolzia guatemalana* (Blume) Wedd. var. *nivea* chi'ini "cotton"
(S. Watson) Friss & Wilm.-Deav.

Luis Valenzuela of Las Rastras tentatively identified this uncommon shrub, so named because the leaves are cottony white underneath. MEDICINE: Luis recalled that people make a bath from a tea brewed from the branches and bark, but he could not recall the infirmity for which it is effective.

*Pouzolzia occidentalis* (Liebm.) Wedd. [*P. palmeri* S. Watson]

We collected this shrub on Cerro Huitchabiri near Basiroa. Ramón Morales thought it had a name and use but could not recall them.

*Urera baccifera* (L.) Guad. (ortiguilla)

This shrub or small tree often acts as a vine. It has tiny white flowers adnate to the vine. It is a tropical plant known in the Mayo region only in tropical deciduous forest near El Rincón Viejo. Francisco Valenzuela pointed out the plant to us and warned that its leaves furnish a potent sting, leaving large welts on the skin. He was not aware of any uses or of a Mayo name.

## VERBENACEAE

*Aloysia sonorensis* Moldenke (mariola)

A small, symmetrical shrub with gray, aromatic leaves. It is common on the coastal plain of extreme southern Sonora and appears to be endemic to coastal thornscrub in southern Sonora and northern Sinaloa. It is common near the

Arroyo Camahuiroa, but is not abundant elsewhere. Vicente Tajia visits the area from time to time to gather piles of the shrub for his pharmacopoeia, for it is not known in his hometown of Teachive, 25 km to the north. The name *mariola* refers to different plants elsewhere in Mexico. In the Chihuahuan Desert and adjacent Texas, it refers to *Parthenium incanum* H.B.K., a shrubby composite with gray, aromatic leaves. MEDICINE: The branches are often brewed into a tea to alleviate colds, grippe, and fever. Vicente Tajia keeps a supply of it at home, and Alejo Bacasegua of Camahuiroa also uses it frequently and prescribes it to people with cold and flu symptoms. Alejo was not aware of a Mayo name, however.

*Avicennia germinans* (L.) L.     ciali "green" (mangle negro; black mangrove)

The roots of this rather drab mangrove grow permanently in the water. It abounds in estuaries in the extreme southwestern part of the region. INDUSTRY: Although the wood is renowned for its hardness, its only use is occasional firewood, for which it is said to be inferior. The bark is steeped and produces a deep blue color good for dyeing, according to Maximiano García of Agiabampo. ARTIFACTS: Balbaneda "Nelo" López of Yavaros uses boat poles made from the straight, thin trunks.

*Citharexylum scabrum*     chaparacoba "cabeza de chachalaca"
Moç. & Sessé ex D. Don

This small tree, rare in the region, was pointed out by Vicente Tajia and identified for us by Felipe Yocupicio Anguamea of Camahuiroa. For several years we knew it from only one specimen growing near the road leading from the International Highway to Masiaca. Since then, with the help of Vicente and Felipe, we have located several more specimens, most of them growing around or near Arroyo Camahuiroa. The tree bears clusters of orange fruits that attract many birds. Felipe believed it had had medicinal uses, but he could not recall them. The chachalaca, for which the tree is named, is a grouselike game bird common in arroyo vegetation in coastal thornscrub and tropical deciduous forest. Natives often locate nests, pilfer the eggs, and place them under a broody hen. The chicks grow into handsome, tame birds. ARTIFACTS: Felipe mentioned that the wood is straight and hard, making it ideal for the choma (string heddle) and sasapayeca (tamping stick) used in weaving blankets. It is also ideal for use as a dibble, or planting stick, for driving holes in the soil into which seeds are dropped.

*Lantana camara* L.     ta'ampisa "molars become numb" (confiturilla negra)

*Confiturilla* is a generic local term for a group of species with showy flat-topped inflorescences. *L. camara* (confiturilla negra), *L. achyranthifolia*, and *L. velutina* (confiturilla blanca) are Verbenaceae, while *Lagascea decipiens* Hemsl. (confiturilla amarilla) is a composite. *L. camara* is a shrub sometimes growing into large thickets. The bright yellow and orange flowers appear year-round. It is common in tropical deciduous forest and foothills thornscrub, tending to proliferate on disturbed soils. MEDICINE: The roots are boiled into a tea taken for kidney pains. The branches are made into a tea and drunk for scorpion stings. The significance of the Mayo name is obscure.

### *Lantana hispida* H.B.K. (confiturilla blanca)

This shrub, similar to *L. camara* but smaller, has white flowers. The fruits are jet black and are eaten by people, contrary to what we had heard from botanists. At the urging of a consultant, one of our party tried them once with discouraging results, but the damage did not appear to be permanent. *L. achyranthifolia* is a similar white-flowered confiturilla found in tropical deciduous forest. It is more herbaceous than *L. hispida*.

### *Lippia alba* (Mill.) N.E. Br. (valeriana)

This is a cultivated aromatic shrub, 1 m tall, with lavender flowers. MEDICINE: The leaves harvested from potted plants are commonly boiled to make a tonic. It is judged especially effective for headaches and for tension and stress, or *los nervios*, detected by the tightening of muscles on the neck. For women who have recently given birth to a baby, it is taken for pasmo, or what might have been diagnosed as hysteria a century ago. Although it is a cultivated plant, we include it because of its widespread use.

### *Lippia graveolens* H.B.K. bura mariola (mariola, orégano de burro)

A nondescript shrub occurring in waste places. It is fairly common northwest of Agiabampo and in Arroyo Las Rastras on the southwest side of the Sierra de Alamos, but not elsewhere in the region. MEDICINE: Vicente Tajia said it is known throughout the region as a good cough remedy. He collects it and stores bunches in his home.

### *Lippia palmeri* S. Watson (orégano)

A small shrub with dense herbage and small white flowers. FOOD: Highly aromatic bushes of the popular seasoning grow wild, especially on calcareous soils in the more arid parts of the region. It is harvested in commercial quantities by the Mayos of San Antonio, north of Masiaca. Villagers fill bags with it and dump them into great piles on the ground. They then gently rake the leaves into a uniform layer perhaps 5 cm thick. After a week of sunlight they

*Orégano* (*Lippia palmeri*) being stirred by Berta Valenzuela, San Antonio. The dried leaves will be sold in city markets. (Photo by T. R. Van Devender)

place the herb in large feed sacks and transport it by bus to the markets of Huatabampo and Navojoa. No home in the region is without it.

**Vitex mollis H.B.K.**　　　　　　　　**júvare "I want that" (igualama, uvalama)**

A spreading tree with rather sparse leafage, sometimes growing to 15 m tall along watercourses. In the tropical deciduous forest near Las Rastras draining the Sierra de Alamos, and near Baimena, the trees grow even taller and nearly equally wide, constituting an important component of the riparian canopy. In April and May the tree flowers delicately, its small, intricate lavender blooms exuding a perfume quite irresistible to pollinators, including hummingbirds. One can detect the smell from many meters' distance. In some canyons, the epiphytic orchid *Oncidium cebolleta* (Jacq.) Sw. covers the branches of the júvares. The fruits are gobbled up by *urracas* (black-throated magpie jays; *Calocitta formosa collei,* Corvidae), cholugos (coatimundis), and many other birds and small mammals. AESTHETICS: Some natives will plant trees in their yards to enjoy the perfume from the flowers as well as the delectable fruits. An Old World species called *carnavalito* (*V. trifolia* L.) is widely planted as a hedge around Mayo yards. It is also called *hazme como quieres* (do what you wish with me), reflecting the ease of transplanting them, according to Elvira Tajia Moroyoqui. FOOD: In August the black, cherry-sized, fleshy fruits are widely eaten. When boiled with sugar, they have a flavor and texture reminiscent of stewed plums. We found them bitter when we first tried them, but Maria Teresa Moroyoqui stewed them with a little sugar, and the result was quite tasty. Don Vicente Tajia, who planted a sapling in his yard, suggests they are tastier eaten with goat's milk. In Baimena the seeds are ground in a metate, or hand mill, and eaten as an atol. LIVESTOCK relish the leaves. MEDICINE: According to Amelia Verdugo Rojo of Los Muertos, the leaves are brewed into a tea and administered to ease labor pains, a practice shared with Guarijíos. She said this potion is made more effective by adding domestic chamomile (*Matricaria chamomilla* L., Asteraceae).

*Vitex pyramidata* B.L. Rob.   jupa'are "Is he the priest?" (negrito)

A small tree with clasping, folding leaves, found in drier and usually higher habitats than the igualama (*V. mollis*). It is much less common than the latter. It often grows on hydrothermally altered soils, along with several other plant species normally found at higher elevations. Otherwise, it is most often found in upper tropical deciduous forest or along with low-elevation oaks on ancient volcanic soils, usually of indurated ash. The Mayo name may be derived from the same origin as El Júpare, the important Mayo town of the Mayo Delta. FOOD: The fruits, though smaller than those of the igualama, are equally popular.

## VIOLACEAE

*Hybanthus mexicanus* Ging.   jépala (jarial)

A nondescript shrub with small leaves in Mayo country, sometimes growing into a small tree up to 6 m tall in moist cajones in tropical deciduous forest. Mayos state that it makes good LIVESTOCK forage.

## VISCACEAE

*Phoradendron californicum* Nutt.   pohótela "phainopepla"
  (toji; desert mistletoe)

*Phoradendron californicum* is a bushy, nearly leafless mistletoe that prefers mezquites and other legumes. The leaves are reduced to juniperlike scales. This is a widespread species in the deserts and desert grasslands of the southwestern United States and northern Mexico, reaching its southern limits in coastal thornscrub. It is a parasite on many species, typically legumes, but also *Larrea divaricata* Cav. in the Sonoran Desert. In the Mayo area, it is most common in coastal thornscrub on palo verdes but has also been collected on ejéa and jócona. It is rare in tropical deciduous forest, appearing mostly on tepeguaje, but also on algarrobo and güinolo. The fruits are eagerly devoured by phainopeplas (*Phainopepla nitens,* Ptilogonatidae), whose Mayo name is the same as that for mistletoe. All species of *Phoradendron* are called *toji* in Spanish. In coastal thornscrub, the broad-leaved *P. brachystachyum* (DC.) Eichler parasitizes mangle dulce, mezquite hormiguillo, sanjuanico, and wolfberry. In tropical deciduous forest, *P. quadrangulare* (Kunth) Krug & Urban is found on various hosts, including Bonpland willow, garambullo, and samo baboso. Other species of *Phoradendron* are found in higher oak woodlands (*P. bolleanum* [Seem.] Eichler, *P. longifolium* Eichler, and *P. tomentosum* [DC.] A. Gray). Mayos distinguish among several species of mistletoe and differentiate their virtues based upon the host species. MEDICINE: Pohótela is widely viewed as brewable into a tea effective for lowering blood pressure. Natives at Yópori on the coastal plain state that the plant growing on mezquite is a better remedy than that growing on other trees. In times of severe drought cowboys will climb trees and knock off branches of the mistletoe, which are then eagerly eaten by LIVESTOCK waiting beneath.

## VITACEAE

*Cissus* cf. *mexicana* DC.          yucohuira "water vine"; coya guani
[*C. mayoensis* Gentry]

*C. trifoliata* (L.)

*C. verticillata* (L.) Nicholson & Jarvis

All these species are conspicuous vines leafing out rapidly after the onset of the summer rains and covering adjacent trees, especially etcho cacti. Their habitats differ, but all three are found in the region and are considered good forage for LIVESTOCK.

## ZYGOPHYLLACEAE

*Guaiacum coulteri* A. Gray.          júyaguo (guayacán)

Guayacán is a large shrub in the Sonoran Desert and thornscrub. In the tropical deciduous forest it becomes a tree of greatly varied habit that grows to 8 m, often with a thick, powerful trunk with maculate bark. The tiny, brilliant emerald green leaves, adnate to the branches, shine through the forest in the dry months. In April, the fragrant dark purple blue blossoms attract pollinators and humans. Some of the trees are valued for shade. Knowledgeable folks do not burn the wood because the smoke has an unpleasant odor. Francisco Valenzuela said the smoke is toxic and if inhaled will cause a person's hair to fall out. MEDICINE: The flower is cooked to make a tea drunk for treating asthma. It is also believed by some to be effective as a wash for psoriasis or skin that has lost its pigmentation. The heartwood is brewed into a tea to cure pujos (bloody stools). CONSTRUCTION and ARTIFACTS: The hard, strong wood (closely related to *lignum vitae* [*Guaiacum officinale* L.]) is used for fence posts and other applications where durable wood is required, such as in constructing looms, for which it is especially valued. Branches are woven and covered with mud to make solid, long-lasting walls. CULTURE: Guarijíos have informed us that they respect the tree as having extraordinary power and refuse to harm it.

*Kallstroemia grandiflora* Torr.    jímuri "chiggers" (baiburín; summer poppy)
*K. parvifolia* Norton
*K. californica* (S. Watson) Vail

These prostrate annuals aggressively fill all available terrestrial niches in thornscrub and some areas of tropical deciduous forest following the onset of summer rains. *K. grandiflora* and *K. parvifolia* flower with showy orange or peach salmon blooms, often illuminating large expanses of land. *K. californica* is a similar plant with smaller leaves and tiny yellow flowers. Mayos attribute no good use to any of these plants but note that they are a host for baiburines, the infernal chiggers that plague those who visit the monte in the wet months.

# APPENDIX A
# Mayo Region Place Names and Their Meanings

| | |
|---|---|
| Agiabampo | Guásima (*Guazuma ulmifolia*) in the water |
| Aquiropo | Pile of *pitahayas* (*Stenocereus thurberi*) |
| Bacabachi | Reed seed |
| Bachoco | Saltwater |
| Bachomojaqui | Arroyo of seepwillow (*Baccharis salicifolia*) |
| Bacobampo | Snake in the water |
| Bacorehuis | Loaned snake |
| Bahuises | Screech owls |
| Baimena | Derived from *baji mela* [he has killed three men] |
| Basiroa | Root in the water |
| Buaysiacobe | He lost it three times in the river |
| Buayums | Loose diapers |
| Buiyacusi | Where the earth sings |
| Buiyagojo | Hole in the ground |
| Camahuiroa | Squash vine |
| Camoa | Without a sprout |
| Capitahuasa | Captain's fence |
| Chichibojoro | Hole full of *estafiate* (*Ambrosia confertiflora*) |
| Choacalle | *Choya* in the road |
| Choquincahui | Starry hill |
| Chorijoa | House that is no good [Place of the quail (?)] |
| Cucajaqui | Arroyo of *vinorama* (*Acacia farnesiana*) |
| Cuchujaqui | Arroyo of fish |
| El Júpare | Is he the priest? [Mezquite over there (?)] |
| El Tábelo | Broken stick |

| | |
|---|---|
| Etchohuaquila | Skinny *etcho* |
| Etchojoa | Place of the *etchos* |
| Etchoropo | Pile of *etchos* |
| Güirocoba | Vulture head |
| Huasaguari | Fence of baskets |
| Huatabampo | Willow in the water |
| Huebampo | *Bledo* (*Amaranthus palmeri*) in the water |
| Jambiolobampo | Old lady in the water |
| Jeobampo | Hiccup in the water |
| Jerocoa | Muddy water [Something appeared in the water (?)] |
| Jopopaco | *Palo blanco* (*Piscidia mollis*) out there |
| Juliantabampo | Julian in the water |
| Las Bocas | Those who are stretched out on the sand (not "the mouths") |
| Macoyahui | The sound of (Pharisees or squashes) being hacked to pieces |
| Masiaca | Centipede hill |
| Mayocahui | *Mauto* (*Lysiloma divaricatum*) hill |
| Mochibampo | Turtle in the water |
| Mochicahui | Turtle hill |
| Mocúzari | Where the owl hoots |
| Moroncárit | Sorcerer's house |
| Nahuibampo | Good for nothing in the water |
| Naopatia | Covered prickly pear fruit |
| Navojoa | Place of the prickly pears |
| Nescotahueca | Upright *nesco* (*Lonchocarpus hermannii*) |
| Piedra Baya | Squash-colored stone (also called Tetabaya) |
| Saneal | Grove of *sanéas* (*Prosopis articulata*) |
| Sebampo | Sand in the water |
| Sinahuisa | Spoon of *sina* (*Stenocereus alamosensis*) |
| Sirebampo | *Sire* (*Malpighia emarginata*) in the water |
| Tapizuela | *Huaraches* (sandals) |
| Tásirogojo | Hole full of *tásiro* (*Jacquinia macrocarpa*) |
| Teachive | Scattered round stones |
| Techobampo | Mud [elbow (?)] in the water |
| Tepahui | He is fat |
| Tesia | The grinding stones found in the stomach of ruminants |

| | |
|---|---|
| Tesopaco | *Teso* (*Acacia occidentalis*) out there |
| Tetajiosa | Paper rock |
| Tóbari | They want to sleep [mullet (?)] |
| Yavaros | Place of mullet or shrimp |
| Yocogigua | Where the jaguar ate |
| Yópori | Pori's place |
| Yucomari | He got struck by lightning |

## APPENDIX B
# Yoreme Consultants

Bacasegua, Alejo; Camahuiroa. Born ca. 1930 in Camahuiroa. A former fisherman and agricultural day laborer, he is knowledgeable about wild plants. He is known for his willingness to prescribe medicinal and spiritual cures for victims of various pathologies.

Badón, Federico; Rincón de Aliso, Sinaloa. Born ca. 1940 in Rincón de Aliso. Federico is an artisan who occasionally works as a day laborer. He weaves blankets on an upright (Spanish-style) loom. He also carves masks and makes Mayo musical instruments.

Baesmo Cota, Ignacio; La Misión, Sinaloa. Born ca. 1920. Ignacio is extremely knowledgeable about the plants of the tropical deciduous forest of the lower Río Fuerte.

Baisegua, Reyes; Sirebampo. Born ca. 1925 in Sirebampo. He is most knowledgeable about plants of the coastal thornscrub. A tall man, he has worked as a cowboy and day laborer, woodcutter, and fisherman. He has woven morrales from ixtle for many years and is probably the last Mayo to produce morrales of the old style.

Buichílame Josaino, Paulino; Las Bocas. Born ca. 1925 in Las Bocas. Paulino has worked as a day laborer, cowboy, and fisherman. He is familiar with herbal cures.

Dórame Buitimea, Mariano; Los Buayums. Born ca. 1940. A well-known curandero, he is recognized as such by the Instituto Nacional Indigenista and provided with a laminated card to that effect. He is knowedgeable about the curative powers of plants, but he is in poor health and reluctant to venture into the monte for fear of criminals.

Estrella, Leonarda; Los Capomos. Born ca. 1935 in Los Capomos. A potter of considerable renown, she also is knowledgeable about Mayo customs of the Río Fuerte region.

Estrella, Patricio; Los Capomos. Born ca. 1928 in Los Capomos. A successful peasant farmer, he is a former pascola dancer who still plays as a musician in Los Capomos fiestas. He is knowledgeable about plants of the monte and studies medicinal herbs as well.

Estrella, Turibio; Los Capomos. Born ca. 1930 in Los Capomos. An ejidatario who for most of his life has raised corn, beans, sesame seed, and watermelons on his ejido plot. From time to time he has worked as a day laborer. He also sings at fiestas.

Félix, Juan; Teachive. Born ca. 1950 in Teachive. Juan is a day laborer and woodcutter. He is intimately familiar with Cerro Terúcuchi, north of Teachive, and other parts of the region.

Gámez, Francisco; San José de Masiaca. Born ca. 1960 in San José. A former pascola, Paco carves superb masks and figurines. He also produces drums and sonajos (rattles).

Gámez Piña, Serapio; San José de Masiaca. Born ca. 1920 in San José. Serapio has worked as a cowboy, woodcutter, and day laborer, and has gleaned from the monte. He is a traditional Mayo whose son became a well-known pascola dancer. He fashions native crafts for sale to outsiders.

García, Maximiano; Agiabampo. Born ca. 1940 in Agiabampo. Maximiano had been a fisherman but now is engaged in miscellaneous enterprises in the ejido of Agiabampo. He is knowledgeable about herbal remedies.

Gonzalez, María; Teachive. Born in 1905 in Teachive. Doña María is blind but capable of discussing many uses of plants.

Ibarra, Domingo; Huasaguari. Born ca. 1930 in San Antonio de Las Ibarras. Domingo tends cows, cuts firewood, and occasionally works as a day laborer. He is remarkably knowledgeable about the history of the Huasaguari area, its plants, and its topography.

Leyva, Francisco "Chico"; Yavaritos. Born ca. 1930 in Yavaritos. Chico has worked as a day laborer, woodcutter, and cow herder. He is knowledgeable about the monte around Yavaritos and about Mayo ceremonials.

Liso Jusaino, Genaro; Nahuibampo. Born ca. 1925 in Nahuibampo. Genaro has been a day laborer and woodcutter. He is a traditional Mayo who is knowledgeable about the monte.

López, Fausto; Nahuibampo. Born ca. 1925 in Nahuibampo. Fausto has worked as an ejidatario, day laborer, cowboy, and woodcutter. A traditional Mayo, he recalls much of the history of the area.

López, José; Nahuibampo. Born ca. 1935 in Nahuibampo. Brother to Fausto, José is also knowledgeable about the monte and the Río Mayo.

López, Leobardo; Chomajibiri. Born ca. 1930 near Yavaros. An ejidatario, he is blessed with a large irrigated parcel of farmland. Leobardo is very knowledgeable about dune and estuarine vegetation and plants.

López Flores, Braulio; Yocogigua. Born in 1935 in Yocogigua. He has several times been president of the Ejido Yocogigua and has worked as a cowboy, woodcutter, agricultural day laborer, farmer, and planter of buffelgrass. Braulio is a good conversationalist.

López García, Balbaneda "Nelo"; Yavaros. Born ca. 1950. Nelo is a fisherman and boatman at Yavaros. He is knowledgeable about plants of the dunes and estuaries.

Mendibles Cota, Wencelao "Mono"; La Cuesta de Basiroa. Born ca. 1930. Mono is a jack-of-all-trades mixing carpentry with native woods and prescribing herbal and supernatural cures. He has worked as a day laborer and cowboy. He is very knowledgeable about the region's natural history.

Mendoza, Buenaventura; Teachive. Born ca. 1905 near Yocogigua. For much of her life she wove fine woolen blankets. She has spent many hours in her later years spinning wool from a malacate (spindle). She is very knowledgeable about the use of native plants. She is the mother of María Soledad Moroyoqui Mendoza.

Morales, Ramón; Los Muertos. Born ca. 1945. An ejidatario of Basiroa, he has been a woodcutter, cowboy, farmer, and day laborer. He operates a small used clothing business. He is knowledgeable about the plants in the region.

Moroyoqui, Aurora; Choacalle. Born ca. 1935. Aurora weaves superb cobijas. She is a traditional Mayo and has been a fiestera for the Masiaca festivals.

Moroyoqui, Gregoria; Sirebampo. Married to Reyes Baisegua. Born ca. 1950 in Chichibojoro, an abandoned village near Las Bocas. She produces tamales made from pitahaya fruits and pitahaya seca.

Moroyoqui, Miguel; Yocogigua. 1920–1999. Born and died in Yocogigua. Miguel was a traditional Mayo familiar with the Yocogigua region.

Moroyoqui Mendoza, María Soledad; Teachive. Born ca. 1940 in Teachive. María is well known as a weaver of fine blankets. She is an expert in preparing chiju.

Moroyoqui Zazueta, María Teresa; Teachive. Married to Vicente Tajia. Born ca. 1938 in Teachive. Teresa is a former weaver still knowledgeable in the art. She is familiar with herbal remedies.

Muñoz Valdés, Jesús; Los Buayums. Born ca. 1934 at Bachoco. Jesús "Chuyón" is an agricultural foreman on ejido lands of Bachoco. He has spent much of his life as a fisherman and knows plants of the dunes and beach exceptionally well.

Navarete Valencia, Filemón; Fernando Solís. Born ca. 1945. Filemón is a day laborer, woodcutter, and cattle herder. He has been a deer dancer and knows a myriad of sones, the old musical verses sung at fiestas. He is a musician at fiestas as well.

Nieblas, Cornelia. Born ca. 1940 in Teachive. She is a cobijera of some renown, as well as a goatherd and baker. Cornelia is knowledgeable in the preparation of vegetable dyes.

Nieblas Moroyoqui, Jesús José "Cheché"; Teachive. Ca. 1940–1997. Cheché worked as an agricultural day laborer, a sand and gravel digger, woodcutter, cowboy, and goat herder.

Nieblas Zazueta, Balbina; Teachive. Born ca. 1960 in Teachive. She and her sons produce katchina-like dolls and pascola masks.

Nieblas Zazueta, Julieta; Teachive. Born ca. 1960 in Teachive. Daughter of Jesús Nieblas and Lidia Zazueta. Julieta is an accomplished weaver knowledgeable in the production of vegetable dyes from local plant materials.

Palomares, Artemisa; Los Capomitos, a tiny village near Los Capomos. Born ca. 1920 in Los Capomitos. Artemisa produces superb baskets and other utensils of woven ta'aco (palm).

Piña, Florentino "Queriño"; Las Bocas. Born ca. 1940 in Las Bocas. He has been *comisario* (chairman) of the Masiaca comunidad and is an authority on local plants.

Tajia Yocupicio, Vicente; Teachive. Born 1934 in Teachive. He has worked as an agricultural day laborer, cowboy, collector of buffelgrass seed, and laborer on highway construction. Vicente is a native botanist with an extensive knowledge of Mayo terminology and plant uses. He is consulted by others on medicinal uses of plants and plant identification.

Valencia Moroyoqui, Seferino; Las Bocas. Born ca. 1940 in Las Bocas. Seferino has been a fisherman most of his life.

Valenzuela, Alvaro; Baimena. Born ca. 1930. He raises corn, beans, squash, and watermelons for his family. In wet years he plants sesame seed. He is knowledgeable about the plants of the region.

Valenzuela, Benjamín; Las Rastras. Born ca. 1973 in San Antonio de Las Ibarras. Benjamín works as a day laborer, cow herder, and woodcutter. He is most familiar with the tropical deciduous forest of the southern reaches of the Sierra de Alamos.

Valenzuela, Luciano "Chano"; Los Muertos. Born ca. 1935 in Los Muertos. Chano is an ejidatario who raises corn and beans, works as a day laborer, and herds cows on the ejido. He is very knowledgeable about the region's topography and plants.

Valenzuela, Luis; Las Rastras. Born ca. 1930 in San Antonio; died 1997. Luis herded cattle and worked as a cowboy on the south side of the Sierra de Alamos. He was a bachelor well known in the region.

Valenzuela, Marcelino; El Rincón Viejo. Born ca. 1945. Marcelino is an expert chair maker and produces other furnishings as well. He is familiar with herbal remedies.

Valenzuela, Marcos; Baimena. Born ca. 1950. Marcos has been a farmer and comisario of the Baimena comunidad. He is politically astute and knowledgeable about the region's fauna and flora.

Valenzuela, Rosario "Chalo"; Baimena. Born ca. 1930 in Baimena. Chalo raises corn, beans, squash, and sesame in his field. He is knowledgeable about the area's plants and topography.

Valenzuela, Sara; El Rincón Viejo. Born ca. 1940. Sara works in various capacities in Alamos: as a teacher, a nurse, and a house cleaner. She is widely viewed as sophisticated in natural and supernatural cures.

Valenzuela Nolasco, Francisco "Don Pancho"; El Rincón. Born ca. 1935 in Los Tanquis, a town some 15 miles north of Alamos. A former pascola dancer, Don Pancho is sophisticated in identifying and classifying plants, and recalling their Mayo names. He is an ejidatario, a cutter of vara blanca, and a day laborer.

Valenzuela Yocupicio, Santiago "El Teco"; Yocogigua. Born ca. 1920 near San Antonio de Las Ibarras. He has worked cutting firewood, producing mezcal in the distillery of Yocogigua (closed in the early 1970s), and as an agricultural day laborer. He is knowledgeable about Mayo traditions.

Verdugo Rojo, Amelia; Los Muertos. Born ca. 1925. Amelia is one of the few Mayo speakers in the area. She has studied herbal medicine much of her life and is knowledgeable about plant remedies.

Yocupicio Anguamea, Felipe; Camahuiroa. Born ca. 1955 in Camahuiroa. Felipe is a fisherman with a vast knowledge of the monte and sea life. His grandmother was a renowned curandera who passed on much of her knowledge to Felipe.

Zazueta, Lidia; Teachive. Born ca. 1940 in Teachive. Widow of Cheché Nieblas. Lidia has been weaving fine blankets since her teenage years. She is knowledgeable about uses of native plants.

Zazueta, María Teresa; Choacalle. Born ca. 1920 in Teachive. She weaves fine blankets and makes ténaborim (cocoon rattles). Doña María Teresa is also expert in making and weaving ixtle. She is most knowledgeable in ancient Mayo customs.

## APPENDIX C
# Gazetteer of the Mayo Region

This list represents localities we have either visited or know to be of importance in understanding the plants of the Río Mayo region, which includes the lower Río Fuerte in Sinaloa. Population estimates are from the 1990 INEGI census unless otherwise noted.

Abaschapa. Settlement of three families, formerly Mayo and now primarily Mestizo, on the Arroyo El Veranito. The arroyo has permanent water (released from the dam above) and handsome cottonwoods. Pop. 16. 26°57.5' N 109°15' W. Elevation 140 m.

Aduana. Gentry wrote in 1942, "Settlement 5 or 6 miles northwest of Alamos and one of the Alamos silver mines active up to about 1914. It lies at the east base of the northern ridge of Sierra de Alamos. Vegetation in the steep canyon above the settlement is rather typical Short-tree Forest type." Aduana languishes below abandoned mills, a village of thirty families among smelter ruins. In spite of its being entirely Mestizo, Aduana remains the site of a November pilgrimage by thousands of Mayos and the most important fiesta in the Mayo region. Many pilgrims walk across the Sierra de Alamos or the Sierra Batayaqui to pay their respects to a *pitahaya* (organpipe cactus; *Stenocereus thurberi*) growing out of the south side of the church. Its shadow, which can be seen when the sun is at a propitious angle, is said to resemble the Virgin Mary. The cactus is possibly the most revered plant in Sonora. Pop. 268. 27°03' N 109°00' W. Elevation 540 m.

Agiabampo. Small seaport in the Agiabampo estuary only a few miles from the border with Sinaloa. Originally a Mayo community, it still has many Mayo families. The surrounding ejido is being developed for commercial agriculture. The development has resulted in the felling of hundreds of thousands of trees, especially pitahayas. The thornscrub just to the north formerly supported the densest stands of pitahayas in the world, and the Arroyo Agiabampo, a small but incised wash, supported a fine mix of trees and shrubs from the south, and others from the north. Pop. 1,671. 26°22.6' N 109°08' W.

Alamos. Gentry wrote in 1942: "Pueblo of about 3000 inhabitants, which in 1910 numbered 6000. A beautiful, quiet old town with a provincial aristocracy, remnant of the mining heyday of thirty years ago. It has many beautiful old houses, two or three primitive hotels, very poor restaurants, and busses coming and going several times a day. The cathedral, the architecture of the whole town, and the salubrious climate are noteworthy. It lies in the Río Fuerte watershed." Alamos is now a historic mining center of 7,000 catering to tourists who find its winter and spring climate most agreeable and its restored colonial architecture enchanting. The surrounding mountains were, and in many places still are, blanketed with rich tropical deciduous forest. The foothills were extensively cleared for pastures and, before the Mexican Revolution, given over to agave cultivation for distilled spirits. There is a sizeable community of North Americans in Alamos, as well as many Mayos and their descendants. 27°01.5' N 108°50' W. Elevation 350–400 m.

Arroyo Camahuiroa. Drainage emptying into the Sea of Cortés near Camahuiroa.

Arroyo Masiaca. Major watercourse draining the south side of the Sierra de Alamos and adjacent smaller ranges, emptying into the Gulf of California at Las Bocas. Contains permanent water over much of its northern portion. Its waters are diverted and its basin's groundwater tapped for agriculture. Its bed is an important source of sand and gravel for Huatabampo and cities and towns of the Mayo Delta.

Ayajcahui. Small, basaltic hill north of Teachive. It has thick groves of *Brongniartia alamosana* growing on its flanks. 160 m at 26°48' N 109°13' W.

Baca (Sinaloa). A former Mayo town on the Río Fuerte. It is now primarily Mestizo, with only three or four Mayo families remaining. The surrounding hills retain patches of luxuriant tropical deciduous forest that seems to flourish in granitic soils and on granitic slopes. Natives manage to raise some crops on *temporales*. Some land is also irrigated by wealthier residents. Pop. (1980) 148. 26°47' N 108°27' W. Elevation 140 m.

Bacabachi. A busy town just east of Mexico Highway 15 south of Navojoa, and the site of a prominent fiesta. The town is notorious for its high rate of crime and violence. Several women weave woolen blankets, but one has ceased weaving because her sheep were stolen. Pop. 1,167. 26°52' N 109°26' W. Elevation 50 m.

Bachoco. Mayo settlement in thornscrub near tidal flats on Yavaros Bay, 10 km east of Yavaros, 16 km east of Huatabampito. The inhabitants gain their livelihood from the sea. Bachoco is a relatively prosperous traditional community, now beset with problems of alcoholism and crime. Pop. 197. 26°44' N 109°25' W. Elevation 20 m.

Bachomojaqui. Village in the Masiaca comunidad located about a kilometer inland from the coast, some 5 km southeast of Las Bocas and just east of Arroyo Bachomojaqui. Most of the houses are constructed primarily of pitahaya ribs, which abound in the region. Nearby are especially rich stands of coastal thornscrub, including inexplicably large numbers of *cuta béjori* (*Schoepfia shreveana*). Pop. 100. 26°32.3' N 109°17' W. Elevation 10 m.

Baimena (Sinaloa). A prosperous village/comunidad backed up to a well-watered series of high hills. The area and surrounding hills support extremely rich vegetation, perhaps the most luxuriant in all of Mayo country. Natives raise *ajonjolí* as a commercial crop, and others raise marijuana in more sheltered fields. Corn, beans, squashes, and watermelons are also abundant in a year with satisfactory rains. Adjacent Arroyo de la Culebra has fine stands of *cedro*. The straight, stately trees are an important resource to the comunidad and are carefully watched. The Arroyo Agua Caliente, which runs through the town, supports huge *sabinos*, the largest we have seen in northwest Mexico. 26°31.2' N 108°18' W. Elevation 200 m.

Basiroa. Former Mayo village on the Río Cuchujaqui, it is now almost entirely Mestizo. The residents raise a few crops on irrigated land and some on temporales. They also raise cows and cut and sell firewood. The area was once the home of a distinct Cáhita-speaking group. More than a dozen homes were washed away in the flooding of the Río Cuchujaqui following Hurricane Isis in 1998. Pop. (1990) 413. 26°43' N 108°54' W. Elevation 170 m.

Batacosa. Village in the municipio of Quiriego located in the transitional thornscrub/tropical deciduous forest of the foothills. It still has an annual fiesta, though few of its inhabitants retain the Mayo language. Pop. 626. 27°32' N 109°25' W. Elevation 200 m.

Buaysiacobe. This ancient village on the Río Mayo is one of the most affluent Mayo towns. Many of the residents own 10 ha parcels of irrigation land. Some have sold them under the government's privatization initiative. For all its (relative) affluence, it is not known as a traditional Mayo center. Pop. 2,874. 27°04' N 109°40' W. Elevation 50 m.

Buayums. Mayo village in coastal thornscrub with large specimens of *Olneya tesota* east of Huatabampo. A few women still weave fine woolen blankets and natives frequent the surrounding monte. Pop. 1,094. 26°52' N 109°26' W. Elevation 20 m.

Buiyacusi. River village of Mayos north of Navojoa. The inhabitants are given to weaving baskets of all sizes from *carrizo* (*Arundo donax*). Pop. (1997) 521. 27°11' N 109°27' W. Elevation 60 m.

Camahuiroa. Resort village on the Sea of Cortés. Originally a Mayo fishing village, it still has a modest Mayo community set away from the long strip of resort homes. Inland are (or were) vast forests of pitahayas, whose fruits the natives gather to consume and sell. In the nearby Arroyo Camahuiroa are three large specimens of *ébano*, and a fine variety of thornscrub and tropical deciduous forest trees. Much of the surrounding countryside has been cleared for planting buffelgrass and is in various stages of succession. Pop. 188. 26°33' N 109°16' W.

Camoa. Ancient village on the Río Mayo upstream from Navojoa. It was an important center at the time of European Contact, but is now a small, somnolent village with a tiny amount of irrigated land. The surrounding hills have remnants of tropical deciduous forest. Pop. 238. 27°14.5 N 109°15' W. Elevation 100 m.

Cerro Bayájuri. Basaltic hill 23 km west of Navojoa in the Río Mayo delta above the Mayo settlement of Bayájuri. Of spiritual significance to Mayos, it preserves one of the last vestiges of coastal/foothills thornscrub in the delta and includes *sahuaro* cacti. 27°05.5' N 109°40' W. Elevation 100 m.

Cerro El Bariste. Summit of a low range well forested with varied tropical deciduous forest 6 km northwest of Yocogigua. 26°50.5' N 109°03' W. Elevation 540 m.

Cerro Mayocahui. Basaltic berg jutting above the vast fields of the Mayo Delta. Although relentlessly grazed by goats and invaded by buffelgrass, it retains original thornscrub, about the only naturally occurring vegetation for many miles in the otherwise sanitized delta region. The Mayo village of Mayocahui skirts its base. The people there are allowed to build their poor huts only where the soil is too shallow and rocky for inclusion in the mega-farms of the delta. 27°04' N 109°44' W. Elevation 25 m at base of hill.

Cerro Terúcuchi. Landmark basaltic hill in arid thornscrub to the north of Teachive. It lies entirely within the Masiaca comunidad. 26°48.8' N 109°14' W. Elevation 220 m.

Chírajobampo. Village adjacent to Mexico Highway 16 some 15 miles south of Navojoa. Nearby is a good amount of intact coastal thornscrub where natives glean firewood and other items. A few women still weave *cobijas* in the village. Most of the people work as agricultural day laborers. Pop. (1997) 692. 26°58' N 109°27' W. Elevation 50 m.

Choacalle. Town in the Masiaca comunidad, immediately adjacent to Masiaca. All of the residents are Mayo, and the language is regularly spoken among older people. Some women weave woolen blankets, and other natives produce arts and crafts. Most, however, work as agricultural day laborers in the Mayo Delta. General Román Yocupicio, Sonoran governor and strongman of the 1930s, was born here. He receives mixed reviews by critical historians. Pop. 257. 26°46.7' N 109°14' W. Elevation 60 m.

Chomajabiri. Agricultural ejido hamlet near Moroncárit. The adjacent relict dune has a rich population of cacti: biznagas, musues, pitahayas, sinas, and tunas. The residents have rich farmland and retain considerable knowledge of estuary vegetation and fishing. In warm weather the place abounds with biting flies, mosquitoes, and no-see-ums. 26°51' N 109°09' W. Elevation 10 m.

Chorijoa. Gentry in 1942 noted: "Old Indian pueblo of Guarijíos or Macoyahuis, now in decay. Eight or ten families of mixed blood remain in residence." The Macoyahuis spoke Cáhita and may have been merely an isolated Mayo group identified by their village, Macoyahui, to the south of San Bernardo. Now a Mestizo/Mayo settlement on the Río Mayo 8 km NNW of San Bernardo, it is surrounded by patches of tropical deciduous forest. In 1998 three Mayo families were in residence. Pop. 65. 27°28' N 108°52' W. Elevation 200 m.

Cohuirimpo. See San Ignacio Cuhirimpo.

Conicárit. Gentry wrote in 1942: "Small pueblo, principally Indian (Mayo, Macoyahui?, and Guarijío), on a mesa at the confluence of the Cedros and Mayo rivers. The site of a Jesuit mission. Population 300." Conicari(t) was one of the original pre-Columbian villages along the Río Mayo, perhaps of a distinct people. It was inundated by the Presa Mocúzari in 1951. Its inhabitants were moved to the present village of Mocúzari.

Cucajaqui. Village in the Masiaca comunidad. The village is spread out over a barren, litter-strewn commons. The natives are for the most part poverty-stricken.

A few women weave blankets, and some men produce various arts and crafts. Pop. 190. 26°46.5' N 109°15' W. Elevation 60 m.

El Júpare. Mayo Indian village and traditional center in the Río Mayo delta. Formerly known as Santa Cruz, it is the location of a sacred cottonwood tree growing on the bank of the old river channel. Often viewed as the most important center of Mayo culture, its Holy Week festivities draw many thousands of people. Pop. (1990) 1,855. 26°48' N 109°43' W. Elevation 25 m.

El Paso. Poor hamlet on the Río Cuchujaqui near Tapizuela. It is one of the few towns in the area that are still predominantly Mayo. Many of the houses are made from handsome cut stone that is quarried nearby. The older people still speak *la lengua*, but its use has disappeared among younger folk. To the east and west are large areas of intact tropical deciduous forest and thornscrub. Pop. 137. 26°41' N 108°49' W. Elevation 150 m.

El Rincón Viejo. Small Mayo settlement of four families on the edge of a mountainous *ejido* 4 km north of Alamos. The surrounding hills host representative vegetation of the tropical deciduous forest. The last fiesta held at El Rincón was in the 1970s. The Mayo language is seldom spoken there. A resident chair maker carries on an old Mayo tradition in fine fashion. Pop. 15. 27°03.55' N 108°56' W. Elevation 410 m.

El Tábelo. Small town of some hundred families, some of Mayo descent, on the arroyo of the same name some 15 km north of Alamos. Graceful sabinos line the watercourse, and rich tropical deciduous forest covers the surrounding hills, while the surrounding valley, probably once cultivated, is heavily grazed and dominated by mesquite. Pop. 291. 27°10' N 108°57.5' W. Elevation 200 m.

Etchohuaquila. Agricultural settlement on the Arroyo Cocoraque. To the east are some remnant patches of thornscrub. Fernando Valenzuela, the famous baseball pitcher, was born here and danced *pascolas*. Pop. 893. 27°17' N 109°44' W. Elevation 25 m.

Etchojoa. Agricultural city in the Mayo Delta and seat of the *municipio*, a bastion of left-wing politics in a right-wing state. Traditional home to many Mayo Indians. It has two important fiestas each year. The Holy Week festivities are second in importance only to those of El Júpare. Agricultural runoff provides flow in the Río Mayo riverbed at the west end of the town, but impenetrable groves of vicious *Mimosa aspera* line the banks, preventing easy access; that is just as well, for the water is frightfully contaminated. Pop. 7,253. 26°52' N 109°38' W. Elevation 30 m.

Etchoropo. Mayo town on the remaining Mayo channel in the Mayo Delta nearly at the edge of the sea. The Holy Week festival is widely thought to be the most traditional in the entire delta. Pop. 1,284. 26°46' N 109°41' W.

Fernando Solís. Poor ejido remotely located among rich coastal thornscrub some 10 km north of Estación Luis on Mexico Highway 15. Several families eke out a tough life from the surrounding *monte*. Pop. 33. 26°36' N 109°06.5' W. Elevation 90 m.

Huasaguari. Mayo settlement in rich thornscrub northeast of Masiaca on the road to San Antonio. Some fine specimens of *júchica* (*Sideroxylon occidentalis*) grow in the yards of two homes. Pop. 80. 26°50.55' N 109°09.05' W. Elevation 140 m.

Huatabampo. Agricultural center in the Distrito de Riego del Río Mayo, the intensely cultivated delta, it is the seat of the municipio of Huatabampo. The local Alvaro Obregón Museum has interesting archives on Mayo affairs, especially concerning the Mayos' role supporting the forces of General Obregón (later president), who was from Huatabampo. It is also a center of Mayo Indian population, although the city itself is Mestizo. Pop. (1995) 26,037. 26°47' N 109°39' W. Elevation 20 m.

Huebampo. Impoverished Mayo village in the Masiaca comunidad. Nearby sits a failed experiment: an attempt at silviculture. In the 1970s villagers planted many hectares of trees of the tropical deciduous forest, hoping to raise them commercially. Forty hectares of jojoba were also planted. The projects were a complete failure, through no fault of the natives. Pop. 111. 26°42.5' N 109°16' W. Elevation 60 m.

Huichabiricahui. Also called Cerro Huichabiri. Prominent hill west of Basiroa with a few yellow *amapas* and interesting native grasses growing thereabouts. It is part of the Basiroa ejido and is well known to the natives of that poor village. 400 m at 26°41.5' N 108°56.5' W.

Jambiolobampo. Village and site of a small comunidad of excellent coastal thornscrub vegetation. The settlement is distinctively Mayo. Some older residents are monolingual Mayo speakers. Pop. 160. 26°43' N 109°19' W. Elevation 30 m.

Jopopaco. Small town in the Masiaca comunidad. Natives plant crops in a limited fashion when rains provide runoff in the Arroyo Masiaca. Although it is a Mayo village, one seldom hears Mayo spoken in the streets. Pop. 613. 26°44' N 109°17' W. Elevation 50 m.

Juliantabampo. Tiny agricultural settlement in the Mayo Delta. It is a nondescript settlement, but with a strong Mayo component. Pop. 80. 26°46' N 109°43' W. Elevation 20 m.

La Cuesta de Basiroa. A hamlet across the Río Cuchujaqui from Basiroa. One or two Mayo families are in residence. One Mayo, nicknamed Mono, ekes out a living carrying on old ways, a bit of *curandismo,* arts and crafts, herbs and potions, and various gleanings from the monte. He is worth visiting. Pop. 101. 26°43' N 108°54' W. Elevation 170 m.

La Gacela. Small Mestizo village with one Mayo family whose influence is extensive, for it has the feel of a Mayo village. It is located on a bluff overlooking the Arroyo Los Cochis, a tributary of the Río Cuchujaqui. Its upper homes offer a commanding view of the Sierra de Alamos. Pop. 168. 26°52.20' N 108°52.10' W. Elevation 250 m.

Las Animas. A hamlet of the Masiaca comunidad. It is situated in the midst of a vast forest of pitahayas 2 km from Las Bocas. The natives have made extensive and effective use of organpipe cactus for constructing homes and fences. Two women weave blankets from handspun wool. M. Kay (1996) carried out some ethnobotanical investigations here. Pop. (1990) 93. 26°37' N 109°20' W. Elevation 20 m.

Las Bocas. A Mayo fishing village where the Arroyo Masiaca empties into the Gulf of California, it is surrounded by a rather seedy beach resort popular with Navojoa weekenders. A Mayo fiesta is held each year on June 28, St. Peter's day. Inland are magnificent stands of pitahaya, ever endangered by the rapid clearing taking place in the region. This forest is unquestionably one of Mexico's natural treasures, but it is in constant jeopardy. The Arroyo Las Bocas, located southeast of the highway, harbors an unexpectedly rich flora. Pop. 424. 26°36' N 109°20' W.

Las Rastras. Mayo *rancho* and abandoned silver mining town on the long arroyo of the same name that drains a large portion of the southwest side of the Sierra de Alamos and empties into the Arroyo Masiaca. The formerly prosperous village is much reduced in size and vitality. The three remaining Mayo families reflect an air of quiet resignation to their poverty and ultimate demise, while the vacation homes of outsiders remain unoccupied nearby. Wealthy ranchers have felled the once rich forests, and the area is now home to endless seas of buffelgrass and rank weeds. Pop. 15. 26°55' N 109°04' W. Elevation 270 m.

Los Capomos (Sinaloa). A traditional Mayo village located in rolling granite uplands whose fertile soils have made the inhabitants relatively prosperous. Los Capomos is surrounded by rich tropical deciduous forest. Two women make good pottery. In the tiny barrio of Los Capomitos, a half kilometer to the south, two or three women weave baskets and hats from *ta'aco* (palm). At least one fellow carves *monos* (figurines) from *capúsari*. The village has an annual festival in June that attracts visitors from the entire region. Mayo is frequently spoken among adults, but the children do not seem to understand it. Fine specimens of *mora* grow on small granitic hills. Pop. ca. 1,000. 26°25.5' N 108°31' W. Elevation 120 m.

Los Muertos. Agricultural village and ejido roughly 4 miles north of Basiroa on the east bank of the Río Cuchujaqui. Pop. (1995) 202. 26°49' N 108°54' W. Elevation 175 m.

Macoyahui. Gentry wrote in 1942, "Small pueblo on the banks of the Río Mayo. A large part of its population is Indian: Mayo, Guarijío, and Macoyahui (?). So-called from the group of Indians formerly inhabiting it, the Macoyahui." It is now a Mestizo village with a few Mayos. The ancient, crumbling church broods over the listless, dying town. Old festivals still take place, but the language is mostly forgotten. Pop. 161. 27°19' N 108°55' W. Elevation 140 m.

Masiaca. Bustling Mayo town of 1,200 on the fringe of the great coastal plain of southern Sonora and one of the most important Mayo towns. Many leather products are produced in the village. Mayo hamlets ring the town. Irrigated pastures of corn, beans, and sesame seed are intermittently harvested, and in wet years or when the arroyo floods, temporales (dry-lands fields) may yield a crop. The town hosts the annual fiesta of San Miguel Arcangel on September 28, the most important time of year for the natives except for Easter week, when the town is also an important festival center. It also celebrates a festival on Mexico's Flag Day on or around February 24. The village church is the spiritual center for several villages in the vicinity and was built by local Mayos. The vegetation around Masiaca is arid coastal thornscrub, although in foothills only a few kilometers to the north one finds foothills thornscrub, and to the east, well-developed tropical deciduous forest. Pop. 1,280. 26°46' N 109°13' W. Elevation 50 m.

Mayocahui. Mayo village at the southern base of Cerro Mayocahui on the arid coastal plain of the Mayo Delta. The uncultivated slopes are the only remaining habitat for woody plants. The area is now heavily agriculturalized, the fields stretching off to the edge of the sea or into the flat horizon. Pop. (1990) 214. 27°07' N 109°44' W. Elevation 25 m.

Mesa Masiaca (Masiacahui; Microondas Masiaca). Basaltic hill between Highway 15 and Masiaca, topped by a microwave installation. The rare *Euphorbia gentryi* sprouts up among the great boulders. On the rocky southern slopes amid thornscrub stand the southernmost sahuaros growing on andesitic basalt, as they do on Tumamoc on the west side of Tucson. Natives of the Masiaca comunidad quarry rocks from the huge basaltic flow, where some large rocks on the flow sing like bells when they are struck. From the top of the mesa the Sea of Cortés is clearly visible. For reasons we have not ascertained, few natives frequent the slopes, and therefore many are unacquainted with the remarkable vegetation, which is decidedly different from that of the surrounding plains. 26°46.30' N 109°13.50' W. Elevation 200 m.

Mesquital (de Buiyacusi). Small town, an extension of Navojoa, located on a canal on the west side of the Río Mayo. Its inhabitants weave utilitarian baskets from carrizo. Pop. 232. 27°07' N 109°28' W. Elevation 75 m.

Mocúzari. Mayo village on the Río Mayo. A local black fly bears the same name. It is the site and service center of a massive dam built on the Mayo in 1951. The resultant reservoir obliterated several historic villages, including Conicárit, whose residents were relocated to Mocúzari. It is still home to Mayos, who perpetuate a fiesta each year. Pop. 731. 27°13' N 109°06' W. Elevation 120 m.

Moroncárit. A hamlet adjacent to the seaside resort of Huatabampito. A few Mayos live in the village. The nearby Loma de Moroncárit is a relict dune with diverse coastal plants. Pop. 1,456. 26°46' N 109°09.05' W. Elevation 10 m.

Nahuibampo. Mayo village and ejido of thirty families on the west bank of the Río Mayo southwest of San Bernardo. It and Vado Cuate on the east bank are the only remaining Mayo villages in the vicinity. To the north and west of Nahuibampo are extensive stands of tropical deciduous forest broken only by plowed *praderas* of buffelgrass. Pop. 127. 27°23' N 108°55' W. Elevation 200 m.

Naopatia. Settlement on Estero Bamocha on the coast of the Gulf of California in the municipio of Huatabampo with a Mayo family in residence. There are fine stands of mangrove in the estuary. Pop. (1990) 21. 26°45' N 109°15' W.

Navojoa. Gentry wrote in 1942, "Old but growing pueblo on the banks of the Río Mayo; population about 9,000. Center of the Río Mayo agricultural district, and the railroad dispersion point for trade up the Mayo Valley and into the Chínipas basin of Chihuahua. An interesting town of some color, containing hotels, restaurants, and garages. Town well water may be contaminated, but the river water is good and is brought in on water carts and peddled from house to house daily." Navojoa is now the agricultural and commercial center of the Río Mayo delta and the most important city in Mayo country. While it has barrios that have sizeable numbers of Mayos, it is now an overwhelmingly Mestizo city. In wet El Niño winters the Río Mayo overtops the spillway at Mocúzari and floods the valley extensively, threatening bridges, closing roads, and driving thousands from their homes. Pop. ca. 100,000. 27°05' N 109°26' W. Elevation 50 m.

Piedra Baya. Hamlet of two families in the Masiaca comunidad. The settlement has a large *represo* (earthen stock tank) that usually contains water. It is somewhat higher than the terrain to the west, with better soils and apparently somewhat higher rainfall. The resultant vegetation is slightly more varied and lush than that below. Pop. 10. 26°41' N 109°9.5' W. Elevation 80 m.

Piedras Verdes. Formerly a Mayo pueblo of a few hundred souls north of Alamos, now largely Mestizo. Residents are said to have been relocated from a town flooded by Mocúzari Lake when it filled in 1951. One or two people apparently speak halting Mayo, but the village is otherwise completely Mestizo in spite of its Mayo appearance. The immediate region is rich in copper deposits and will someday become an open pit mine, thus necessitating another relocation of the inhabitants. Residents produce utensils of carved mezquite (*Prosopis glandulosa*) and *teso* (*Acacia occidentalis*). Tropical deciduous forest and foothills thornscrub dominate the area. Pop. (1990) 234. 27°10' N 109°01' W. Elevation 200 m.

Pisicahui. A small mesa connected to the larger Mesa Masiaca (Masiacahui). The name means "hill of papache borracho" (*Randia obcordata*). Its slopes and summit are covered with an interesting mix of foothills and coastal thornscrub. 26°47' N 109°17.7' W. Elevation 160 m.

Puerto el Candelario. Some maps refer to it as Sierra Natochis. Landmark rhyolite peak to the north of Teachive. 26°52.5' N 109°11.5' W. Elevation 480 m.

Quiriego. Ancient seat of the municipio of Quiriego. It lies in rich thornscrub/tropical deciduous forest on the banks of the Río de los Cedros. A few Mayos inhabit the town. Pop. (1995) 1,154. 27°31' N 109°15' W. Elevation 200 m.

Rincón de Aliso (Sinaloa). Mestizo/Mayo agricultural village on the Río Fuerte near San Blas. Two Mayo brothers weave woolen blankets using Spanish-style foot-operated looms. They also produce other arts and crafts. Pop. ca. 500. 26°20' N 108°43' W. Elevation ca. 100 m.

San Antonio (de Las Ibarras). Ancient and traditional Mayo village adjacent to hacienda of the historically powerful Almada (now Ibarra) family on Arroyo Las Rastras on the southwest side of the Sierra de Alamos. The Ibarras have cleared much of the surrounding ranchland of forest and have planted it with buffelgrass, perhaps the largest such planting in Sonora. In 1995, although the poverty in the village was disheartening, other Mayos viewed San Antonio as a repository of Mayo culture. The families have close ties with those of Las Rastras. The residents gather *orégano* (*Lippia palmeri*) from shrubs growing wild on nearby calcareous soils, sun-dry the leaves on large mats, and market it in Navojoa. A fiesta is held each year in late February. Pop. 104. 26°52' N 109°07' W. Elevation 150 m.

Saneal. Mayo village and ejido in dire poverty, located in arid coastal thornscrub just east of Mexico Highway 15 south of Navojoa. The natives raise cattle and a few crops. Some women weave woolen blankets. Pop. (1995) 893. 26°55' N 109°22' W. Elevation 40 m.

San Ignacio (San Ignacio Cuhuirimpo). Ancient Mayo town, one of the seven traditional Mayo villages, on the Río Mayo 2 km south of Navojoa. It is now largely Mestizo. 27°04' N 109°30' W. Elevation 30 m.

San José de Masiaca. Mayo village of the Masiaca comunidad on the banks of the Arroyo Masiaca, 1 km south of Masiaca. The road connecting the villages is lined with *josos*. Some women weave woolen blankets, and men produce arts and crafts. A few men raise garbanzo beans, corn, and beans when water is available from the arroyo. Most, however, work as day laborers in the fields of the Mayo Delta. Pop. 596. 26°45' N 109°14.7' W. Elevation 70 m.

San Miguel Zapotitlán (Sinaloa). Former Mayo town on the southeast bank of the Río Fuerte, now largely Mestizo. It is home to an important fiesta and is viewed as a Mayo center by many Sinaloan Mayos. Pop. ca. 2,500. 25°55' N 109°05' W. Elevation 100 m.

San Pedrito. Impoverished hamlet in the Masiaca comunidad, especially known for its musicians. Pop. 196. 26°45.7' N 109°15.3' W. Elevation 50 m.

San Pedro Viejo. Strongly Mayo agricultural town on the Río Mayo some 15 km southwest of Navojoa. The residents have planted many citrus trees and other plants that make their humble town most attractive. Pop. (1995) 302. 27°01' N 109°37' W. Elevation 30 m.

Sebampo. Predominantly Mayo agricultural village in the Mayo Delta. Pop. 734. 26°45' N 109°32' W. Elevation 20 m.

Sibiricahui. Prominent mountain on private property to the north of Teachive covered with dense foothills thornscrub. 26°49.5' N 109°13' W. Elevation 440 m.

Sierra Batayaqui. Low, prominent range east of Navojoa. It contains good stands of foothills thornscrub. Batayaqui means "dressed as a Yaqui." Highest point 600 m at 26°54' N 109°15.5' W.

Sierra de Alamos. The tallest mountain in the Mayo region. The Mayo name is *Nojme Cahui* (highest point). West of the main mass of the Sierra Madre, this "sky island" mountain range rises abruptly from foothills at 300 m. Its higher slopes are mostly granites, contrasting with the eruptive volcanics of the Sierra Madre to the north and east. The highest point, Cerro Aduana at 1,760 m (26°58.5' N 108°59.4' W), is 8 km southwest of Alamos. Foothills thornscrub, tropical deciduous forest, oak woodland, pine-oak forest, and semideciduous tropical forest are all found on the mountain. Natives warn against visiting certain remote canyons due to the overzealous and aggressive protection of illicit plants by their caretakers. In 1996 the Sierra de Alamos was designated an ecologically protected area by the Sonoran government. A tiny staff is working to prevent further clearing and destruction of the native vegetation.

Sinahuisa. Mayo village on the coastal plains south of Navojoa and south of Mexico Highway 15. The village is home to several very large ironwood trees and is surrounded by rich coastal thornscrub. A few women of the village still weave fine wool blankets. Pop. 844. 26°52' N 109°20' W. Elevation 50 m.

Sirebampo. Traditional village 1 km south of Mexico Highway 15, south of Navojoa on the route to Agiabampo. The nearby coastal thornscrub is rich in cacti, especially pitahaya. The village is nearly purely Mayo and has an annual fiesta. The residents survive by selling loads of sand to builders in Huatabampo and, during the pitahaya harvest, selling the delectable fruits in local markets. A traditional fellow still makes *morrales* (shoulder bags) from *ixtle* (fiber from *Agave vivipara*). Pop. 428. 26°38' N 109°15' W. Elevation 40 m.

Tapizuela. Mestizo/Mayo village on the Río Cuchujaqui not far from the Sinaloa border. It is known to produce talented pascola dancers. The countryside retains considerable monte, a transition between foothills thornscrub and tropical deciduous forest. The name apparently refers to the soles of *huaraches*. Pop. 715. 26°40' N 108°52' W. Elevation 160 m.

Teachive. Ancient Mayo hamlet associated with the Masiaca comunidad. At the village the Arroyo Masiaca trickles permanently, and an *aguaje* (pool) upstream is permanently full. The water gives rise to varied riparian vegetation and bird life. Residents produce native arts and crafts of high quality, including hand-loomed woolen rugs and pascola masks. Most of the inhabitants work as day laborers in the fields of the Mayo Delta. Pop. 583. 26°47' N 109°14' W. Elevation 75 m.

Tepahui. Gentry in 1942 wrote, "Small pueblo of mixed inhabitants. Jesuits reported a tribe of Indians under that name as inhabiting the upper Cedros." Tepahui is a charming, quiet village in tropical deciduous forest on the banks of the Río Cedros southeast of Quiriego. A few Mayo families still inhabit the town and make chairs from guásima wood. Pop. 282. 27°23' N 109°13' W. Elevation 150 m.

Tesia. An agricultural town, partly Mayo, some 10 km north of Navojoa. While it has considerable historical significance, it has now become a center of ongoing violence associated with the drug trade. Pop. 422. 27°12' N 109°23' W. Elevation 50 m.

Tesopaco (Rosario de Tesopaco). The center of the upper Río Cedros country and commercial center for most of the region between Ciudad Obregón and the east-west Mexico Highway 16. The village is nondescript, rather barren of plants. Although it is often referred to on maps as Rosario, natives still use its aboriginal name, Tesopaco. Until recently, villages nearby retained Mayo customs. Pop. 2,735. 27°50' N 109°21' W. Elevation 450 m.

Tetajiosa. Settlement on the west slopes of the Sierra de Alamos with only a single Mayo family remaining. A small reservoir built many decades ago provides permanent water. From Tetajiosa a decent dirt road leads to the Navojoa-Alamos highway. A far worse road, recently washed out, connects to Las Rastras and San Antonio, then to Masiaca. Pop. 38. 27°00.02' N 109°04' W. Elevation 360 m.

Tobarito. Mestizo/Mayo settlement on the coast north of the mouth of the Río Mayo. The bay of the same name is formed by the delta of the Río Cocoraque, which traditionally separates Mayo country from Yaqui country. 27°05' N 109°55' W.

Vado Cuate. A poor and depressed but neat village south of San Bernardo on the east bank of the richly vegetated Río Mayo. Many of the inhabitants speak Mayo and perpetuate religious festivals. Some inhabitants occasionally produce pascola masks from the wood of *palo mulato* (*Bursera grandifolia*) and other Burseras. Vado Cuate and Nahuibampo are the only strongly Mayo villages remaining on the Río Mayo upstream from Mocúzari Dam. Pop. 80. 27°21' N 108°55.30' W. Elevation 100 m.

Yavaritos. Traditional Mayo village surrounded by rich coastal thornscrub but only a few kilometers southeast of the vast irrigated complex of the Mayo Delta. The handsome houses are of traditional Mayo construction—wattle and daub, dirt roofs, porches with beams of *mezquite*, ironwood, or, occasionally, *mauto*. Yavaritos is known to produce excellent pascola dancers who are in great demand for fiestas all over the region. Pop. 321. 26°45' N 109°20' W. Elevation 30 m.

Yavaros. Shallow-water port on an important bay, south of the delta of the Río Mayo, probably formed from the silts and currents of the river. Originally a Mayo village, it has become overwhelmingly Mestizo since the construction of canning and fish-processing plants. The bay is now contaminated beyond description and reeks of offal. Pop. 3,506. 26°42' N 109°32' W.

Yococahui. A saddle and protuberance on Cerro Terúcuchi, north of Teachive. It has some highly unusual tree growth and is a fine viewpoint for lands to the north. The name means "jaguar hill." 26°48.55' N 109°13.5' W. Elevation 160 m.

Yocogigua. Mayo settlement of some forty families south of Alamos, 18 km east of Masiaca. The village is set in tropical deciduous forest. Immediately adjacent to the east is the *yori* (non-Mayo) part of town, inhabited by an equal number of non-Mayos. The name was derived from a nearby rock whose colors present a likeness of a jaguar eating. It is the site of an abandoned *mezcal* factory, which closed in the mid-1970s. The Mayo language is still spoken by older townsfolk. Numerous escaped cultivars of agaves are still found in the nearby tropical deciduous forest. The natives gather piles of *chiltepines* each year. They sell some of the fiery chiles and keep the rest for their own stocks. The ejido of Yocogigua has seen considerable strife between Mayo and non-Mayo members. Still, they have together managed to clear many fields and plant buffelgrass. The town well is an interesting, ancient rock-lined affair with steps leading to the water. Pop. 205. 26°47.40' N 109°01.30' W. Elevation 240 m.

Yópori. A settlement of four Mayo families on an ejido, some 8 km north of Teachive. The families speak Mayo among themselves. A small, hilly promontory, the Cerro de Yópori, immediately behind the settlement supports an interesting population of sahuaro cacti. 26°51.5' N 109°14.3' W. Elevation 80 m.

# APPENDIX D
# Mayo Plants Listed by Spanish Name

| SPANISH NAME | MAYO NAME | SCIENTIFIC NAME |
| --- | --- | --- |
| aguama | chó'ocora | *Bromelia alsodes* |
| aguamita | huítbori | *Hechtia montana* |
| aguaro | tan cócochi | *Martynia annua* |
| aguaro | aguaro | *Proboscidea altheaefolia, P. parviflora* |
| ajonjolí | — | *Sesamum orientale* |
| álamo | aba'aso | *Populus mexicana* |
| algarrobo | — | *Acacia pennatula* |
| algodoncillo | chi'ini | *Wimmeria mexicana* |
| amapa | to'bo | *Tabebuia impetiginosa* |
| amapa amarilla | to'bo saguali | *Tabebuia chrysantha* |
| amole | aimuri | *Agave vilmoriniana* |
| amolillo | tupchi | *Sapindus saponaria* |
| añil | chiju | *Indigofera suffruticosa* |
| arrellán | — | *Psidium sartorianum* |
| ayal | choca'ari | *Crescentia alata* |
| bachata | jutuqui | *Opuntia* sp. |
| baiburilla | na'atoria | *Dorstenia drakena* |
| baiburín | jímuri | *Kallstroemia californica, K. grandiflora, K. parvifolia* |
| batamote | bachomo | *Baccharis salicifolia* |
| batayaqui | batayaqui | *Montanoa rosei* |
| bebelama | bebelama | *Sideroxylon persimile* |
| bellísima | — | *Solanum seaforthianum* |

| | | |
|---|---|---|
| biznaga | ónore | *Ferocactus herrerae* |
| biznaguita | chicul ónore | *Mammillaria grahamii, M. mainiae, M. yaquensis* |
| bledo | hué | *Amaranthus palmeri* |
| bola colorada | síquiri tájcara | *Trichilia americana, T. hirta* |
| brasil | júchajco | *Haematoxylum brasiletto* |
| brasil chino | tebcho | *Haematoxylum* sp. |
| brea | choy | *Parkinsonia praecox* |
| brincador | túbucti | *Sebastiania pavoniana* |
| bugambilia | — | *Bougainvillea spectabilis* |
| cabeza de violín | laven co'oba | *Heliotropium angiospermum* |
| cacachila | aroyoguo | *Karwinskia humboldtiana* |
| candelilla | cantela oguo | *Pedilanthus macrocarpus* |
| candelilla | — | *Euphorbia colletioides* |
| cardo | táchino | *Argemone ochroleuca* |
| carrizito de huitlacoche | baca cupojome jicto | *Lasiacis ruscifolia* |
| carrizo | ba'aca nagua | *Arundo donax* |
| cascalosúchil | — | *Plumeria rubra* |
| cebolla de coyote | guoy cebolla | *Hymenocallis sonorensis* |
| cedro | — | *Cedrella odorata* |
| chalate | tchuna | *Ficus insipida* |
| chalate | — | *Ficus trigonata* |
| chamisa del mar | bahue mo'oco | *Lycium brevipes, Suaeda moquinii* |
| chanate pusi | chana pusi | *Rhynchosia precatoria* |
| chani | comba'ari | *Hyptis suaveolens* |
| chapote | ja'apa ahuim | *Casimiroa edulis* |
| chichicoyota | goy minole, hayá huíchibo | *Cucumis melo* |
| chichicoyota | tetaraca | *Cucurbita argyrosperma* |
| chichiquelite | mambia | *Solanum americanum, S. nigrescens* |
| chicura | jiogo | *Ambrosia ambrosioides* |
| chicurilla | cau nachuqui | *Ambrosia cordifolia* |
| chilicote | jévero | *Erythrina flabelliformis* |
| chiltepín | có'cori | *Capsicum annuum* |
| chírahui | chírajo | *Acacia cochliacantha* |

| | | |
|---|---|---|
| chócola | chócola, tónaso | *Jarilla chocola* |
| chopo | cho'opo | *Mimosa palmeri* |
| choya | choya | *Opuntia fulgida* |
| choya | jijí'ica | *Opuntia leptocaulis* |
| choya | to'otori huita | *Opuntia pubescens* |
| choya | sibiri | *Opuntia thurberi* |
| chuparosa | semalucu | *Justicia californica, J. candicans* |
| ciática | — | *Bonplandia geminiflora* |
| ciruela del monte | jó'otoro | *Ziziphus obtusifolia* |
| cola de la zorra | hayas guasia | *Setaria liebmannii* |
| cola de zorrillo | jupachumi | *Petiveria alliacea* |
| compio | — | *Combretum fruticosum, Heteropteris palmeri* |
| confiturilla blanca | — | *Lantana hispida* |
| confiturilla negra | ta'ampisa | *Lantana camara* |
| copalquín | tapichogua | *Hintonia latiflora* |
| coquillo | coni saquera | *Cyperus compressus* |
| coquillo | coni saca | *Cyperus perennis* |
| coralillo cimarrón | bacot mútica | *Rivina humilis* |
| corcho | — | *Diphysa suberosa* |
| cordoncillo | — | *Elytraria imbricata* |
| cordoncillo | — | *Siphonoglossa mexicana* |
| cornetón | — | *Nicotiana glauca* |
| cresta de gallo | toto'ora | *Plumbago scandens* |
| cuajilote | — | *Pseudobombax palmeri* |
| cuernito | aguaro | *Proboscidea parviflora* |
| cúmero | cumbro | *Celtis iguanaea, C. pallida, C. reticulata* |
| cuna de niño | osi | *Commelina erecta* |
| dais | yore cotéteca | *Dalea scandens* |
| dais | siteporo, jíchiquia | *Desmanthus bicornutus, D. covillei* |
| damiana | — | *Turnera diffusa* |
| diente de culebra | bacot tami | *Brickellia coulteri* |
| diente de culebra | — | *Cupania dentata* |
| duraznilla | navo | *Opuntia gosseliniana* |
| ébano | ébano | *Caesalpinia sclerocarpa* |
| ejotillo | baiquillo | *Sesbania herbacea* |

| | | |
|---|---|---|
| ejotillo cafecillo | — | *Senna obtusifolia, S. occidentalis* |
| encino | cusi | *Quercus albocincta* |
| encino, encino roble | — | *Quercus chihuahuensis* |
| epazote | lipazote | *Chenopodium ambrosioides* |
| escoba | to'oro cojuya, jíchiquia | *Abutilon incanum* |
| escoba | — | *Bastardiastrum cinctum* |
| escoba | siteporo, jíchiquia | *Desmanthus bicornutus* |
| estafiate | chíchibo | *Ambrosia confertiflora, Parthenium hysterophorus* |
| etcho | etcho | *Pachycereus pecten-aboriginum* |
| flor de capomo | capo segua | *Nymphaea elegans* |
| flor de iguana | huícori bísaro | *Senna pallida* |
| flor de mayo | jévero | *Erythrina flabelliformis* |
| flor de piedra | teta segua | *Selaginella novoleonensis, S. pallescens* |
| frijol de codorniz | subuai muni | *Desmodium scorpiurus* |
| garambullo | baijuo | *Pisonia capitata* |
| garbanzo de coyote | goy carabanzo, tóroco | *Bunchosia sonorensis* |
| gloria | — | *Tecoma stans* |
| golondrina | cuépari | *Euphorbia abramsiana, Euphorbia* spp. |
| golondrina del mar | bahue cuépari | *Palafoxia linearis* |
| gordolobo | talcampacate | *Pseudognaphalium leucocephalum* |
| granadilla | sire | *Malpighia emarginata* |
| guachapori | na'a chú'uqui | *Cenchrus* spp. |
| guachaporito | — | *Triumfetta semitriloba* |
| guaje | huique | *Leucaena lanceolata* |
| guamúchil | maco'otchini | *Pithecellobium dulce* |
| guamuchilillo | — | *Calliandra emarginata* |
| guamuchilillo | jodimaco'otchini | *Pithecellobium unguis-cati* |
| guásima | agia | *Guazuma ulmifolia* |
| guasimilla | — | *Aphananthe monoica* |
| guayacán | júyaguo | *Guaiacum coulteri* |
| guayavilla | cuey oguo | *Salpianthus arenarius* |
| guayavillo | baihuío | *Acacia coulteri* |

| | | |
|---|---|---|
| guayparín | caguorara | *Diospyros sonorae* |
| güerequi | choya huani | *Ibervillea sonorae* |
| güinolo (chírahui) | chírajo | *Acacia cochliacantha* |
| güirote de culebra | caamugia | *Serjania mexicana, S. palmeri* |
| hierba de la víbora | jamyo olouama | *Stegnosperma halimifólium* |
| hierba del indio | guasana jibuari | *Aristolochia watsonii* |
| hierba del venado | cuchu pusi | *Porophyllum gracile* |
| hierba de pasmo | — | *Baccharis sarothroides, Hymenoclea monogyra* |
| hierba la flecha | túbucti | *Sebastiania pavoniana* |
| hierba lanuda | mochis | *Tidestromia lanuginosa* |
| higuera | tchuna | *Ficus pertusa* |
| higuera | — | *Ficus trigonata* |
| higuerilla | — | *Ricinus communis* |
| huevos del toro | berraco | *Stemmadenia tomentosa* |
| huichilame | jutuqui, jó'otoro | *Ziziphus obtusifolia* |
| huiloche | huoquihuo, huicobo | *Diphysa occidentalis* |
| huisache | huanaca | *Caesalpinia cacalaco* |
| igualama | júvare | *Vitex mollis* |
| incienso | choyoguo | *Encelia farinosa* |
| jaboncillo | murue | *Fouquieria diguetii, F. macdougalii* |
| jarial | jépala | *Hybanthus mexicanus* |
| jeco | jeco | *Hymenoclea monogyra, Xylothamia diffusa* |
| jícama | tosa huira | *Ipomoea bracteata* |
| jócona | jócona | *Havardia sonorae* |
| jojoba | — | *Simmondsia chinensis* |
| joso de la sierra | joso | *Conzattia multiflora* |
| lechuguilla | huítbori | *Agave aktites* |
| limoncillo | o'ouse suctu | *Zanthoxylum fagara* |
| lluvia de oro | — | *Tecoma stans* |
| maguey | cu'u | *Agave vivipara* |
| mala mujer | — | *Heliotropium curassavicum* |
| malva | to'oro cojuya | *Abutilon incanum* |
| malva | to'oro cojuya, jíchiquia | *Abutilon mucronatum* |

| | | |
|---|---|---|
| malva | — | *Bastardiastrum cinctum* |
| malva | to'oro jíchiquia | *Malvastrum biscuspidatum* |
| mamoa | momo ogua | *Erythroxylon mexicanum* |
| mangle | — | *Conocarpus erecta* |
| mangle | pasio tosa, moyet | *Laguncularia racemosa* |
| mangle dulce | pasio | *Maytenus phyllanthoides* |
| mangle negro | ciali | *Avicennia germinans* |
| mangle rojo | cánari, pasio síquiri | *Rhizophora mangle* |
| mariola | — | *Aloysia sonorensis* |
| mariola | bura mariola | *Lippia graveolens* |
| mataneni | sanarogua | *Callaeum macropterum* |
| mauto | mayo | *Lysiloma divaricatum* |
| melón de coyote | hayá huíchibo | *Cucumis melo* |
| mescalito | huítbori | *Hechtia montana* |
| mescalito (de huitlacoche) | huírivis cu'u | *Tillandsia exserta, T. recurvata* |
| mezcal | baogoa | *Agave* cf. *colorata* |
| mezcal | aiboli | *Agave* cf. *jaiboli* |
| mezcal | cu'u | *Agave vivipara* |
| mezquite | sanéa | *Prosopis articulata* |
| mezquite | juupa | *Prosopis glandulosa* |
| mezquite hormiguillo | juupa quecara | *Condalia globosa* |
| mimbro | — | *Cephalanthus occidentalis* |
| mochis | mochis | *Boerhavia coccinea, B. spicata, Tidestromia lanuginosa* |
| mora | mora | *Chlorophora tinctoria* |
| morita | sire | *Malpighia emarginata* |
| mostaza | ma'aca | *Drypetalon runcinatum* |
| nacapuli | nacapul | *Ficus cotinifolia, F. pertusa* |
| negrito | jupa'are | *Vitex pyramidata* |
| nesco | nesco | *Lonchocarpus hermannii* |
| nopal | navo | *Opuntia wilcoxii* |
| ocotillo, ocotillo macho | murue | *Fouquieria diguetii, F. macdougalii* |
| orégano | — | *Lippia palmeri* |
| orégano de burro | bura mariola | *Lippia graveolens* |
| oreja de ratón | tori naca | *Chiococca alba, Malvella lepidota* |

| Spanish | Mayo | Scientific |
|---|---|---|
| ortiga | tatio | *Croton ciliatoglandulifer* |
| ortiga | nata'ari | *Tragia jonesii* |
| ortiguilla | — | *Urera baccifera* |
| otate | ba'aca | *Otatea acuminata* |
| palma | ta'aco | *Brahea aculeata* |
| palma, palma real | ta'aco | *Sabal uresana* |
| palo adán | murue | *Fouquieria diguetti* |
| palo amarillo | jójona, momoguo | *Esenbeckia hartmanii* |
| palo barril | ciánori | *Cochlospermum vitifolium* |
| palo blanco (shrub) | ánima ogua | *Carlowrightia arizonica* |
| palo blanco (tree) | joopo | *Piscidia mollis* |
| palo bofo | huique | *Leucaena lanceolata* |
| palo cachora | cuta béjori | *Schoepfia schreberi, S. shreveana* |
| palo chino | chino | *Havardia mexicana* |
| palo colorado | mapáo, cuta síquiri | *Caesalpinia platyloba* |
| palo de asta | pómajo | *Cordia sonorae* |
| palo dulce | baijguo | *Eysenhardtia orthocarpa* |
| palo fierro | ejéa | *Olneya tesota* |
| palo fierro (palo pinto) | ooseo suctu | *Chloroleucon mangense* |
| palo huaco | úsim yuera | *Agonandra racemosa* |
| palo jito | jito | *Forchhammeria watsonii* |
| palo joso | joso | *Albizia sinaloensis* |
| palo limón | úsim yuera | *Amyris balsamifer* |
| palo limón | — | *Zanthoxylum* sp. |
| palo mulato | to'oro mulato | *Bursera grandifolia (B. simaruba)* |
| palo piojo | cuta nahuila | *Brongniartia alamosana* |
| palo piojo | jícamuchi | *Caesalpinia palmeri* |
| palo piojo (blanco) | jícamuchi | *Caesalpinia caladenia* |
| palo pitillo | murue | *Fouquieria diguetii, F. macdougalii* |
| palo santo | jútuguo | *Ipomoea arborescens* |
| palo torsal | guaji, paenepa | *Alvaradoa amorphoides* |
| palo verde | caaro | *Parkinsonia florida* |
| palo verde | úsim yuera | *Agonandra racemosa* |
| palo zorillo | jupachumi | *Senna atomaria* |

| | | |
|---|---|---|
| pamita | — | *Descurainia pinnata* |
| papache | jósoina | *Randia echinocarpa* |
| papache borracho | pisi | *Randia obcordata, R. thurberi* |
| papachili | — | *Ruellia nudiflora* |
| papelío | sato'oro | *Jatropha cordata* |
| peonía | jévero | *Erythrina flabelliformis* |
| pintapán | jíchiquia jéroca, ta'ri sorogua | *Abutilon abutiloides* |
| piocha | síquiri tájcara | *Trichilia americana, T. hirta* |
| pitahaya | sahuira | *Stenocereus montanus* |
| pitahaya | aaqui | *Stenocereus thurberi* |
| pitahaya barbona | aaqui jímsera | *Pilosocereus alensis* |
| pitahayita | chicul ónore | *Mammillaria* spp. |
| pochote | baogua | *Ceiba acuminata* |
| quelite | choal | *Chenopodium neomexicanum* |
| rama amarga | juya chibu | *Brickellia brandegei* |
| rama blanca | choyoguo | *Encelia farinosa* |
| rama del toro | — | *Dicliptera resupinata, Henrya insularis, Tetramerium abditum* |
| rama de venado | ma'aso o'ota | *Justicia candicans* |
| rama dormilona | yete ogua | *Mimosa asperata* |
| rama lisita | to'oro jíchiquia | *Sida aggregata* |
| rama quemadora | nata'ari | *Tragia jonesii* |
| retama | guacaporo, bacaporo | *Parkinsonia aculeata* |
| rocía | júptera | *Helenium laciniatum* |
| romerillo, jécota | jeco | *Hymenoclea monogyra* |
| sábila | — | *Aloe barbadensis* |
| sabino | ahuehuete | *Taxodium distichum* |
| sacamanteca | paros pusi | *Solanum azureum, S. tridynamum* |
| sacamatraca | nómom | *Peniocereus striatus* |
| sahuaro | saguo | *Carnegiea gigantea* |
| saitilla | yemsa ba'aso | *Bouteloua aristidoides* |
| saituna | báis cápora | *Ziziphus amole* |
| salvia | bíbino | *Hyptis albida* |
| samo, samóta | causamo | *Coursetia glandulosa* |

| | | |
|---|---|---|
| samo baboso | sa'amo | *Heliocarpus attenuatus* |
| sangrengado | sa'apo | *Jatropha cinerea, J. malacophylla* |
| sangrengado azul | sa'apo | *Jatropha cardiophylla* |
| sanjuanico | tásiro | *Jacquinia macrocarpa* |
| sanmiguelito | masasari | *Antigonon leptopus* |
| sapuchi | sapuchi | *Randia laevigata* |
| sauce, saúz | huata | *Salix bonplandiana, S. gooddingii* |
| saya | saya, saya mome | *Amoreuxia gonzalezii, A. palmatifida* |
| sayla de pintapán | — | *Sida aggregata* |
| sina | nómom | *Peniocereus marianus, P. striatus* |
| sina | sina | *Stenocereus alamosensis* |
| sina volador | sina cuenoji | *Selenicereus vagans* |
| sinita | musue | *Lophocereus schottii* |
| sitavaro | sitavaro | *Vallesia glabra* |
| tabaco de coyote | goy tabaco | *Nama hispidum* |
| tabaco de coyote | goy biba | *Nicotiana obtusifolia* |
| tajuí | tajimsi | *Krameria erecta* |
| tajuí | cósahui | *Krameria sonorae* |
| talayote | ma'aso alócossim | *Passiflora arida* |
| tampicerán | tampicerán | *Platymiscium trifoliolatum* |
| tasajo | tasajo | *Acanthocereus occidentalis* |
| tatachinole | tatachinole | *Tournefortia hartwegiana* |
| tavachín | ta'áboaca | *Caesalpinia pulcherrima* |
| té de lila | — | *Melochia speciosa, M. tomentosa* |
| tempisque | ca'ja | *Sideroxylon tepicense* |
| tepeguaje | macha'aguo | *Lysiloma watsonii* |
| tepozana | cocolmeca | *Eupatorium quadrangulare* |
| tescalama | báisaguo | *Ficus petiolaris* |
| teso | teso | *Acacia occidentalis* |
| teta del diablo | jo'osi pipi | *Matelea altatensis* |
| tetas del venado | ma'aso pipi | *Asclepias subulata* |
| toji | pohótela | *Phoradendron californicum* |
| toji | chíchel | *Struthanthus palmeri* |

| | | |
|---|---|---|
| toloache | tebue | *Datura discolor, D. lanosa* |
| tomatillo | tombrisi | *Physalis maxima* |
| tonchi | mabe | *Marsdenia edulis* |
| torote, torote de vaca | to'oro, to'oro sahuali | *Bursera fagaroides, B. microphylla* |
| torote acensio, torote de incienso | to'oro chutama | *Bursera penicillata* |
| torote copal | to'oro chutama, to'oro síquiri | *Bursera lancifolia* |
| torote copal | to'oro sajo | *Bursera stenophylla* |
| torote lechoso | — | *Euphorbia gentryi* |
| torote panalero | sato'oro | *Jatropha cordata* |
| torote prieto | to'oro chucuri | *Bursera laxiflora* |
| torote puntagruesa | to'oro chutama | *Bursera penicillata* |
| trompillo | jícure | *Ipomoea pedicellaris* |
| tronador | — | *Ruellia nudiflora* |
| trucha | tatio | *Croton ciliatoglandulifer* |
| tule | — | *Scirpus americanus, Typha domingensis* |
| tullidora | aroyoguo | *Karwinskia humboldtiana* |
| uña de gato | nésuquera | *Mimosa distachya* |
| uvalama | juvaré | *Vitex mollis* |
| vainoro | cumbro | *Celtis iguanaea* |
| vainoro | baijuo | *Pisonia capitata* |
| valeriana | — | *Lippia alba* |
| vara blanca | cuta tósari | *Croton fantzianus* |
| vara prieta | huo'ótobo | *Cordia parvifolia* |
| vara prieta | cuta chicuri | *Croton alamosanus* |
| vara prieta | júsairo | *Croton flavescens, C. sonorae* |
| verdolaga | huaro | *Portulaca oleracea* |
| verdolaga | mochis | *Trianthema portulacastrum* |
| vinorama | cu'uca | *Acacia farnesiana* |
| yerbanís | — | *Tagetes filifolia* |
| zacate | ba'aso | *Aristida ternipes* |
| zacate buffel | — | *Pennisetum ciliare* |
| zacate de lana | — | *Cynodon dactylon* |
| zapote (chapote) | ja'apa ahuim | *Casimiroa edulis* |

## APPENDIX E
# Mayo Plants Listed by Mayo Name

| MAYO NAME | SPANISH NAME | SCIENTIFIC NAME |
|---|---|---|
| aaqui | pitahaya | *Stenocereus thurberi* |
| aaqui jímsera | pitahaya barbona | *Pilosocereus alensis* |
| aba'aso | álamo | *Populus mexicana* |
| agia | guásima | *Guazuma ulmifolia* |
| aguaro | aguaro, cuernito | *Proboscidea altheaefolia, P. parviflora* |
| ahuehuete | sabino | *Taxodium distichum* |
| aiboli | mezcal | *Agave* cf. *jaiboli* |
| aimuri | amole | *Agave vilmoriniana* |
| ánima ogua | palo blanco | *Carlowrightia arizonica* |
| aroyoguo | cacachila, tullidora | *Karwinskia humboldtiana* |
| ba'aca | otate, bambú | *Otatea acuminata* |
| ba'aca nagua | carrizo | *Arundo donax* |
| ba'aco | — | *Phaulothamnus spinescens* |
| ba'aso | zacate | *Aristida ternipes* |
| ba'aso síquiri | — | *Gouinia virgata* |
| baca cupojome jicto | carrizito de huitlacoche | *Lasiacis ruscifolia* |
| bacaporo | retama | *Parkinsonia aculeata* |
| bacot mútica | — | *Byttneria aculeata* |
| bacot mútica | coralillo cimarrón | *Rivina humilis* |
| bachomo | batamote | *Baccharis salicifolia* |
| bacot tami | diente de culebra | *Brickellia coulteri* |
| bahue cuépari | golondrina del mar | *Palafoxia linearis* |

| | | |
|---|---|---|
| bahue huitcha | — | *Monanthochloë littoralis* |
| bahue mo'oco | — | *Allenrolfea occidentalis* |
| bahue mo'oco | chamisa del mar | *Lycium brevipes, Suaeda moquinii* |
| bahuepo juatóroco | — | *Atriplex barclayana* |
| baihuío | guayavillo | *Acacia coulteri* |
| baijguo | palo dulce | *Eysenhardtia orthocarpa* |
| baijuo | garambullo | *Pisonia capitata* |
| baiquillo | ejotillo | *Sesbania herbacea* |
| báisaguo | tescalama | *Ficus petiolaris* |
| báis cápora | saituna | *Ziziphus amole* |
| baogoa | mezcal | *Agave* cf. *colorata* |
| baogua | pochote | *Ceiba acuminata* |
| batayaqui | batayaqui | *Montanoa rosei* |
| bebelama | bebelama | *Sideroxylon persimile* |
| berraco | huevos del toro | *Stemmadenia tomentosa* |
| bíbino | salvia | *Hyptis albida* |
| bura mariola | mariola, orégano de burro | *Lippia graveolens* |
| caamugia | güirote de culebra | *Serjania mexicana, S. palmeri* |
| caaro | palo verde | *Parkinsonia florida* |
| caguorara | guayparín | *Diospyros sonorae* |
| ca'ja | tempisque | *Sideroxylon tepicense* |
| cánari | mangle rojo | *Rhizophora mangle* |
| cantela oguo | candelilla | *Pedilanthus macrocarpus* |
| capo segua | flor de capomo | *Nymphaea elegans* |
| capúsari | — | *Crataeva palmeri* |
| cau nachuqui | chicurilla | *Ambrosia cordifolia* |
| causamo | samo | *Coursetia glandulosa* |
| chana pusi | chanate pusi | *Rhynchosia precatoria* |
| chaparacoba | — | *Citharexylum scabrum* |
| chíchel | toji | *Struthanthus palmeri* |
| chíchibo | estafiate | *Ambrosia confertiflora, Parthenium hysterophorus* |
| chicul ónore | biznaguita, pitahayita | *Mammillaria grahamii, M. mainiae, M. mazatlanensis, M. yaquensis* |

| | | |
|---|---|---|
| chi'ini | — | *Pouzolzia guatemalana* |
| chi'ini | algodoncillo | *Wimmeria mexicana* |
| chiju | añil | *Indigofera suffruticosa* |
| chino | palo chino | *Havardia mexicana* |
| chírajo | chírahui, güinolo | *Acacia cochliacantha* |
| choal | quelite | *Chenopodium neomexicanum* |
| choca'ari | ayal | *Crescentia alata* |
| chócola, tónaso | chócola | *Jarilla chocola* |
| chó'ocora | aguama | *Bromelia alsodes* |
| cho'opo | chopo | *Mimosa palmeri* |
| choy | brea | *Parkinsonia praecox* |
| choya | choya | *Opuntia fulgida* |
| choya huani | güerequi | *Ibervillea sonorae* |
| choyoguo | rama blanca, incienso | *Encelia farinosa* |
| choyoguo | — | *Encelia halimifolia* |
| ciali | mangle negro | *Avicennia germinans* |
| ciánori | palo barril | *Cochlospermum vitifolium* |
| cochibachomo | — | *Alvordia congesta* |
| cochirepa | — | *Acacia* sp. |
| cochizayam | — | *Talinum paniculatum, T. triangulare* |
| cocolmeca | tepozana | *Eupatorium quadrangulare* |
| có'cori | chiltepín | *Capsicum annuum* |
| comba'ari | chani | *Hyptis suaveolens* |
| coni saca | coquillo | *Cyperus perennis* |
| coni saquera | coquillo | *Cyperus compressus* |
| cósahui | tajuí | *Krameria sonorae* |
| coya guani | — | *Cissus* spp. |
| cuchu pusi | hierba del venado | *Porophyllum gracile* |
| cuépari | golondrina | *Euphorbia abramsiana, Euphorbia* spp. |
| cuey oguo | guayavilla | *Salpianthus arenarius* |
| cumbro | cúmero | *Celtis iguanaea, C. pallida, C. reticulata* |
| cumbro | vainoro | *Celtis iguanaea* |
| cusi | encino | *Quercus albocincta* |
| cuta béjori | — | *Colubrina triflora* |

| | | |
|---|---|---|
| cuta béjori | palo cachora | *Schoepfia schreberi,*<br>*S. shreveana* |
| cuta chicuri | vara prieta | *Croton alamosanus* |
| cuta nahuila | palo piojo | *Brongniartia alamosana* |
| cuta síquiri | palo colorado | *Caesalpinia platyloba* |
| cuta tósari | vara blanca | *Croton fantzianus* |
| cu'u | mezcal, maguey | *Agave vivipara* |
| cu'uca | vinorama | *Acacia farnesiana* |
| ébano | ébano | *Caesalpinia sclerocarpa* |
| ejéa | palo fierro | *Olneya tesota* |
| etcho | etcho | *Pachycereus pecten-*<br>*aboriginum* |
| goy biba | tabaco de coyote | *Nicotiana obtusifolia* |
| goy carabanzo | garbanzo de coyote | *Bunchosia sonorensis* |
| goy minole | chichicoyota, melón<br>de coyote | *Cucumis melo* |
| goy sisi | — | *Pectis coulteri* |
| goy tabaco | tabaco de coyote | *Nama hispidum* |
| guacaporo | retama | *Parkinsonia aculeata* |
| guaji, paenepa | palo torsal | *Alvaradoa amorphoides* |
| guasana jibuari | hierba del indio | *Aristolochia watsonii* |
| guoy cebolla | cebolla de coyote | *Hymenocallis sonorensis* |
| hayá huíchibo | chichicoyota | *Cucumis melo* |
| hayas guasia | cola de la zorra | *Setaria liebmannii* |
| huanaca | huisache | *Caesalpinia cacalaco* |
| huaro | verdolaga | *Portulaca oleracea* |
| huata | sauce, saúz | *Salix bonplandiana,*<br>*S. gooddingii* |
| hué | bledo, quelite | *Amaranthus palmeri* |
| huichori | — | *Sarcostemma cynanchoides* |
| huicobo | huiloche | *Diphysa occidentalis* |
| huícori bísaro | flor de iguana | *Senna pallida* |
| huilanchi | — | *Digitaria bicornis* |
| huique | guaje, palo bofo | *Leucaena lanceolata* |
| huírivis cu'u | mescalito<br>(de huitlacoche) | *Tillandsia exserta,*<br>*T. recurvata* |
| huítbori | lechuguilla | *Agave aktites* |
| huítbori | mescalito, aguamita | *Hechtia montana* |

| | | |
|---|---|---|
| huo'ótobo | vara prieta | *Cordia parvifolia* |
| huoquihuo | huiloche | *Diphysa occidentalis* |
| ja'apa ahuim | zapote (chapote) | *Casimiroa edulis* |
| jamyo olouama | hierba de la víbora | *Stegnosperma halimifolium* |
| jeco | romerillo, hierba de pasmo, jécota | *Hymenoclea monogyra* |
| jeco | jeco | *Xylothamia diffusa* |
| jejeri | — | *Pereskiopsis porteri* |
| jejiri, jotónari | — | *Lycium* aff. *berlandieri* |
| jépala | jarial | *Hybanthus mexicanus* |
| jévero | chilicote, flor de mayo, peonia | *Erythrina flabelliformis* |
| jícamuchi | palo piojo blanco | *Caesalpinia caladenia* |
| jícamuchi | palo piojo | *Caesalpinia palmeri* |
| jíchiquia | escoba | *Abutilon incanum* |
| jíchiquia | malva | *Abutilon mucronatum* |
| jíchiquia | dais | *Desmanthus bicornutus, D. covillei* |
| jíchiquia jéroca | pintapán | *Abutilon abutiloides* |
| jícure | trompillo | *Ipomoea pedicellaris* |
| jijí'ica | choya | *Opuntia leptocaulis* |
| jímaro | — | *Euphorbia californica, E. gentryi* |
| jímuri | baiburín | *Kallstroemia californica, K. grandiflora, K. parvifolia* |
| jiogo | chicura | *Ambrosia ambrosioides* |
| jito | palo jito | *Forchhammeria watsonii* |
| jócona | jócona | *Havardia sonorae* |
| jodimaco'otchini | guamuchilillo | *Pithecellobium unguis-cati* |
| jójona | palo amarillo | *Esenbeckia hartmanii* |
| joopo | palo blanco | *Piscidia mollis* |
| jo'osi pipi | teta del diablo | *Matelea altatensis* |
| jó'otoro | ciruela del monte, huichilame | *Ziziphus obtusifolia* |
| joso | palo joso | *Albizia sinaloensis* |
| joso | joso de la sierra | *Conzattia multiflora* |
| jósoina | papache | *Randia echinocarpa* |
| juana huipili | — | *Boerhavia* spp. |

| | | |
|---|---|---|
| júchajco | brasil | *Haematoxylum brasiletto* |
| júchica | — | *Sideroxylon occidentale* |
| jupa'are | negrito | *Vitex pyramidata* |
| jupachumi | cola de zorillo | *Petiveria alliacea* |
| jupachumi | palo zorillo | *Senna atomaria* |
| júptera | rocía | *Helenium laciniatum* |
| jusa'iro | — | *Cordia curassavica* |
| júsairo | vara prieta | *Croton flavescens, C. sonorae* |
| jútuguo | palo santo | *Ipomoea arborescens* |
| jutuqui | bachata | *Opuntia* sp. |
| jutuqui | huichilame | *Ziziphus obtusifolia* |
| juupa | mezquite | *Prosopis glandulosa* |
| juupa quecara | mezquite hormiguillo | *Condalia globosa* |
| júvare | igualama, uvalama | *Vitex mollis* |
| juvavena | — | *Capparis atamisquea* |
| juya chibu | rama amarga | *Brickellia brandegei* |
| júyaguo | guayacán | *Guaiacum coulteri* |
| juya jotoro | — | *Bernardia viridis* |
| laven co'oba | cabeza de violín | *Heliotropium angiospermum* |
| lipazote | epazote | *Chenopodium ambrosioides* |
| ma'aca | mostaza | *Drypetalon runcinatum* |
| ma'as ba'aso | — | *Bouteloua barbata* |
| ma'aso alócossim | talayote | *Passiflora arida* |
| ma'aso arócosi | — | *Eucnide hypomalaca* |
| ma'aso o'ota | rama de venado | *Justicia candicans* |
| ma'aso pipi | tetas del venado | *Asclepias subulata* |
| mabe | tonchi | *Marsdenia edulis* |
| macha'aguo | tepeguaje | *Lysiloma watsonii* |
| maco'otchini | guamúchil | *Pithecellobium dulce* |
| mambia | chichiquelite | *Solanum americanum, S. nigrescens* |
| mapáo | palo colorado | *Caesalpinia platyloba* |
| masasari | sanmiguelito | *Antigonon leptopus* |
| mayo | mauto | *Lysiloma divaricatum* |
| miona | — | *Commicarpus scandens* |

| | | |
|---|---|---|
| mochis | mochis | *Boerhavia coccinea, B. spicata* |
| mochis | hierba lanuda | *Tidestromia lanuginosa* |
| mochis | verdolaga | *Trianthema portulacastrum* |
| momoguo | palo amarillo | *Esenbeckia hartmanii* |
| momo ogua | mamoa | *Erythroxylon mexicanum* |
| mo'oso | — | *Bebbia juncea, Machaeranthera tagetina* |
| mora | mora | *Chlorophora tinctoria* |
| moyet | mangle | *Laguncularia racemosa* |
| murue | ocotillo, palo adán, palo pitillo | *Fouquieria diguetti* |
| murue | ocotillo macho, jaboncillo, palo pitillo | *Fouquieria macdougalii* |
| musue | sinita | *Lophocereus schottii* |
| na'a chú'uqui | guachapori | *Cenchrus* spp. |
| na'atoria | baiburilla | *Dorstenia drakena* |
| nacapul | nacapuli | *Ficus cotinifolia, F. pertusa* |
| nata'ari | rama quemadora, ortiga | *Tragia jonesii* |
| navo | nopal, duraznilla | *Opuntia gosseliniana* |
| navo | nopal | *Opuntia wilcoxii* |
| nesco | nesco | *Lonchocarpus hermannii* |
| nésuquera | uña de gato | *Mimosa distachya* |
| nómom | sina, sacamatraca | *Peniocereus marianus, P. striatus* |
| ona jújugo | — | *Adelia virgata* |
| ona jújugo | — | *Colubrina viridis* |
| ónore | biznaga | *Ferocactus herrerae* |
| ooseo suctu | palo fierro, palo pinto | *Chloroleucon mangense* |
| o'ouse suctu | limoncillo | *Zanthoxylum fagara* |
| oseu nácata | — | *Perityle cordifolia* |
| osi | cuna de niño | *Commelina erecta* |
| otate | otate, bambú | *Otatea acuminata* |
| paros pusi | sacamanteca | *Solanum tridynamum* |
| pasio | mangle dulce | *Maytenus phyllanthoides* |
| pasio síquiri | mangle rojo | *Rhizophora mangle* |
| pasio tosa | mangle | *Laguncularia racemosa* |

| | | |
|---|---|---|
| pisi | papache borracho | *Randia obcordata, R. thurberi* |
| pohótela | toji | *Phoradendron californicum* |
| pómajo | palo de asta | *Cordia sonorae* |
| sa'amo | samo baboso | *Heliocarpus attenuatus* |
| sa'apo | sangrengado azul | *Jatropha cardiophylla* |
| sa'apo | sangrengado | *Jatropha cinerea, J. malacophylla* |
| saguo | sahuaro | *Carnegiea gigantea* |
| sahuira | pitahaya | *Stenocereus montanus* |
| samo | samo, sámota | *Coursetia glandulosa* |
| sanarogua | mataneni | *Callaeum macropterum* |
| sanéa | mezquite | *Prosopis articulata* |
| sapuchi | sapuchi | *Randia laevigata* |
| sato'oro | torote panalero, papelío | *Jatropha cordata* |
| saya, saya mome | saya | *Amoreuxia gonzalezii, A. palmatifida* |
| semalucu | chuparosa | *Justicia californica* |
| sibiri | choya | *Opuntia thurberi* |
| sigropo | — | *Lycium andersonii* |
| sina | sina | *Stenocereus alamosensis* |
| sinaaqui | — | *Stenocereus alamosensis* ×, *S. thurberi* |
| sina cuenoji | sina volador | *Selenicereus vagans* |
| síquiri tájcara | piocha, bola colorada | *Trichilia americana, T. hirta* |
| sire | granadilla, morita | *Malpighia emarginata, M. glabra* |
| sitavaro | sitavaro | *Vallesia glabra* |
| siteporo | dais, escoba | *Desmanthus cornutus, D. covillei* |
| subuai muni | frijol de codorniz | *Desmodium scorpiurus* |
| ta'áboaca | tavachín | *Caesalpinia pulcherrima* |
| ta'aco | palma | *Brahea aculeata* |
| ta'aco | palma, palma real | *Sabal uresana* |
| ta'ampisa | confiturilla negra | *Lantana camara* |
| tábelojeca | — | *Capparis flexuosa* |
| táchino | cardo | *Argemone ochroleuca* |
| tajimsi | tajuí | *Krameria erecta* |

| | | |
|---|---|---|
| talcampacate | gordolobo | *Pseudognaphalium leucocephalum* |
| tampicerán | tampicerán | *Platymiscium trifoliolatum* |
| tan cócochi | aguaro | *Martynia annua* |
| tapichogua | copalquín | *Hintonia latiflora* |
| ta'ri sorogua | pintapán | *Abutilon abutiloides* |
| tasajo | tasajo | *Acanthocereus occidentalis* |
| tásiro | sanjuanico | *Jacquinia macrocarpa* |
| tatachinole | tatachinole | *Tournefortia hartwegiana* |
| tatio | trucha, ortiga | *Croton ciliatoglandulifer* |
| tchuna | chalate, higuera | *Ficus insipida* |
| tchuna | nacapuli, higuera | *Ficus pertusa* |
| tebcho | brasil chino | *Haematoxylum* sp. |
| tebue | toloache | *Datura discolor, D. lanosa* |
| teso | teso | *Acacia occidentalis* |
| tetaraca | chichicoyota | *Cucurbita argyrosperma* |
| teta segua | flor de piedra | *Selaginella novoleonensis, S. pallescens* |
| to'bo | amapa | *Tabebuia impetiginosa* |
| to'bo saguali | amapa amarilla | *Tabebuia chrysantha* |
| tombrisi | tomatillo | *Physalis maxima* |
| to'oro | torote, torote de vaca | *Bursera fagaroides, B. microphylla* |
| to'oro, to'oro sahuali | torote copal | *Bursera stenophylla* |
| to'oro bichu | — | *Lycium* aff. *fremontii*, *Mammillaria* spp. |
| to'oro chucuri | torote prieto | *Bursera laxiflora* |
| to'oro chutama | torote copal | *Bursera lancifolia* |
| to'oro chutama | torote acensio, torote de incienso, torote puntagruesa | *Bursera penicillata* |
| to'oro cojuya | malva, escoba | *Abutilon incanum* |
| to'oro jíchiquia | malva, escoba | *Abutilon mucronatum* |
| to'oro jíchiquia | malva | *Malvastrum biscuspidatum* |
| to'oro jíchiquia | rama lisita, sayla de pintapán | *Sida aggregata* |
| to'oro mulato | palo mulato | *Bursera grandifolia (B. simaruba)* |
| to'oro sajo | torote copal | *Bursera stenophylla* |

| | | |
|---|---|---|
| to'oro síquiri | torote copal | *Bursera lancifolia* |
| to'otori huita | choya | *Opuntia pubescens* |
| tori naca | oreja de ratón | *Chiococca alba, Malvella lepidota* |
| tóroco | garbanzo de coyote | *Bunchosia sonorensis* |
| tosa huira | jícama | *Ipomoea bracteata* |
| toto'ora | cresta de gallo | *Plumbago scandens* |
| túbucti | brincador | *Sebastiania pavoniana* |
| tupchi | abolillo, amolillo | *Sapindus saponaria* |
| úsim yuera | palo verde, palo huaco | *Agonandra racemosa* |
| úsim yuera | palo limón | *Amyris balsamifer* |
| yemsa ba'aso | saitilla | *Bouteloua aristidoides* |
| yete ogua | rama dormilona | *Mimosa asperata* |
| yocutsuctu | — | *Zanthoxylum mazatlanum* |
| yore cotéteca | dais | *Dalea scandens* |
| yucohuira | — | *Cissus* cf. *mexicana, C. trifoliata, C. verticillata* |

## APPENDIX F
# Glossary of Mayo and Spanish Terms

Mayo entries are labeled (M). All other entries are in Spanish.

| | |
|---|---|
| aaqui begua (M) | husks or peels of organpipe fruits |
| aaqui nábera (M) | crestate organpipe cactus |
| aaqui nójim (M) | tamales of organpipe fruits |
| aaqui tej'ua (M) | ripe organpipe fruits |
| aca'ari (M) | basket for gathering organpipe fruits; made of organpipe ribs |
| aguates | glochids; tiny spines on cactus of genus *Opuntia* and others |
| alaguássim (M) | fiesta director |
| albóndigas | meatballs |
| alferecía | a condition in which the mouth or face is twisted; Bell's palsy |
| atol | porridge |
| babatuco (M) | indigo snake (*Drymarchon corais*) |
| baboso (adj.) | drooling |
| bacote | pole (usually an agave shoot) for gathering cactus fruit |
| bajada | alluvial fan; gradually sloping base of mountains |
| batea | carved wooden bowl |
| birria | stewed meat |
| brujo | witch |
| buásim (M) | sweet, ripe organpipe fruits |
| buli | gourd canteen |
| cabeza | head |
| cachora | lizard |

| | |
|---|---|
| cacique | local chieftain |
| Cáhita (M) | the linguistic group to which Mayo and Yaqui belong (translated it means "There is nothing") |
| cajón | box canyon |
| calzones | formerly the loose white trousers worn by Mexican peasants; now undershorts |
| camote | tuber; sweet potato |
| carrizo | any canelike grass; usually *Arundo donax* in the Mayo region |
| chapa (M) | a pole with a wedged open end used as a tool for gathering prickly pear fruits |
| chinami (M) | fence constructed from organpipe ribs |
| cholugo | coatimundi (*Nasua narica*), a long-tailed raccoon relative |
| choma (M) | string heddle (weaving) |
| chúcata (M) | gum from mesquite (*Prosopis glandulosa*) |
| cobata (M) | roasted agave head |
| cobija | blanket |
| cobijera | blanket weaver |
| cogollos | tips of leaf shoots or fronds (agave or palm) |
| colono | colonist, settler |
| comal | griddle, often with a concave surface |
| comunero | member of a *comunidad* |
| comunidad | communally owned lands controlled by indigenous people, similar to reservations in the United States |
| conti (M) | ceremonial procession through a village during Lent |
| corcho | cork |
| costal | feed sack used to transport materials |
| culebra | snake |
| curandero | curer, healer |
| desmonte | former forest that has been cleared |
| dulce | sweet |
| ejidatario,-a | member of an *ejido* |
| ejido | communally owned lands with individual parcels |
| empacho | chronic indigestion, constipation |
| equipatas | Sonoran term for cyclonic (winter) rains |
| escorpión | Gila monster, beaded lizard (*Heloderma* spp.) |
| fariseo | Pharisee in Semana Santa procession |

| | |
|---|---|
| fiestero | festival official |
| gajo | piece or section of a ribbed plant or citrus fruit |
| gloria | procession and rituals of Holy Saturday morning before Easter |
| gobernador | governor |
| guacabaqui (M) | stew of beef and vegetables eaten during fiestas |
| guari (M) | handwoven palm or sotol basket |
| hacendado | owner of a hacienda |
| hacienda | landed feudal estate with workers attached to it |
| hechicero | sorcerer |
| hermano de la fe | brother of the faith; evangelical Protestant |
| horcón | post for holding up a roof |
| huaraches | leather sandals, often with soles made from tires |
| huo'ótobo (M) | shrub (*Cordia parvifolia*) or smooth stick made from the branches |
| ixtle | agave fiber |
| jabalí | javelina, peccary (*Tayassu tayacu*) |
| jamut (M) | female, woman |
| jejenes | no-see-ums, biting gnats |
| jíchiquia (M) | broom |
| jíconi (M) | fruit of *etcho* cactus (*Pachycereus pecten-aboriginum*) |
| jiote | spots on the skin with no pigment |
| jornalero | day laborer |
| juqui (M) | semi-subterranean hut in which women weave palm hats |
| juya ania (M) | the world of living nature; everything that is alive |
| ladrillo | fired adobe brick |
| la lengua | the Mayo language, according to Mayos |
| las aguas | summer rains, primarily thunderstorms |
| lata | cross-hatching; the bottom layer of a roof |
| latifundista | owner of large tract of land |
| ma'aso (M) | deer; deer dancer |
| malacate | spinning shaft for producing yarn from fleece |
| mal de orín | urinary disorder |
| malezas | weeds |
| malpuesto | evil spell |
| manglares | mangrove swamps |

| | |
|---|---|
| manteca | fat, lard |
| matachines | ritual dance society of men wearing elaborate headdresses and gaudy clothing |
| matríz | womb, uterus |
| maya | roasting pit |
| mecate | rope |
| mestizo | person of mixed indigenous and Spanish origin |
| miel | sweet syrup; honey |
| milpa | rain-fed cornfield |
| monte | thornscrub; native vegetation; unurbanized and uncleared land |
| móriac (M) | sorcerer |
| morral | handbag of woven ixtle |
| nahuila (adj.) (M) | worthless, homosexual |
| navo (-im, pl.) (M) | prickly pear fruits |
| nopalitos | edible prickly pear pads |
| olla | pot; water jug |
| onobachia (M) | seeds, primarily of barrel cactus (*Ferocactus herrerae*) |
| o'o (M) | male; man |
| panal | wild beehive; wild honeycomb |
| panela | homemade (cottage) cheese pressed into forms |
| pascola | ceremonial dancer wearing mask and rattles |
| pasmo | a vague illness characterized by feeling bad, having fever, being fearful, etc. |
| péchita | pod of mesquite |
| penca | large, long leaf (of palm or agave) or cactus pad |
| petate | sleeping mat of woven palm or reed |
| piloncillo | brown sugar cone |
| piojo | louse |
| pitahaya | fruit of organpipe cactus (*Stenocereus thurberi*) |
| poposahui (M) | between green and ripe (said of cactus fruits) |
| porfiriato | the period when Porfirio Díaz was dictator of Mexico, roughly 1880–1910 |
| portal | porch |
| pradera | pasture |
| pujos | bloody stools |
| punzada | throbbing headache, perhaps migraine |
| quiote | agave stalk, shoot |

| | |
|---|---|
| ramada | a small, shaded pavilion, usually constructed of mesquite log uprights and branches for roofing |
| ramadón | the principal ramada of a fiesta |
| ranchería | settlement |
| repartamiento | system of sharing of native Indians' labor between their fields and mines |
| resfrío | the common cold; also the rash associated with colds in children |
| ropa de manta | calico, the humble cloth with which Mayo women formerly made their clothing |
| sasapayeca (M) | tamping stick for loom |
| sasonar | to ripen |
| sebua (M) | glochids; tiny spines on cacti of genus *Opuntia* and others |
| Semana Santa | Holy Week; the week before Easter |
| serrano (adj.) | mountain |
| sitori (M) | syrup |
| son | old song, folk song |
| sonaja | rattle-type instrument; gourd rattle |
| susto | a pathological condition brought on by sudden fright |
| tabaco rústico | wild or native tobacco |
| taburete (M) | stool with cross-hatched sides and animal skin seat |
| tájcarim (M) | tortillas |
| tapanco | raised bed of organpipe ribs for drying fruits |
| tapeste (M) | traditional wooden Mayo bed, with a frame of *etcho* wood and leather or poles for mattress |
| tarime | bed with numerous thin poles for suspension (same as *tapeste*) |
| tásic (M) | agave fiber |
| tatemado | roasted |
| telar (n.) | loom |
| temporales | rain-fed agricultural fields |
| ténaborim (M) | gravel-filled cocoons sewn together in long strands and wrapped around the legs of pascola and deer dancers |
| terreno baldío | "empty terrain"; land available to anyone who claims it |
| teteraca (adj.) (M) | worthless |
| tienda | store |
| tinajera | tripod for supporting a water jug |

| | |
|---|---|
| tiricia | chronic sadness, depression |
| tótosi (M) | white |
| tullidora | crippler |
| vara | pole, stick |
| varero | cutter of poles, usually from *Croton fantzianus* |
| venadito | wild deer |
| venado | deer; for Mayos it often means a deer dancer, who performs an elaborate ritual at Mayo ceremonies |
| viga | roof beam |
| yoreme (-em, pl.) (M) | term used by Mayos to refer to themselves |
| zarza (adj.) | white, cloudy |
| zarzo (n.) | drying rack suspended from ceiling |

# Notes

CHAPTER 1

1. Etchohuaquila lies on the west bank of the Arroyo Cocoraqui, an ephemeral streambed with its own entry into the Sea of Cortés.
2. For a comprehensive analysis of the partitioning of Mayo and Yaqui lands, see Calderón Valdés 1985, vol. 5.
3. In 1996 it was possible to drive along rural highways from near Huatabampo to slightly west of Ciudad Obregón, a distance of nearly sixty miles, without leaving the fields.
4. Martin et al. (1998) noted 2,865 species in an area 2 degrees square, roughly one-fifth of which is the land of the Mayos. At least 50 more species have been added to that number as of the year 2001.
5. We use the terms "arid" and "semiarid" with caution. For our purposes, semiarid conditions permit the planting of rain-fed crops of corn and beans, while arid climates do not.
6. Unless otherwise noted, all data are from Hastings and Humphrey 1969; however, we advise against relying heavily on these and other published rainfall statistics because of systemic problems with reporting and recording data.
7. This figure is cited from Calderón Valdés 1985, 1:168
8. On first mention of a plant, we will give the Mayo name, if available, with the Spanish name and scientific name and family. Thereafter we will use the more familiar name, usually Spanish. All plants are listed in appendixes D and E.
9. In December 1997, freezing temperatures were reported in various locations, including parts of the Mayo and Fuerte Valleys. Crop and vegetation damage in the Mayo Delta and southward in coastal thornscrub was minimal, far less than the disastrous losses predicted by government officials. Ornamentals in Alamos were heavily damaged. *Cumbro* (*vainoro; Celtis iguanaea*), *nacapul* (*nacapulis; Ficus pertusa*), and *agia* (*guásima; Guazuma ulmifolia*) in Alamos and adjacent arroyos were also damaged. Other areas nearby escaped damage. This was an unusual climatic phenomenon in which a bubble of very cold, dense air spilled over

the Sierra Madre from the Mexican Plateau and descended onto the lowlands of extreme southern Sonora.

10. We did, however, note crop damage in the lower Fuerte Valley to the south in Sinaloa. The huge valley's steep canyon walls upstream appear to act as a funnel for cold air masses.

11. Political factors cannot be ruled out, however. See, for example, Yetman 1995.

12. Even so, crops at Los Capomos produced poorly in 1995 and 1999. After a disastrously dry spring and early summer in 1996, the rains were adequate, producing bumper crops of sesame and corn and an adequate yield of beans. At Nahuibampo the 1996 harvest was below average, and that of 1999 was barely worth the effort.

13. Barranca Candemeña, the deep, sheer-walled gorge below Basaseáchic Falls on the upper Mayo drainage, is apparently the rim of a massive caldera (Robert Schmidt, pers. comm., 1998).

CHAPTER 2

1. Conicárit only later came to be identified as a Mayo town. It was inundated by the waters of Mocúzari Dam in 1951. Its dislocated inhabitants moved to Mocúzari or the now mestizo village of Piedras Verdes.

2. Thomas Robertson (1964), writing in 1947 about a utopian experiment begun by Americans on the Río Fuerte, makes no distinction between Mayos of the Río Fuerte and those of the Río Mayo.

3. Almada (1990) claims that the mission of Masiaca was founded by the Jesuit Pedro Méndez in 1612. He provides no archival sources for this claim, which is probably incorrect.

4. That Mayo comunidades indígenas exist is proof that whatever their origin, the *comuneros'* ties to their villages can be traced back for generations.

5. O'Connor (1989) lists four indicators of "Mayo-ness": having a Mayo last name, participating in a fiesta, speaking the Mayo language, and displaying a house cross. The last seems to be dying out, and the first is ambiguous because many Mayos have the same last names as mestizos (although apparently all Moroyoquis, Buitimeas, and Yocupicios are Mayos). Without disagreeing with O'Connor, I have reduced the list to two indicators, use of the Mayo language and participation in fiestas, but even this list must not be applied too diligently. These criteria are more rigorous than in the United States, where tribal membership is determined by a percentage of indigenous ancestry, often as low as one-sixteenth.

6. Because the INEGI (Instituto Nacional de Estadísticas, Geografía, e Informática) census data do not break down the numbers of indigenous speakers by language, it is impossible to ascertain more precise figures. The total was arrived at by totaling the indigenous speakers over five years of age listed in the municipios of Etchojoa, Huatabampo, and Navojoa, where all or nearly all the indigenous residents are Mayos, and adding the native language speakers in the municipio of Alamos locality by locality based on the aboriginal language spoken in each. The municipio of Alamos includes several hundred Guarijío speakers, but in only

one known instance (Chorijoa) does the same village contain both Mayo- and Guarijío-speaking residents.

7. Of 468 Seris living in Sonora, 43 were considered monolingual Seri speakers. Seris are considered the least assimilated indigenous group in Sonora, perhaps in the entire nation. Roughly 5 percent of Yaquis are monolingual Yaqui speakers, while less than 1 percent of Mayos are monolingual.

8. The actual number of Mayo speakers is probably higher than those revealed by the census. During our research, we repeatedly encountered individuals who initially denied their ability to speak la lengua when, in fact, they do. Their fear of being looked down upon and considered backward leads them to deny their linguistic abilities. In addition to those identified in the census as Mayo speakers, an equal number of individuals can be considered Mayos who are unable to speak the Mayo language. Our investigations suggest that most people under thirty are unable to converse in Mayo.

9. At Los Capomos the dances are associated with delicious intrigues whereby young people "elope" and run off to a new home, and are thus married. The match is known to most residents of the village in advance, and the party that follows is anticipated by all. Such a union is soon consecrated by marriage.

10. For a contemporary description of Semana Santa, see Melchor C. 1991.

11. Disputes occasionally arise between traditional Mayos and the clergy over control of local churches. In Masiaca, for example, the local priest, who found the fiestas too heavily laden with pagan elements, installed benches in the formerly open, seatless church deliberately to impede the progress of fariseos and *matachines* (members of a dance society with highly decorated costumes) during the fiestas. This caused deep resentments and divisions within the Mayos. Other clergy accept the Mayo traditions and incorporate them into Catholic liturgy.

12. Don Benito Alcaraz once walked from his home in Choacalle de Masiaca, Sonora, to the fiesta de San Miguel in Los Capomos, Sinaloa, a distance of nearly 100 kilometers.

CHAPTER 3

1. Legend has it that Sinaloa from Culiacán south was settled by Aztec forerunners who founded the empire of Aztlán. See Zavala C. 1989 and Secretaría de Educación Pública 1991.

2. Beals (1943) compiled an impressive description of Cáhitas aboriginal culture, but most of his studies are limited to Yaqui culture and pertain little to Mayos and other Cáhitas.

3. Pfefferkorn's *Sonora: A Description of the Province* was not written until 140 years after the arrival of the Jesuits in Sonora.

4. Our primary source of Pérez de Ribas is an original manuscript in Special Collections at the University of Arizona Library. The handwritten manuscript contains no sections or page numbers. Other versions have divided his works into sections and chapters.

5. Kirchhoff argues that societies of central Mexico (Mesoamerica) were class

or state societies, those of the river country of the northwest (Oasisamerica) were tribal societies, and those of the northwest desert (Aridamerica) were nomadic societies. This is an interesting hypothesis, but probably an overgeneralization.

6. Pérez de Ribas (1645) notes of the explorer Francisco de Ibarra that in 1565 he "salió a Culiacán y de allí (con buen número de soldados) entró por la provincia de Sinaloa. Andúvola toda y visitó sus naciones . . . y viendola poblada de tanta gente . . . y que los colores con que se emijaban y pintaban los indios, daban señales de minas (porque esos colores los sacan de ellas) determinó dejar poblada una villa, en al río que llaman Zuaque [Fuerte]." [Ibarra "set out from Culiacán and from there (with a good number of soldiers) entered the province of Sinaloa. He traveled all over it and visited its peoples . . . and seeing that the colors with which they decorated and painted themselves were signs of mining (because the colors were derived from minerals), he decided to found a villa on the river they call Zuaque (Fuerte)."]

Pérez de Ribas also attributes the arrival in 1605 of Captain Martínez de Hurdaide to the Crown's interest in metals. "El conde de Monterrey, virrey de la Nueva España, habiéndole dado noticias los que las tenían, que en la provincia de Sinaloa había veneros de minas, que prometían much riqueza, cuyo descrubrimiento le estaba muy bien al rey y a sus vasallos . . . despachó con estas noticias e informes, su excelencia orden y mandato al capitán de Sinaloa [Hurdaide]." ["The count of Monterrey, viceroy of New Spain, having received word of what things were there, and that in the province of Sinaloa there were veins of minerals that promised to be rich, whose discovery would sit well with the king and his vassals, . . . dispatched an order to the captain of Sinaloa (Hurdaide) to verify this."]

Navarro García (1967:29) notes that "Las minas de Copala [near Mazatlán] fueron destruidas con ocasión del mismo levantamiento de los tepehuanes de 1616." ["The mines of Copala were destroyed during the very uprising of the Tepehuanes in 1616."] Hurdaide's expedition against the Chínipas was motivated by rumors of silver.

7. Frank (1972) notes that the most underdeveloped parts of Latin America are the areas that had the greatest riches at the time of contact; in contrast, the areas where development took place were those with little to export. The Mayo region fell in between. Their agricultural resources were exploited by Spaniards prior to the discovery of silver at Alamos.

8. This point is made forcefully by Figueroa (1994:62).

9. Contemporary descriptions of Hurdaide suggest that he was sufficiently astute militarily that he would not recruit local warriors merely to be used as cannon fodder.

10. Pérez de Ribas notes: "La palabra mayo en su lengua significa término, por ventura por estar este río entre otros dos de gentes encontradas y que traían guerras contínuas con los mayos y no les daban lugar a salir de sus términos" (Robertson 1968, book 4, chap. 1). ["The word Mayo in the Mayo language means 'boundary' because the river lies between two others who are constantly at war with the Mayos and hardly allow them to leave their lands."] Our studies indicate that the

name Mayo is derived from the name for the tree *Lysiloma divaricatum* (*mauto* in Spanish). Thus *Mayo* means "mauto people."

11. Historians rely almost exclusively on Pérez de Ribas's account of the Yaqui campaign. He was not present for the battles, however, and relied on others' accounts. In exalting the achievements of Hurdaide, he relates an incident in which the Yaquis purportedly set fire to the brush around a hill on which Spaniards had established a defensive position. Hurdaide, he says, craftily ignited a counter-fire, which blew back at the Yaquis and saved the Spaniards from being incinerated. The problem with this account is that the coastal thornscrub in which the fighting took place does not burn, even during the scorching dry season, and cannot be ignited except for scattered individual shrubs. (In order to plant buffelgrass, ranchers must clear the pastures by hand and wait for the cut brush to dry before they can burn it.) The account is either a great exaggeration or a fabrication, if not by Pérez de Ribas, then by one of his informants.

12. Pérez de Ribas says of the arrival of Pedro de Méndez: "Porque los mayos recibieron a su padre Méndez, con singulares demonstraciones de alegría, saliendole a recibir dos o tres leguas antes que llegara a sus pueblos, y por todas ellas levantaron sus arcos triunfales de ramos de árboles, que son sus tapices." ["The Mayos received Padre Méndez with unusual demonstrations of joy, venturing out to meet him two or three leagues from their village, and they erected triumphal arches made of branches and covered the ground as well."]

13. Although pleased with their piety, Pérez de Ribas was distressed to find "an abuse among the Mayos, that was difficult to correct: this was, that it was quite easy for pregnant women to obtain abortions." Divine intervention assisted in ending this blasphemy, however. Of an old midwife, he records that "the old woman had the devil in her, to do abortions for pregnant women, exhorting them and giving them remedies to carry it out. God took mercy on this old woman, because he moved her heart to request baptism, which she received and the next morning she awoke dead" (Robertson 1968, book 4, chap. 3). No inquest was recorded.

14. The "reduced" pueblos were Batacosa, Camoa, Cuhuirimpo, Conicárit, El Júpare, Navojoa, Tepahui, and Tesia. Jesuit sources (Pérez de Ribas) divided the Mayos into three greater subdivisions: Camoa (the upstream Mayos), Navojoa (midstream), and Etchojoa (delta).

15. In describing the arrival and activity of Padre Pedro Méndez, Bannon (1947) unwittingly reveals his allegiance to the long list of writers who evoke the cultural superiority of Europeans. He characterizes the Mayo spiritual and medicinal practitioners who objected to the Jesuits presence as sorcerers (hechiceros), dismissing their objections as mere jealousy. (Ambrose Bierce would have defined a sorcerer as a practitioner from someone else's religion.) Bannon also credits the Jesuits with vastly improving agriculture almost overnight, thus alleviating perpetual impending starvation among Mayos, a claim that cannot be substantiated by the historical record. It is perhaps difficult for a Jesuit to be objective about the exploits of those in his own order, but such cultural chauvinism by a sophisticated scholar demonstrates how deeply some cultural myths have penetrated.

16. Reff (1991:256–57) successfully debunks the myth of Jesuit-spawned technological revolution among New World natives. In spite of claims to the contrary, wheat, cattle, and the metal plow were scarcely used among indigenous people of the Northwest until well into the nineteenth century. Pfefferkorn acknowledges that Sonorans preferred wild game, horse, and mule meat to beef, which they appeared to resist for aesthetic (and, perhaps, cultural) reasons. Indeed, among natives of northwest Mexico, only the horse seems quickly to have gained wide acceptance, and that by Comanches and Apaches, much to the dismay of Spaniards. Likewise, cattle were highly valued only by the Seris, who rustled them at will, thus providing an excuse for Spanish and Mexican extermination campaigns. Today corn tortillas are said in Sonora to be the food of Indians, while flour tortillas are food of *gente de razón*, indicating that wheat never made a great impact on indigenous Sonorans (with the possible exception of Yaquis), and that the introduction of wheat benefited only colonists.

Pfefferkorn grudgingly admits that the European diet was less wholesome than the native Sonoran diet: "the food supplies of the Sonorans . . . are in part bad, in part insipid and nauseating. And yet these people live contentedly, reach a great age, and are much more healthy on such a fare than are others whose daily board consists only of artificial and highly spiced dishes" (1949:200). This admission seems to undercut the common claim that European horticulture and husbandry benefited the natives of the New World.

17. In return for the myriad diseases introduced by Europeans, the best the New World natives could send to Europe was syphilis. The exchange was highly unequal.

18. Reff (1991) notes that the smallpox virus can be communicated through clothing and can remain virulent on items such as stored blankets for more than a year. Thus, in addition to the danger of direct contamination through nasal discharge and saliva, the Indians were also exposed to disease when they accepted gifts of European textiles.

19. Reff (1991) notes epidemics of 1616–17, 1619–23, 1623–25, and several between 1636 and 1660. That any natives of Mexico survived at all seems miraculous and a tribute to their genetic toughness (or, perhaps, to whatever "primitive" rituals they practiced, including bathing, which was anathema to Europeans).

20. It is also clear from the narrative that natives quickly learned to invoke the Devil to explain any perceived deviance. Pérez de Ribas goes to great lengths to detail the descriptions of The Adversary provided by various Indian groups. These garish portraits amuse those who have worked at length among native peoples and learned how they often take great delight in pulling the legs of supposedly highly trained observers. We once spent several weeks doing an ethnobotanical study with a Mayo, dutifully jotting down all of his comments, only to learn later from his daughter that he amused himself from time to time by inventing preposterous tales about uses of plants, wondering when we would catch on. When we confronted him with one especially egregious exaggeration, he chuckled happily for some time.

21. Brading (1971:146), however, notes that "Mexican mine-workers, far from being the oppressed peons of legend, constituted a free, well-paid geographically

mobile labor force which in many areas acted as the virtual partners of the owners.... The vast majority of Mexican miners—they did not number more than 45,000 individuals—worked voluntarily."

22. For an excellent discussion of the machinations of miners, missionaries, and Indians, see Calderón Valdés 1985, 2:102–10.

23. If Brading (1971) is correct, the Jesuits may have protested too loudly at the recruitment of indigenous men to work in the mines. The Indians may well have viewed mine work as desirable, while the Jesuits probably resented the loss of influence over the men, inflating their "concerns" about the effects of mining on the Indians.

24. During the late 1940s and 1950s the Mexican government poured vast amounts of capital into developing the agricultural potential of the Mayo and Yaqui Valleys. These were viewed as the potential breadbaskets of Mexico, which they have indeed become. The Yaqui Valley is today Mexico's most important wheat-producing center.

25. A good portion of these profits was invested in the construction of ornate churches in Sonora. See, for example, Roca 1967.

26. In 1996 Sonora was Mexico's principal producer of copper, and newly reopened gold mines promise to make it the leading producer of gold as well. The state's mining industry still draws southerners in search of work.

27. The view of indigenous people as subhuman is still current in Sonora, where mestizos often refer to themselves as gente or gente de razón, in contrast to Indians, who are simply referred to as *indios*, a pejorative term.

28. Sheridan (1979) has documented the campaign of genocide directed by the Crown and the Jesuits against the Seris in reprisal for their refusal to accept subjugation and their unwillingness to cease their raids against intruders in their territory.

29. There were exceptions, of course. The Pimas expressed genuine affection for Father Kino, who repeatedly acted as their advocate, even in the face of rebellions.

30. In 1767 the Spanish Crown, under the influence of liberal Bourbon reformers, expelled the Jesuits from the New World. The intrigue surrounding the expulsion is the subject of much literature. For a Sonoran perspective, see Calderón Valdés 1985, 2:193–210.

31. Cáhitas had no ax to grind with Apaches, who primarily (though not exclusively) plundered mestizo settlements, a policy Cáhitas could readily understand. Apaches later raided Opata settlements after Opatas accepted assimilation by Europeans.

32. *Latifundistas* are owners of large estates but are not necessarily hereditary landowners, nor are those who work their lands traditionally in peonage.

33. A mestizo historian blames Juan Banderas for the ongoing conflicts between the Cáhitas and the government. "Having been dormant for so many years, the hatred that Jusacamea [Juan Banderas] had aroused in his people continued to plague Sonora into the present century" (Stagg 1978:40).

34. Federalists favored states' rights and a weak central government, while centralists preferred a strong central government and relatively weak states. These dis-

tinctions were frequently glossed over, however, as warlords tended to attract followers on the basis of their personalities and family connections rather than their ideologies.

35. For a brief history of the invasion, see Chisem 1990:157–68. The French occupation of Mexico followed the announcement by President Benito Juárez of a two-year moratorium on foreign debt payments, which enraged the French. The Mexican treasury was empty, and tax revenues were paltry. England and Spain, also indignant, made their separate peace. The United States was too involved in the Civil War to intervene, and France seized the opportunity to expand its empire. See Calderón Valdés 1985, 3:175–78.

36. Yaquis also supported the French. One of the most important non-French military leaders of the imperial forces in Sonora was an Opata named Refugio Tánori, who supported the empire for the same reasons.

The French invaded Sonora primarily because Napoleon III believed reports that Sonora possessed enormous mineral wealth that could be readily exploited. Revenues from the mines were to be used to offset the costs of the invasion and occupation (Calderón Valdés 1985, 3:177). When it became clear that the costs of exploiting the minerals would be exorbitant and that the French would face unending hostility in the state, Napoleon's interest faded. Guerrilla actions by patriotic Sonorans became increasingly effective, and the resistance grew. On September 4, 1866, nationalist forces decisively defeated occupying forces (which consisted primarily of pro-imperialist Sonorans) near Ures. A week later the French withdrew from Sonora, and six months later, from Mexico.

37. Pesqueira probably bore a festering resentment against indigenous people. Born in 1820 to Creole parents, he was educated in Spain and France. He insinuated himself into Sonoran politics and rose quickly to command a garrison at Baviácora on the Río Sonora, ultimately becoming governor of the state. His promising career suffered a severe setback at the battle of the Pozo Hediondo (Stinking Well) on January 20, 1851, when his dashing troops were handed a crushing defeat at the hands of a small army of Apaches. Twenty-six of Pesqueira's troops died in the contest, and forty-three, including Pesqueira, were wounded. Knocked from the saddle, Pesqueira was forced to run for his life, an ignominious taint for the vaunted Sonoran. The Apache force proceeded to steal 1,300 horses and raid the defenseless town of Bacoachi, which they sacked and pillaged (Almada 1952:507–9).

Pesqueira recovered quickly, managing to be elected to the state legislature the following year and ultimately to become one of Sonora's most famous and successful strongmen. As governor his programs included the colonization and agricultural development of the Río Mayo and Río Yaqui lands, both programs contingent upon the submission of the Mayo and Yaqui peoples (Hu DeHart 1984:58). In 1872 he also proposed the establishment of Pesqueira Colony, a model mestizo town to be built according to the most modern plans between Navojoa and El Júpare. It was never built.

38. The favorite tool used by government representatives to take over Yaqui

and Mayo lands was the constitutional right of an individual in Mexico to claim *terrenos baldíos,* idle or unoccupied lands. Unless land was specifically being put to beneficial use, it was "denouncable"—that is, claimable—as terreno baldío. Because a large proportion of Yaqui and Mayo lands was common land used for grazing, gathering, hunting, and religious purposes, they were not "worked" in any normal sense and hence were considered baldío. Colonos were encouraged with the full backing of the law to claim these lands for themselves as terrenos baldíos in spite of their critical importance to indigenous peoples. Thus untold thousands of hectares of communal lands were expropriated for the benefit of mestizo settlers and, more often still, of wealthy landowners and "survey" companies. This provision still exists, and the Mexican government has usually interpreted the term "baldío" to benefit the wealthy. And, of course, what some people view as a use, others view as useless. The communal lands were to Mayos and Yaquis what bison were to the Plains Indians.

39. For a fictionalized but intriguing account of the life of Teresa Urrea, see Domecq 1990.

40. Teresa's influence did not end with her exile. In 1896 an armed group organized by Pomposo Ramos Rojo launched from the United States an attack on the Nogales, Sonora, customshouse with the hope of establishing a base of operations to reinstate Teresa to her former position of influence. Ramos Rojo was defeated, and most of his men were arrested by U.S. agents for violations of neutrality laws (Almada 1990:22).

41. Teresa's prophetic movement may have been short-lived, but Erasmus (1967:102–3) cites the outbreak of a new movement in 1957 when "God" appeared to a young Mayo, demanding a resurgence of the fiestas. Word of this revelation quickly spread through the region, resulting in an increase in fiestas and embellished tales about the appearance of "God."

42. At the forefront of the enterprise to make the Yaqui and Mayo Valleys into one huge agricultural enterprise was a remarkably resilient entrepreneur and opportunist named Carlos Conant. A native of Guaymas, he rose rapidly in the military but proved adept at siding with losers. At one point he was exiled from Sonora, using his enforced vacation to open a mine near Ocampo on the Río Mayo in Chihuahua, where in 1883 he conscripted a small army to put down a strike. Returning to Sonora in 1888, he was given a concession by the Porfirio Díaz regime to survey and develop the Mayo and Yaqui Deltas as a first step toward "opening" the deltas to modern agriculture. He went broke but sold out to others, ultimately the Richardson Construction Company, which purchased the delta scheme in 1905 and went on to dominate delta economy and politics for nearly forty years. For an account of Conant's schemes, see Dabdoub 1964:267–91.

43. Robertson (1968:21–22) describes parenthetically how Mayos from the Río Fuerte stormed the Navojoa garrison in 1915 during the height of the Mexican Revolution.

44. Hacendados—who owned large, usually hereditary estates that included peons or semi-indentured laborers who worked the lands—were invariably aligned

with the oligarchy, and usually the clergy as well. They opposed the porfiristas (followers of Porfirio Díaz) primarily because they refused to share wealth and power with the more traditional elite. Obregón's role in the revolution vacillated between that of a populist hero and an opportunistic elitist.

45. Obregón also managed to parley his military success into personal financial success. His farm on the Río Mayo increased from 180 hectares in 1912 to 3,500 in 1917. In 1918 the price of chickpeas doubled. He profited handsomely from crops sown on lands legitimately claimed by Mayos (Krauze 1997:388).

46. The Yaquis, whom he had lured into his army with the same lying promises he made to Mayos, fared even worse. In 1927 Obregón led a military assault on Yaqui lands and delivered their final defeat (Krauze 1997:401).

47. Calles made a practice of eliminating his enemies or those with whom he merely disagreed. He developed a secret police that became notorious for using torture on political prisoners. The opposition in the late 1920s, including the Cristeros (a rabidly pro-Catholic military movement centered in west-central Mexico), responded in kind to his violence. It was not a good period for Mexico.

48. The porfiriato refers to the period during which Porfirio Díaz was dictator in Mexico, roughly 1880–1910.

49. The most prominent of evangelical groups is La Iglesia de Dios Completo.

50. For an excellent description of Yocupicio's activities, see Bantjes 1998.

51. See Reynolds 1996 for a description of Owen's beliefs and an interesting documentary history of the scheme. Owen (1975) describes his own principles in a pompous book. A class analysis of Owen's efforts and their place in Mexican history is provided by Ortega (1978). Reynolds, typical of North American writers, makes no mention of the Mayo society that had used the lands appropriated by Owen and his followers. His description of the region's indigenous people (variously Ahomes, Zuaques, and Huites, now all considered Mayos) is as follows: "Most of the local population lived along the river. Thousands of Mayo Indians scratched an easy living from the river bottom lands, some working tiny plots to which the government had given them squatter's right, others sharecropping for some 20 well-to-do hacienda owners, called hacendados. The hacendados took at least half of the crop, but the apathetic Indians had as much as they wanted" (Reynolds 1996:11). Reynolds, in the Owen and North American tradition, looked upon unexploited lands as vacant and ripe for the plucking.

52. Owen or someone from his enterprise was also responsible for the founding of Los Mochis, the principal city on the lower Río Fuerte, now an urban center of more than a half million.

CHAPTER 4

1. Estero Tóbari is associated with the mouth of Arroyo Cocoraque, not with any major river. Estero Aquiropo is associated with the delta of the Río Mayo. Bahía Yavaros lies considerably south of the Mayo Delta, although it may have been formed by the Mayo in recent geological times. Similarly, Bahía Agiabampo is not associated with a major river but may have been formed by the delta of the Río

Fuerte. Both bays appear to have their geological origins in the rifting of Baja California in the late Miocene.

2. Shell fishing has given rise to an associated industry, lime production. Tons of spent shells are heaped into piles and left for at least two years to dry and remove organic matter. They are then roasted in ovens and smashed and ground into lime for use in local construction and for making whitewash.

3. This change in species density is also reflected ethnobotanically. For Guarijíos, who occupy tropical deciduous forest on the middle Río Mayo, pitahayas decrease in importance while etchos, which flourish in tropical deciduous forest, greatly increase in importance, thus reversing the roles of the cacti among Mayos.

4. Gentry (1962:73) found that the indigenous people of the Río Mayo region occupied different ecological habitats. "These three tribes are stratified ecologically. The Mayo are Thorn Forest [thornscrub] people, the Warihío Short-tree Forest [tropical deciduous forest] and the Tarahumare largely Pine Forest. It is rare that tribes and environments are so closely correlated." Gentry's analysis was accurate for the Río Mayo of his day. South of the Río Mayo, however, thousands of Mayos live in tropical deciduous forest as well as in coastal and foothills thornscrub.

## CHAPTER 5

1. In southern Sinaloa the common name is *cardón,* the same Spanish name given to numerous arborescent cacti. In Baja California it is referred to as *pitahaya barbona.* In northwest Argentina at least four species of columnar cacti are referred to as *cardón.*

## CHAPTER 6

1. Richard Felger informs us that this mist is actually insect excrement.

2. For a more general discussion of the relation between the Mayos' use of herbs and symptomatology, see Bañuelos 1999.

3. Numerous studies have shown that Arizona Pimas, who adopted a Western diet high in sugar and alcohol, have astronomical rates of diabetes, while Mexican Pimas, genetically similar but still traditional farmers with a traditional diet of corn, beans, and squash, have very low rates. See also Nabhan 1985.

4. Even with a heightened awareness of the etiology of disease, however, the resources to cope with requirements for sanitary food preparation are greatly limited in the region. Water must be boiled using precious or expensive firewood or prohibitively expensive propane gas. Commercial detergents are also beyond the budgets of many families, and water for washing and rinsing dishes is in short supply in many communities. Finally, refrigerators are a quasi luxury. In most villages the majority of homes do not have one. Food-handling strategies for preventing the transmission of disease must take these factors into consideration. Indeed, even in the region's cities, few public restaurants would be in compliance with standard public health and sanitation requirements of the United States.

5. *Curandismo,* a tradition dating from the Jesuit period in Mayo history, is practiced throughout Mexico.

6. The Jesuit Ignaz Pfefferkorn devotes considerable discussion in *Sonora: A Description of the Province* to healing practices and substances of the eighteenth century, but all in the context of European conception of disease and health. By his time, most vestiges of pre-Columbian healers had been eliminated.

# Works Cited

Acosta, R. 1949. *Apuntes históricos sonorenses.* Hermosillo: Gobierno del Estado de Sonora.

Alegre, F. J. 1888. *La historia de la compañía de Jesús en Nueva España.* N.p. Special Collections, University of Arizona Library, Tucson.

Almada, F. R. 1937. *Apuntes históricos de la región de Chínipas.* Chihuahua: Gobierno del Estado.

———. 1952. *Diccionario de historia, geografía, y biografía sonorense.* Hermosillo: Gobierno del Estado.

———. 1990. *La revolución en el estado de Sonora.* Hermosillo: Gobierno del Estado.

Almada Bay, I. 1993. La conexión Yocupicio: Soberanía estatal, tradición cívico-liberal, y resistencia al reemplazo de las lealtades en Sonora, 1913–1939. Ph.D. thesis, El Colegio de México.

Arbelaez, M. S. 1991. The Sonoran missions and Indian raids of the eighteenth century. *Journal of the Southwest* 33:366–77.

Axelrod, D. 1979. *Age and origin of the Sonoran Desert.* California Academy of Sciences Occasional Paper 132:1–74.

Bannon, J. 1947. Black robe frontiersman: Pedro Méndez, S.J. *Hispanic American Historical Review* 27:61–86.

Bantjes, A. 1998. *As If Jesus Walked on Earth: Cardenismo, Sonora, and the Mexican Revolution.* Wilmington: Scholarly Resources.

Bañuelos F., N. 1999. *De plantas, mujeres, y salud.* Hermosillo: Centro de Investigación en Alimentación y Desarrollo.

Beals, R. 1943. *The aboriginal culture of the Cáhita Indians.* Berkeley: University of California Press.

Berlin, B. 1992. *Ethnobiological classification.* Princeton: Princeton University Press.

Berlin, B., D. Breedlove, and P. Raven. 1974. *Principles of Tzeltal plant classification: An introduction to the botanical ethnography of a Mayan-speaking people of highland Chiapas.* New York: Academic Press.

Bolton, H. 1936. *Rim of Christendom.* Tucson: University of Arizona Press.

Brading, D. 1971. *Miners and merchants in Bourbon Mexico, 1763–1810.* Cambridge: Cambridge University Press.

Brunk, S. 1995. *Emiliano Zapata: Revolution and betrayal in Mexico.* Albuquerque: University of New Mexico Press.

Bye, R. 1995. Ethnobotany of the Mexican tropical dry forests. Pp. 423–38 in S. Bullock, H. A. Mooney, and E. Medina, eds., *Seasonally dry tropical forests.* New York: Cambridge University Press.

Cabeza de Vaca, A. 1972. *The narrative of Alvar Núñez Cabeza de Vaca.* Translated by F. Bandelier. Barre, Mass.: Imprint Society.

Calderón Valdés, S., ed. 1985. *Historia general de Sonora.* 5 vols. Hermosillo: Gobierno del Estado.

Chisem, J. 1990. *Apuntes para la historia de Guaymas.* Hermosillo: Gobierno del Estado de Sonora.

Cochemé, J., and D. Dement. 1991. *Geology of the Yécora area, northern Sierra Madre Occidental, Mexico.* Geological Society of America Special Paper 254.

Collard, D., and E. Collard. 1962. *Castellano-mayo, mayo-castellano.* Mexico City: Summer Institute of Linguistics.

Crosby, A. 1986. *Ecological imperialism: The biological expansion of Europe, 900–1900.* Cambridge: Cambridge University Press.

Crosby, H. 1994. *Antigua California.* Albuquerque: University of New Mexico Press.

Crumrine, R. 1977. *The Mayo Indians of Sonora: A people who refuse to die.* Tucson: University of Arizona Press.

———. 1983. Mayo. Pp. 264–75 in A. Ortiz, ed., *Handbook of North American Indians.* Washington: Smithsonian Institution.

Dabdoub, C. 1964. *La historia de el valle del Yaqui.* Hermosillo: Gobierno del Estado.

Domecq, B. 1990. *La insólita historia de la Santa de Cabora.* Mexico City: Planeta.

Erasmus, C. 1967. Culture change in northwest Mexico. In J. Steward, ed., *Contemporary change in traditional societies,* vol. 3. Urbana: University of Illinois Press.

Everitt, B. L. 1968. Use of the cottonwood in an investigation of the recent history of a flood plain. *American Journal of Science* 266:417–39.

Felger, R. S. 1999. *The flora of Cañón de Nacapule: A desert-bound tropical canyon near Guaymas, Sonora, Mexico.* Proceedings of the San Diego Society of Natural History no. 35.

Felger, R. S., and E. Joyal. 1999. The palms (Arecaceae) of Sonora, Mexico. *Aliso* 18:1–18.

Felger, R. S., and M. B. Moser. 1985. *The people of the desert and sea: An ethnobotany of the Seri Indians.* Tucson: University of Arizona Press.

Figueroa, A. 1994. *Por la tierra y por los santos: Identidad y persistencia cultural entre yaquis y mayos.* Mexico City: Culturas Populares.

Frank, A. G. 1972. *Lumpenbourgeoisie: Lumpendevelopment.* New York: Monthly Review Press.

Freeze, R. 1989. *Mayo de Los Capomos.* Mexico City: El Colegio de México.

Friedman, S. L. 1996. Vegetation and flora of the coastal plains of the Río Mayo region, southern Sonora, Mexico. M.S. thesis, Arizona State University, Tempe.

Friedrich, P. 1970. *Agrarian reform in a Mexican village.* Englewood Cliffs, N.J.: Prentice-Hall.
Gentry, H. S. 1942. *Río Mayo Plants.* Washington, D.C.: Carnegie Institution Publication 527.
———. 1963. *The Warihío Indians of Sonora-Chihuahua: An ethnographic survey.* Washington, D.C.: Smithsonian Institution Anthropological Papers no. 65.
Gerhard, P. 1982. *The northern frontier of New Spain.* Princeton: Princeton University Press.
Germán E., J. L., L. Ríos R., C. E. Flores G., and O. S. Ayala P. 1987. *Génesis y desarrollo de la cultura mayo de Sonora.* Ciudad Obregón: Instituto Tecnológico de Ciudad Obregón, Sonora.
Hall, L. 1981. *Alvaro Obregón: Power and revolution in Mexico 1911–1920.* College Station: Texas A & M University Press.
Hastings, J. R., and R. R. Humphrey. 1969. *Climatological data and statistics for Sonora and northern Sinaloa.* University of Arizona Technical Reports on the Meteorology and Climatology of Arid Regions no. 19.
Hu De-Hart, E. 1984. *Yaqui resistance and survival.* Madison: University of Wisconsin Press.
Kay, M. A. 1996. *Healing with plants in the American and Mexican West.* University of Arizona Press, Tucson.
Kearney, T. H., and R. H. Peebles. 1969. *Arizona flora.* Berkeley: University of California Press.
Kirchhoff, P. 1953. Mesoamerica. *Acta Americana* 1:92–107.
Krauze, E. 1997. *Mexico: Biography of power.* New York: HarperCollins.
Krizman, R. D. 1972. Environment and season in a tropical deciduous forest in northwestern Mexico. Ph.D. dissertation, University of Arizona.
Lindquist, C. 2000. Dimensions of sustainability: The use of *vara blanca* as a natural resource in the tropical deciduous forest of Sonora, Mexico. Ph.D. dissertation, University of Arizona.
Lonsdale, P. 1989. Geology and tectonic history of the Gulf of California. In E. Winterer, D. Hussong, and R. Decker, eds., *The Eastern Pacific Ocean and Hawaii, Geological Society of America, Geology of North America,* N:499–521.
Martin, P. S., D. Yetman, M. Fishbein, P. Jenkins, T. R. Van Devender, and R. K. Wilson. 1998. *Gentry's Río Mayo Plants: The tropical deciduous forest and environs of northwest Mexico.* Tucson: University of Arizona Press.
McAuliffe, J. R. 1995. Landscape evolution, soil formation, and Arizona's desert grasslands. Pp. 100–129 in M. McClaran and T. R. Van Devender, eds., *The desert grassland.* Tucson: University of Arizona Press.
Melchor C., R. L. 1991. Organización de la fiesta de la Santísima Trinidad en El Júpare. In D. Gutiérrez and J. Gutiérrez T., eds., *El noroeste de Mexico: Sus culturas étnicas.* Mexico City: National Museum of Anthropology.
Nabhan, G. P. 1985. *Gathering the desert.* Tucson: University of Arizona Press.
Navarro García, L. 1967. *Sonora and Sinaloa in the seventeenth century.* Sevilla: University of Seville.
Nentvig, J. 1951. *Rudo ensayo.* Tucson: Arizona Silhouette.

Obregón, A. 1959. *Ocho mil kilómetros en campaña*. Mexico City: Fondo de Cultura Económica.

Obregón, B. 1988. *Historia de los descubrimientos antiguous y modernos de la Nueva España escrita por el conquistador en el año de 1584*. Mexico City: Editorial Porrúa.

O'Connor, M. 1989. *Descendants of Totoliguoqui*. University of California Publications in Anthropology. Berkeley: University of California Press.

Ortega N., S. 1978. *El eden subvertido: La colonización de Topolobampo, 1886–1896*. Mexico City: Departamento de Investigaciones Históricas, Instituto Nacional de Antropología e Historia.

Owen, A. K. 1975. *Integral co-operation: Its practical application*. Philadelphia: Porcupine Press.

Padden, R. 1967. *The hummingbird and the hawk*. Columbus: Ohio State University Press.

Pérez de Ribas, A. 1645. *Crónicos de los triunfos de nuestra santa fe*. Original copy in Special Collections, University of Arizona Library, Tucson.

Pfefferkorn, I. 1949. *Sonora: A description of the province*. T. Treutlein, editor and translator. Albuquerque: University of New Mexico Press.

Reff, D. 1991. *Disease, depopulation, and culture changes in northwestern New Spain, 1518–1764*. Salt Lake City: University of Utah Press.

Reynolds, R. 1996. *Cats' paw utopia: Albert K. Owen, the adventurer of Topolobampo Bay, and the last great utopian scheme*. San Bernardino: Borgo Press.

Robertson, T. 1964. *Southwestern utopia*. Los Angeles: Ward Ritchie Press.

———. 1968. *My life among the savage natives of New Spain*. Los Angeles: Ward Ritchie Press.

Roca, P. 1967. *Paths of the padres through Sonora: An illustrated history and guide to its Spanish churches*. Tucson: Arizona Pioneers Historical Society.

Sauer, C. 1932. *Road to Cíbola*. Berkeley: University of California Press.

———. 1935. *Aboriginal populations of northwest Mexico*. Berkeley: University of California Press.

Secretaría de Educación Público. 1991. *Sinaloa: Tierra fértil entre la costa y la sierra*. Mexico City: Secretaría de Educación Pública.

Sellers, W. D., and R. H. Hill. 1974. *Arizona climate, 1931–1972*. Tucson: University of Arizona Press.

Sheridan, T. 1979. The cross or the arrow? The breakdown in Spanish-Seri relations, 1729–1750. *Arizona and the West* 21:221–334.

Shreve, F., and I. Wiggins. 1964. *Vegetation and flora of the Sonoran Desert*. Palo Alto: Stanford University Press.

Sobarzo, H. 1991. *Vocabulario sonorense*. Hermosillo: Gobierno del Estado.

Spicer, E. 1962. *Cycles of conquest*. Tucson: University of Arizona Press.

———. 1980. *The Yaquis: A cultural history*. Tucson: University of Arizona Press.

Stagg, A. 1978. *The Almadas and Alamos, 1783–1867*. Tucson: University of Arizona Press.

Troncoso, F. P. 1905. *Las guerras con las tribus yaqui y mayo del estado de Sonora*. Mexico City: Departamento del Estado Mayor.

Turner, R. M., J. E. Bowers, and T. L. Burgess. 1995. *Sonoran Desert plants, an ecological atlas.* Tucson: University of Arizona Press.

Valenzuela Y., H. 1984. *Utilización de las plantas en la comunidad mayo de los Buayums, municipio de Navojoa.* Dirección General de Culturas Populares. N.p.: Secretaría de Educación Popular. Manuscript in Special Collections, University of Arizona, Tucson.

Van Devender, T. R., A. C. Sanders, R. K. Wilson, and S. A. Meyer. 2000. Flora and vegetation of the Río Cuchujaqui, a tropical deciduous forest near Alamos, Sonora. Pp. 36–101 in R. H. Robichaux and D. Yetman, eds., *The tropical deciduous forest of Alamos: Biodiversity of a threatened ecosystem.* Tucson: University of Arizona Press.

Vázquez, R. E. 1955. *Geografía del estado de Sonora.* Mexico City: Pluma y Lápiz de México.

Villa, E. 1951. *La historia del estado de Sonora.* Hermosillo: Editorial Sonora.

Villalpando, E. 1985. Los grupos agrícolas de Sonora en la época prehispánica. In S. Calderón Valdés, ed., *Historia general de Sonora* 1:225–34. Hermosillo: Gobierno del Estado.

West, R. C. 1993. *Sonora: Its geographical personality.* Austin: University of Texas Press.

Yetman, D. 1995. San Bernardo revisited. *Journal of the Southwest* 37:142–56.

———. 1998. *Scattered round stones: A Mayo village in Sonora.* Albuquerque: University of New Mexico Press.

Yetman, D., and A. Búrquez. 1994. Buffelgrass—Sonoran Desert nightmare. *Arizona Riparian Council Newsletter* 7:8–10.

Yetman, D., T. R. Van Devender, P. Jenkins, and M. Fishbein. 1995. The Río Mayo: A history of studies. *Journal of the Southwest* 37:294–345.

Zavala C., P. 1989. *Apuntes sobre el dialecto yaqui.* Hermosillo: Gobierno del Estado.

Zuñiga, I. 1835. *Rápida ojeada al estado de Sonora, territorios de California, y Arizona.* Reprint, Mexico City: Editor Vargas, 1948.

# Index

All species are indexed by scientific name, with common names in Mayo, Spanish, and English given in parentheses and roman type. Partial listings also appear under the common names most frequently used in the text. (See appendixes D and E for a complete list of Mayo and Spanish names, including the corresponding scientific names.) Page numbers in italic denote illustrations.

Abolillo (*Sapindus saponaria*), 74, 125, 251
*Abutilon* spp., 115, 130; *abutiloides* (jíchiquia jéroca; pintapán; Indian mallow), 224; *incanum* (to'oro cojuya; escoba; Indian mallow), 224–25; *mucronatum* [*A. trisulcatum*], 225
*Acacia: cochliacantha* (chírahui; güinolo; boat-spine acacia), 12, 65, 124, 191; *coulteri* (baihuío; guayavillo), 12, 67, 68, 191–92; *farnesiana* (cu'uca; vinorama; sweet acacia), 65, 191, 192; *occidentalis* (teso; Sonoran catclaw), 114, 192; *pennatula* (algarrobo), 193
Acanthaceae, 112, 123, 130; species descriptions, 133–35
*Acanthocereus occidentalus* (tasajo), 161
Achatocarpaceae, 65, 135
Acosta, Benigno, 160, 204
*Adelia virgata* (ono jújugo), 65, 184
Aduana, 153
Aesthetics, 109–10; *Amyris balsamifer* for, 248; Bignoniaceae for, 151, 152; Cactaceae for, 167; *Caesalpinia* spp. for, 194, 195–96; *Conzattia multiflora* for, 198; *Hyptis albida* for, 221; Moraceae, for, 229, 230; *Olneya tesota* for, 210; *Plumeria rubra* for, 138; *Solanum seaforthianum* for, 260; *Vitex mollis* for, 268–69

Agavaceae, 115; species descriptions for, 135–37
*Agave*, 125, 160, 213; *aktites* (huítbori; lechuguilla), 135; *colorata* (baugoa; mezcal), 136; *jaiboli* (aiboli; mezcal), 136–37; *vilmoriniana* (aimuri; amole; octopus agave), 115, 137; *vivipara* (cu'u; mezcal, maguey), 87, 90–95, 137
Agave, false (*Hechtia montana*), 157
Agiabampo, 6, 63, 66
Agiabampo, Estero, 62, 63, 328–29n.1
*Agkistrodon bilineatus* (pichicuate), 117
*Agonandra racemosa* (úsim yuera; palo verde, palo huaco), 234–35
Agribusiness, 19, 66
Agriculture, 5, 19, 33, 50, 58, 319n.5, 320nn.10, 12, 322n.7, 328n.51; commercial, 58–59, 319n.3; Jesuits and, 43–44, 323n.15; land clearing for, 84–85; on Mayo Delta, 4, 57–58; milpa, 12–13; on Río Mayo, 4–5, 53; Spanish era, 31, 33, 42; water allocation and, 7–8
Aguama (*Bromelia alsodes*), 157
Aguamita (*Hechtia montana*), 157
Aguaro (Martyniaceae), 125, 226–27
Aguas, las, 11, 12, 13, 65
Ahome, 9
Ahomes, 16, 17, 35, 36, 45

337

# Index

Aizoaceae, 137
Ajonjolí (*Sesamum orientale*), 13, 235
Álamo (*Populus mexicana* subsp. *dimorpha*), 74, 115, 249–50
Alamos, 10, 42, 49, 70
Alamos, Sierra de, 9, 73
*Albizia sinaloensis* (joso; palo joso), 74, 115, 193
Alcohol, 92
Alfalfa (*Medicago sativa*), 132
Algarrobo (*Acacia pennatula*), 193
Algodoncillo (*Wimmeria mexicana*), 71, 112, 113, 175
*Allenrolfea occidentalis* (bahue mo'oco), 176
Almada, José María "El Chato," 49
*Aloe barbadensis* (sábila; aloe vera [*A. vera*]), 142
*Aloysia sonorensis* (mariola), 120, 265–66
*Alvaradoa amorphoides* (guaji; palo torsal), 253–54
*Alvordia congesta* (cochibachomo), 143
Amapa (*Tabebuia impetiginosa*), 12, 71, 74, 110, 113, 151–52
Amapa amarillo (*Tabebuia chrysantha*), 150–51
Amaranthaceae, 123, 125, 137–38
*Amaranthus palmeri* (hué; bledo; pigweed), 125, 137–38
Amaryllidaceae, 138
*Amazona* spp., 77, 243
*Ambrosia*: *ambrosiodes* (chicura; canyon ragweed), 119, 143; *confertiflora* (chíchibo; estafiate; slimleaf bursage), 121, 143; *cordifolia* (cau nachuqui; chicurilla; Sonoran bursage), 143, 144
Amole (*Agave vilmoriniana*), 115, 137
Amolillo (*Sapindus saponaria*), 74, 125, 251
*Amoreuxia gonzalezii* and *palmatifida* (saya, saya mome), 103–4, 125, 152
*Amyris balsamifer* (úsim yuera; palo limón), 76, 110, 248
Añil (*Indigofera suffruticosa*), 115, 133, 205–6
Animal products, 77, 118
*Anoda* spp., 130
*Antigonon leptopus* (masasari; sanmiguelito; queen's wreath), 125, 241
Antimalarial plants, 119, 133, 134
Apaches, 45, 325nn.31, 37
*Aphananthe monoica* (guasimilla), 175
Apocynaceae, 68, 74, 109, 138–39

Aquiropo, Estero, 62
Arecaceae, 113, 139–40
*Argemone ochroleuca* subsp. *ochroleuca* (táchino; cardo; prickly poppy), 235
*Arisida ternipes* var. *ternipes* (ba'aso; zacate; spider grass), 236
*Aristolochia watsonii* (guasana jibuari; hierba del indio; pipevine), 140–41
Aristolochiaceae, 140–41
Arrellán (*Psidium sartorianum*), 231–32
Arrows, 144, 198
Arroyos, plant communities in, 74, 76
Arthropod stings and bites, treatment of, 116, 117, 142, 201, 233, 267
Artifacts: of *Agave*, 92–94, 137; of Arecaceae, 139–40; of Asteraceae, 146; of *Cordia parvifolia*, 81; of Fabaceae, 208, 210, 215; of *Jacquinia macrocarpa*, 263–64; of *Jatropha cordata*, 101–2; of *Lycium andersonii*, 258; of *Lysiloma watsonii*, 208; of *Martynia annua*, 226; of *Pachycereus pectenaboriginum*, 95; of *Phaulothamnus spinescens*, 135; of *Randia* spp., 247; of *Rhizophora mangle*, 244; of *Salix*, 250; of *Sapindus saponaria*, 251; of Verbenaceae, 266. *See also by type*
*Arundo donax* (ba'aca nagua; carrizo; giant cane), 132, 170, 237, 239
Asclepiadaceae, 141–42
*Asclepias subulata* (ma'aso pipi; tetas del venado), 141
Asphodelaceae, 141
Assimilation, 47
Asteraceae (Compositae), 71, 119, 121, 123, 129, 130, 269; species descriptions of, 143–49
Asthma, 121, 212, 217, 270
*Atriplex barclayana* (bahuepo juatóroco; saltbush), 177
*Avicennia germinans* (ciali; mangle negro; black mangrove), 62, 130, 266
Ayal (*Crescentia alata*), 110, 149–50
Aztecs, 16, 30–31, 321n.1

Ba'aco (*Phaulothamnus spinescens*), 65, 135
Baby's breath, pink (*Talinum paniculatum*), 241
Baca, 17
Bacabachis, 17
Bacanora, 95
Bacasegua, Alejo, 118–19, 143, 145, 163, 165,

173, 175, 187, 196, 242, 243, 248, 249, 262, 266, 276
*Baccharis: salicifolia* (bachomo; batamote; seepwillow), 144; *sarothroides* (hierba de pasmo, romerillo; desert broom), 144
Bachata (*Opuntia* sp.), 167
Bachoco, 19, 29
Bachomohaqui, *31*, 110
Bacot mútica (*Byttneria aculeata*), 261
Badón, Federico, 204, 250, 276
Baesmo Cota, Ignacio, 249, 276
Bahía Yavaros, 62, 328n.1
Baiburilla (*Dorstenia drakena*), 123, 228–29
Baiburín (*Kallstroemia* spp.), 271
Baimena, 17, 18, 19, 110, 161, 168, 175
Báis cápora (*Ziziphus amole*), 12
Baisegua, José Luis, 20
Baisegua, Reyes, 23, *24*, 87, 138, 214, 220, 224, 276; on ixtle manufacture and weaving, 92–95
Ballmoss (*Tillandsia* spp.), 11, 157
Bamboo (*Otatea acuminata*), 113, 239
Banderas, Juan (Juan Ignacio Jusacamea), 48, 325n.33
Baroyeca, 42, 45
Baroyeca, Sierra, 17
Barrel cactus (*Ferocactus herrerae*), 11, 65, 120, 125, 162–63
Basin and Range Province, 13, 14–15
Basiroans, 16, 17
Baskets, 139, 140, 223, 237, 265
*Bastardiastrum cinctum* (malva, escoba), 130, 225
Batamote (*Baccharis salicifolia*), 144
Batayaqui (*Montanoa rosei*), 71, 113, 146–47
Batopilas, 43
Bats: fruit-eating, 66, 230, 231; lesser long-nosed (*Leptonycterus curasoae*), 66
Bayájuri, Cerro, 3
Bazán, Hernando de, 35
Beans, 13
*Bebbia juncea* (mo'oso; sweetbush), 124, 144
Bebelama (*Sideroxylon* spp.), 252
Bees, 102, 111, 112, 117
Bellísima (*Solanum seaforthianum*), 110, 260
Benito Juárez, 21
*Bernardia viridis* (juya jotoro), 184
Bignoniaceae, 12, 110; species descriptions for, 149–52
Bilingualism, 19
Biological diversity, 8

Bird-of-paradise, red (*Caesalpinia pulcherrima*), 72, 195–96
Birds, 77
Bixaceae, 71, 152–53
Biznaga (*Ferocactus herrerae*), 11, 65, 120, 125, 162–63
Biznaguita (*Mammillaria* spp.), 11, 65, 122, 164
Bledo (*Amaranthus palmeri*), 125, 137–38
*Boa constrictor* (boa constrictor), 76
*Boerhavia* (mochis), 123, 130, 232; *coccinea* (juana huipili; red spiderling [*B. diffusa*]), 232; *spicata* (spiderling), 232
Boidae, 76
Bola colorada (*Trichilia* spp.), 228
Bombacaceae, 71, 153–54
*Bonplandia geminiflora* (ciática), 240–41
Boraginaceae, 65, 67, 72, 124, 154–56
*Bougainvillea spectabilis* (bugambilia), 232
*Bouteloua: aristidoides* (yemsa ba'aso; saitilla; six-weeks needle grama), 237–38; *barbata* (ma'as ba'aso; six-weeks grama, Mayo grama), 238
*Brahea aculeata* (ta'aco; palma; fan palm), 113, 139
Brasil (*Haematoxylum brasiletto*), 67, *112*, 113, 115, 120, 122–23, 126, 202–3, 221
Brasil chino (*Haematoxylum* sp.), 203–4
Brassicaceae, 132, 156–57
Brea (*Parkinsonia praecox*), 65, 111, 112, 115, 126, 211–13, *212*
*Brickellia* (brickel-bush): *brandegei* (juya chibu; rama amarga), 144; *coulteri* (bacot tami; diente de culebra), 145
Brincador (*Sebastiania pavoniana*), 110, 190
Brittlebush (*Encelia farinosa*), 145
*Bromelia alsodes* (chó'ocora; aguama), 157
Bromeliaceae, 11, 157
Bronchitis, 121
*Brongniartia alamosana* (cuta nahuila; palo piojo), 12, 194
Brooms, 155, 199, 224–25
Brujos (witches), 119
Buaysiacobe, 19
Buayums, 118
Buckets, made of pitahaya, 87, *88*
Buffelgrass, 58–59, 69, 240
Bugambilia (*Bougainvillea spectabilis*), 232
Buichílame Josaino, Paulino, 148, 226, 258, 276
Buitimea, Benigno, 235

Bulrush (*Scirpus americanus*), 183
Bumelia occidentalis. See *Sideroxylon occidentale*
*Bunchosia sonorensis* (goy carabanzo; garbanzo de coyote), 222
Burros, 118; forage for, 143, 144, 177, 216, 233, 239, 240
Bursage: slimleaf (*Ambrosia confertiflora*), 121, 143; Sonoran (*A. cordifolia*), 144
*Bursera* (to'oro; torote), 113, 114, 130; *fagaroides* (to'oro sahuali; torote de vaca), 65, 158; *grandifolia* (to'oro mulato; palo mulato), 71, 123, 125, 158, *159*; *lancifolia* (to'oro chutama; torote copal), 158–59; *laxiflora* (to'oro chucuri; torote prieto), 12, 65–66, 67, 68, 69, 120, 159–60; *microphylla*, 65, 160; *penicillata* (to'oro chutama; torote acensio), 71, 160; *simaruba* (to'oro mulato; gumbo limbo), 161; *stenophylla* (torote copal), 71, 113, 161
Burseraceae, 12; species descriptions for, 158–61
Buttercup tree (*Cochlospermum vitifolium*), 71, 152–53
Button bush (*Cephalanthus occidentalis*), 244
*Byttneria aculeata* (bacot mútica), 261

Cabeza de Vaca, Alvar Nuñez, 33–34
Cabora, 52
Cacachila (*Karwinskia humboldtiana*), 67, 121, 122, 123, 242–43
Cactaceae, cacti, 11, 65, 125; species descriptions for, 161–73
Caesalpinia: cacalaco (huanaca; huisache), 110, *194*; caladenia (jícamuchi; palo piojo blanco), 68, 194; *palmeri* (jícamuchi; palo piojo), 65, 194–95; *platyloba* (mapáo; palo colorado), 67, 113, 195; *pulcherrima* (ta'aboaca; tavachín; red bird-of-paradise), 72, 110, 133, 195–96; *sclerocarpa* (ébano), *76*, 123, 196, *197*
Cáhitas, 16, 17, 26, 27, 30–31; Mexican persecution of, 49–50; rebellions by, 44–45, 51
Cajeme (José María Leyva), 50–51
Calixto, Juan, 45
*Callaeum macropterum* (sanarogua; mataneni), 124, 223
Calles, Plutarco Elías, 54–55, 328n.47
*Calliandra emarginata* (guamuchilillo [*C. rupestris*]), 196
*Callipepla douglasii* (choli; elegant quail), 199

*Calocitta formosa collei* (black-throated magpie jays), 268
Camahuiroa, 6, 19, 63, 118
Camahuiroa, Arroyo, *76,* 196
Camoa, 6, 48, 323n.14
Cancer, treatments for, 120, 167, 182, 244
Candelilla: *Pedilanthus macrocarpus*, 65, 189–90; *Euphorbia colletioides*, 68, 132, 187
Cane, giant (*Arundo donax*), 132, 170, 237, 239
Canes, 200
Capetamaya, 51
Capparaceae, 65, 76, 173–74
*Capparis atamisquea* (juvavena [*Atamisquea emarginata*]), 65, 124, 173
*Capparis flexuosa* (tábelojeca), 173
*Capsicum annuum* (có'cori; chiltepín), 71, 125, 254–57
Capúsari (*Crataeva palmeri*), 76, 173–74
Cardo (*Argemone ochroleuca* subsp. *ochroleuca*), 235
Caricaceae, 125, 174
*Carlowrightia arizonica* (ánima ogua; palo blanco), 133
*Carnegiea gigantea* (saguo; sahuaro; saguaro), 62, 161–62
Carranza, Venustiano, 54
Carrizito de huitlacoche (*Lasiacis ruscifolia*), 239
Carrizo (*Arundo donax*), 132, 170, 237, 239
Carved objects, 153, 158, 160, 161, 174, 192, 193, 201, 204, 228, 231, 242, 250, 251, 252
*Casimiroa edulis* (ja'apa ahuim; chapote, zapote), 112, 248
Castor bean (*Ricinus communis*), 132, 190
Catclaw (*Mimosa distachya*), 65, 209; Sonoran (*Acacia occidentalis*), 114, 192
Catholicism, Catholic Church, 325n.25; and Calles, 54–55; Mayos and, 20–21, 36, 38, 39, 41–42, 321n.11, 324n.20; plants important in, 125–26
Cattle, 43, 58, 69, *111*, 324n.16; fodder and forage for, 112, 148, 157, 163, 165, 166, 187, 202, 216, 240; medical treatment of, 92, 251, 264
Cebolla de coyote (*Hymenocallis sonorensis*), 138
*Cedrella odorata* (cedro), 113, 227
Cedro (*Cedrella odorata*), 113, 227
*Ceiba acuminata* (baogua; pochote; kapok), *70*, 71, 125, 153, *154*
Celastraceae, 63, 71, 174–75

Celtidaceae, 12, 175–76
Celtis, 71, 124, 125; *iguanaea* (cumbro; vainoro; tropical hackberry), 175–76; *pallida* (cumbro; cúmero; desert hackberry), 176; *reticulata* (cumbro; cúmero; netleaf hackberry), 176
Cenchrus spp. (na'a chú'uqui; guachapori; sandbur), 238. See also *Pennisetum ciliare*
Cenzontle (*Mimus polyglottos*), 135
Cephalanthus occidentalis var. *salicifolius* (mimbro; button bush), 244
Chacalacas (*Ortalis wagleri*), 77
Chalate (*Ficus* spp.), 71, 229, 231
Chamisa del mar: *Lycium brevipes*, 258; *Suaeda moquinii*, 177
Chamomile (*Matricaria chamomilla*), 269
Chanate pusi (*Rhynchosia precatoria*), 121, 216–17
Chani (*Hyptis suaveolens*), 122, 125, 221
Chaparacoba (*Citharexylum scabrum*), 76, 266
Chapote (*Casimiroa edulis*), 112, 248
Charcoal, 209
Cheesebush (*Hymenoclea monogyra*), 146
Cheesemaking, 115, 146, 252
Chenopodiaceae, 176–77
Chenopodium: ambrosioides (lipazote; epazote; Mexican tea), 121, 177; *neomexicanum* (choal; quelite; fish goosefoot), 177
Chichicoyota (Cucurbitaceae), 180
Chichiquelite (*Solanum* spp.), 125, 259–60
Chickens, 112–13, 216
Chicotera (*Masticophis flagellum*), 233
Chicura (*Ambrosia ambrosioides*), 119, 143
Chicurillo (*Ambrosia cordifolia*), 144
Chiggers, 271
Chi'ini (*Pouzolzia guatemalana*), 265
Childbirth, 119, 267, 269
Chilicote (*Erythrina flabelliformis*), 12, 113, 115, 200–201
Chiltepín (*Capsicum annuum*), 71, 125, 254–57
Chinchweed (*Pectis coulteri*), 148
Chiococca alba (tori naca; oreja de ratón), 244
Chírahui (*Acacia cochliacantha*), 12, 65, 124, 191
Chloroleucon mangense (ooseo suctu; palo fierro), 67–68, 197
Chlorophora tinctoria (mora), 71, 113, 228
Chócola (*Jarilla chocola*), 125, 174

Choix, 16, 19
Choix (people), 17
Cholugos (*Nasua narica*), 118, 268
Chopo (*Mimosa palmeri*), 74, 209
choya, cholla (*Opuntia*), 123–24; chainfruit (*fulgida*), 65, 164–65; desert Christmas (*leptocaulis* var. *britonii*), 11, 165; hybrid (*leptocaulis* × *thurberi*), 165–66; *pubescens*, 166; *thurberi*, 65, 166
Christianity, 56. See also Catholicism
Christmas cactus, desert (*Opuntia leptocaulis* var. *brittoni*), 11, 165
Chuparosa (*Justicia californica*), 134
Ciática (*Bonplandia geminiflora*), 240–41
Ciconiiformes, 77
Circulatory problems, 120
Ciruela del monte (*Ziziphus obtusifolia* var. *canescens*), 120, 243–44
Cissus spp. (yucohuira), 270
Citharexylum scabrum (chaparacoba), 76, 266
Ciudad Obregón, 8
Classification scheme, Mayo, 129–30
Climate, 8–13, 319–20n.9; rainfall, 7, 9–10; seasons, 10–11
Clover, sour (*Melilotus indica*), 132
Clubs, 200
Coastal flats, 9
Coastal plain, 13
Coastal thornscrub, 5, 9, 64–67, 77, 131
Coastal vegetation, 62–63
Coatimundis (*Nasua narica*), 118, 268
Cochibachomo (*Alvordia congesta*), 143
Cochlospermum vitifolium (ciánori; palo barril; buttercup tree), 71, 152–53
Cocoraque, Arroyo, 16–17, 62, 328n.1
Cola de la zorra (*Setaria liebmannii*), 240
Cola de zorillo (*Petiveria alliacea*), 121, 235–36
Colds, 120–21
Coleonyx variegatus, 117–18
Colorados, Cerros, 73, 74
Colubridae, 77
Colubrina: triflora (cuta béjori), 241–42; viridis (ono jújugo), 242
Combretaceae, 62, 130, 177–78
Combretum fructicosum (compio), 177
Combs, 95
Commelina erecta (osi; cuna de niño; dayflower), 178
Commelinaceae, 178

*Commicarpus scandens* (miona), 123, 232
Compio (*Combretum fructicosum*), 177
Compio (*Heteropteris palmeri*), 223
Comunidades (communities), 18, 19, 48, 49, 60–61
*Condalia globosa* (juupa quecara; mezquite hormiguillo; bitter condalia), 242
Confiturilla: amarilla (*Lagacea decipiens*), 130; blanca (*Lantana achyranthifolia, L. hispida*), 130, 267; negra (*L. camara*), 130, 266–67
Conicaris, 17
Conicárit, 34, 320n.1
*Conocarpus erecta* (mangle; button mangrove), 130, 177–78
Construction, 113; *Agonandra racemosa* used in, 235; *Alvaradoa amorphoides* used in, 254; *Ambrosia confertiflora* used in, 143; Bignoniaceae used in, 151–52; Cactaceae used in, 85, 96, 167, 171; Capparaceae used for, 173; *Cedrella odorata*, 227; *Ceiba acuminata* as, 153; Celtidaceae used in, 175, 176; *Cordia sonorae* used in, 156; *Erythroxylon mexicanum* used in, 184; Euphorbiaceae used in, 184, 186; Fabaceae used in, 192, 194, 195, 196, 200, 201, 203, 204, 207–8, 209, 210, 213, 214, 215, 216, 217; *Guaiacum coulteri* used in, 271; *Heliocarpus attenuatus* used in, 264; *Hintonia latiflora* in, 245; *Ipomoea arborescens* used in, 179; *Montanoa rosei* used in, 146–47; Moraceae used in, 228; palms used for, 139; Poaceae used in, 236, 237, 239; Rhamnaceae used in, 242, 243; Rutaceae used in, 248, 249; Salicaceae in, 250; *Sapindus saponaria* in, 251; *Schoepfia schreberi* as, 234; *Scirpus americanus* used in, 183; *Wimmeria mexicana* used for, 175
Consultants, 127–28
Convolvulaceae, 123, 178–80
*Conzattia multiflora* (joso; joso de la sierra), 71, 198
Cooperatives, 19
Copalquín (*Hintonia latiflora*), 67, 113, 119, 120, 122, 123, 124, 245
Coquillo (*Cyperus* spp.), 183
Coralbean (*Erythrina flabelliformis*), 12, 113, 115, 200–201
Coralillo cimarrón (*Rivina humilis*), 236
Coralillos. See Snakes, coral
Corcho (*Diphysa suberosa*), 74, 200

*Cordia: curassavica* (jusa'iro), 154–55; *parvifolia* (huo'ótobo; vara prieta), 65, 79–83, 115, 126, 155; *sonorae* (pómajo; palo de asta), 67, 114, 155–56
Cordoncillo: *Elytraria imbricata*, 123, 190; *Siphonoglossa mexicana*, 134–35
Corks, 200, 201
Corn, 13
Cornetón (*Nicotiana glauca*), 132, 258
Coronado, Francisco Vásquez de, 34
Corral, Ramón, 50
Corrals, 112, 113
Cortés, Sea of, 13
Corua, la (*Boa constrictor*), 76
Corvidae, 77, 268
Cottonwood, Mexican (*Populus mexicana* subsp. *dimorpha*), 74, 115, 126, 249–50
*Coursetia glandulosa* (samo), 68, 198–99
Coyotes, 78
Cracidae, 77
*Crataeva palmeri* (capúsari), 76, 173–74
Creosotebush (*Larrea divaricata*), 116
*Crescentia alata* (choca'ari; ayal), 110, 149–50
Cresta de gallo (*Plumbago scandens*), 124, 236
*Crotalus* spp., 77
*Croton: alamosanus* (cuta chicuri; vara prieta), 185, 186; *ciliatoglandulifer* (tatio; trucha), 185; *fantzianus* (cuta tósari; vara blanca), 71, 113, 185–86; *flavescens* (júsairo; vara prieta), 68, 186; *sonorae* (júsairo; vara prieta), 186–87
Cruz, Diego de la, 39
*Ctenosaura hemilopha* (spiny-tailed iguana), 77, 118
Cuajilote (*Pseudobombax palmeri*), 153–54, 155
Cucajaqui de Masiaca, 95
Cuchujaqui, Río, 15, 57, 75, 132, 182, 183
Cuckoo, squirrel (*Piaya cayana*), 77
Cuculidae, 77
*Cucumis melo* (goy minole; chichicoyota; coyote melon), 180
*Cucurbita argyrosperma* subsp. *argyrosperma* var. *palmeri* (tetaraca; chichicoyota; wild gourd [*C. palmeri*]), 180
Cucurbitaceae, 65, 180–82
Cudweed, white (*Pseudognaphalium leucocephalum*), 149
Culebra, Arroyo de la, 137
Cultivars, 95, 138, 149, 195, 232, 235, 254
Culture, plants important to, 81, 100, 125–26, 133, 142, 196, 209, 217, 233

Cumbro (*Celtis* spp.), 71, 124, 125, 175–76
Cuna de niño (*Commelina erecta*), 178
*Cupania dentata* (diente de culebra), 250–51
Cupressaceae, 182–83
Curanderos, curandismo, 115, 118–19, 126, 243, 330n.5
Cuta béjori (*Colubrina triflora*), 241–42
*Cyanocorax beecheii*, 77
*Cydia deshaisiana*, 190
*Cynodon dactylon* (zacate de lana; Bermuda grass), 238
Cyperaceae, 183
*Cyperus* (coquillo; flatsedge): *compressus* (coni saquera), 183; *perennis* (coni saca), 183
Cypress, Mexican bald (*Taxodium distichum*), 74, 75, 110, 114, 132, 182–83

*Dactyloctenium aegyptium*, 132
Dais: *Dalea scandens* var. *occidentalis*, 199; *Desmanthus covillei*, 68, 115, 199
Daisy, rock (*Perityle cordifolia*), 148
*Dalea scandens* var. *occidentalis* (jíchiquia; dais, escoba), 199
Damiana (*Turnera diffusa*), 265
Dams, 5, 8, 9, 15
*Datura* (tebue; toloache): *discolor* (desert thorn apple), 257; *lanosa* (sacred datura), 121, 257
Dayflower (*Commelina erecta*), 178
Deafness, treatment for, 261
Deer, 78, 111; forage for, 134, 179, 187, 235; Sinaloan whitetail (*Odocoileus virginianus* subsp. *sinaloense*), 174
Deer dancers (venados), 6, 23–24; rattles for, 150; songs of, 90, 104, 199, 260
Deer weed (*Porophyllum gracile*), 123, 148
Deportations, 55
*Descurainia pinnata* subsp. *halictorum* (pamita; tansy mustard), 156
Desert broom (*Baccharis sarothroides*), 144
*Desmanthus* spp. (siteporo; dais, escoba), 68, 115, 199
*Desmodium scorpiurus* (subuai muni; frijol de codorniz), 199
Devil's claw (*Proboscidea parviflora* subsp. *parviflora*), 227
Diabetes, 116, 119, 124, 182, 212, 217, 220, 226, 235, 244, 253, 329n.3
Díaz, Porfirio, 50, 57, 328n.48
*Dichanthium annulatum*, 132

*Dicliptera resupinata* (alfalfilla, rama del toro), 130, 133
Diéguez, M. M., 55
Diente de culebra: *Brickellia coulteri*, 145; *Cupania dentata*, 250–51
Digestive ailments, medicines for, 116, 117, 121–22, 133, 141, 143, 156, 165, 166, 177, 186, 187, 190, 199, 201, 206, 207, 208, 214, 221, 223, 224, 228, 244, 246
*Digitaria bicornis* (huilanchi), 238
*Diospyros sonorae* (caguorara; guayparín; Sonoran persimmon), 74, 125, 133, 183–84
*Diphysa*: *occidentalis* (huoquihuo; huiloche), 68, 200; *suberosa* (corcho), 74, 200
Disease, 324nn.17, 18; etiology of, 116, 329n.4, 330n.6; during Spanish era, 39, 40–41, 324n.19
Distrito de Riego del Río Mayo, 60
Domínguez Dam, Josefa Ortiz, 57
Dórame Buitimea, Mariano, 118, 276
*Dorstenia drakena* (na'atoria; baiburilla), 123, 228–29
Doves, 77; white-winged (*Zenaida asiatica*), 87
Dress, of Mayos, 25–26
Drought, 11, 131; and fodder collection, 111–12; and pitahaya fruit production, 89–90
Drug trafficking, 74, 127
*Drymarchon corais* (babatuco; indigo snake), 77, 233
*Dryopetalon runcinatum* (ma'aca; mostaza), 157
Dune and beach communities, 62, 63
Dyes, 115; *Acacia* spp. as, 192, 193; *Ambrosia confertiflora*, 143; *Avicennia germinans* as, 266; *Chlorophora tinctoria*, 228; Fabaceae used in, 201, 202, 206; *Jacquinia macrocarpa* as, 121, 264; *Jatropha cinerea* for, 189; *Krameria erecta* used as, 220; *Vallesia glabra* used as, 100

Ear and eye problems, 121
Ébano (*Caesalpinia sclerocarpa*), 76, 123, 196, 197
Ebenaceae, 183–84
Economy, Mayo participation in, 59–61
Ejidos, 8, 18, 19, 60
Ejotillo: *Sesbania herbacea*, 218; *Senna* spp., 217–18
El Fuerte, 38

## 344 / Index

El Jitón, 105, 106, 110
El Júpare, 19, 48
El Máhone (Presa Miguel Hidalgo), 15, 57
El Paso, 113
El Rincón Viejo, 134
El Sabino, 74
*Eleusine indica,* 132
*Elytraria imbricata* (cordoncillo; scaly-stem), 123, 133, 190
*Encelia: farinosa* (choyoguo; rama blanca, incienso; brittlebush), 145; *halmifolia* (choyoguo), 145
Encinales, 131
Encino (*Quercus* spp.), 73, 218–19
Encino roble (*Quercus tuberculata*), 74
Endoparasites, 116
Epazote (*Chenopodium ambrosioides*), 121, 177
Epidemics, Spanish-era, 39, 40–41
Epiphytes, 11
Equipatas, 11, 131
*Eragrostis cilianensis,* 132
*Erythrina flabelliformis* (jévero; chilicote; coralbean), 12, 113, 115, 200–201
Erythroxylaceae, 72, 184
*Erythroxylon mexicanum* (momo ogua; mamoa), 72, 74, 184
Escoba: *Abutilon incanum,* 224–25; *Bastardiastrum cinctum,* 225; *Dalea scandens* var. *occidentalis,* 199; *Desmanthus bicornutus, D. covillei,* 199
Escorpiones, 77, 117
*Esenbeckia hartmanii* (jójona; palo amarillo), 248
Estafiate: *Ambrosia confertiflora,* 121, 143; *Parthenium hysterophorus,* 147–48
Estrella, Leonardo, 143, 158, 238, 276
Estrella, Patricio, 128, 149, 157, 176, 180, 199, 223, 243, 248, 276
Estrella, Turibio, 149, 276–77
Estuaries, 62–63
Etcho (*Pachycereus pecten-aboriginum*), 64, 66–67, 70; uses of, 95–99, 113, 124, 125, 167
Etchohuaquila, 3, 4, 319n.1
Etchojoa, 22, 323n.14
*Eucnide hypomalaca* (ma'aso arócosi; rock nettle), 222
*Eupatorium quadrangulare* (cocolmeca; tepozana), 145
*Euphorbia,* 124; *abramsiana* (cuépari; golondrina; spurge), 187; *californica* (jímaro), 187; *colletioides* (candelilla), 68, 132, 187; *gentryi* (torote lechosa), 68
Euphorbiaceae, 12, 65, 110, 123, 129; species descriptions for, 184–91
Evil, warding off, 119
Exotics, 132, 232, 235, 237, 238, 258; buffelgrass, 58–59, 69, 240
*Eysenhardtia* (baijguo; palo dulce; kidneywood), 68, 114, 115, 126, 221; *orthocarpa,* 201–2

Fabaceae, 12, 62, 65, 67, 68, 71, 72, 74, 76, 110, 114, 129, 132; species descriptions for, 191–218
Fagaceae, 73, 218–19
Farming. *See* Agriculture
Fatigue, 122
Fats, in medicines, 118, 217, 223
Félix, Juan, 110, 136, 147, 148, 277
Female disorders, 116, 143, 148, 177, 208, 267
Fences, 113; of brasil, *112;* of *Bursera,* 158; Cactaceae, *85, 98,* 99; of *Caesalpinia platyloba,* 195; of *Condalia globosa,* 242; of *Erythrina flabelliformis,* 201; of *Erythroxylon mexicanum,* 184; of Fabaceae, 206, 209, 211, 215, 217; of *Fouquieria,* 219; of *Guaiacum coulteri,* 271; of *Jatropha cordata,* 102; of Nyctaginaceae, 233; of *Schoepfia schreberi,* 234; of *Zanthoxylum fagara,* 249
*Ferocactus herrerae* (ónore; biznaga; barrel cactus), 11, 65, 120, 125, 162–63
Fiber, agave, 92–94, 137
*Ficus,* 110, 125; *cotinifolia* (nacapuli; strangler fig), 229; *insipida* (tchuna; chalate), 71, 229; *pertusa,* 230; *petiolaris* (báisaguo; tescalama; rock fig), 72, 124, 230–31; *trigonata* (higuera, chalate [*F. goldmanii*]), 231
Fiestas, Mayo, 20–25, 227, 237, 321nn.11, 12. *See also* Holy Week ceremonies
Fiesteros, 22–23
Figs (*Ficus*), 110, 125; rock (*petiolaris*), 72, 230–31; strangler (*cotinifolia*), 229
Firewood, 115, 158, 174, 176, 178, 184, 218, 219, 234, 242, 248, 254, 266; Fabaceae as, 191, 192, 193, 194, 197, 202, 204, 205, 206, 207, 211, 216, 217
Fish, stunning, 207, 213
Fishermen, fisheries, 6, 19, 63, 329n.2
Flatsedge (*Cyperus* spp.), 183
Flax (*Linum* sp.), 217
Flor de capomo (*Nymphaea elegans*), 233

Flor de iguana (*Senna pallida*), 71, 218
Flor de mayo (*Erythrina flabelliformis*), 12, 113, 115, 200–201
Flor de piedra (*Selaginella* spp.), 123, 253
Fodder, forage, 111–12, 233; Acanthaceae as, 133, 134, 135; Achatocarpaceae as, 135; Aizoaceae as, 137; Asclepiadaceae as, 141; Asteraceae as, 143, 144, 145, 148; Bromeliaceae as, 157; Cactaceae as, 163, 165, 166, 167; Capparaceae as, 173; Chenopodiaceae as, 176, 177; Convolvulaceae as, 179, 180; Euphorbiaceae as, 187; Fabaceae as, 191, 195, 196, 197, 202, 205, 209, 210, 211, 212, 214, 217; *Hybanthus mexicanus* as, 269; Malvaceae as, 225, 226; *Maytenus phyllanthoides* as, 174; *Phoradendron californicum* as, 270; Poaceae as, 238, 239, 240; Rhamnaceae as, 242, 243, 244; *Rivina humilis* as, 236; Solanaceae as, 258; *Strutanthus palmeri* as, 222; *Talinum paniculatum* as, 241; Tiliaceae as, 264; Verbenaceae as, 269; *Vitex mollis* as, 269
Folk diseases, 122–23
Food, 125; *Agave* spp. as, 90, 92, 135, 136–37; *Amaranthus palmeri* as, 137; *Amoreuxia* spp. as, 103–4, 152; *Antigonon leptopus* as, 241; Bombacaceae as, 153, 154; Brassicaceae as, 156, 157; Bromeliaceae as, 157; Burseraceae as, 158, 160; Cactaceae as, 161, 162, 163, 164–65, 166–67, 168, 169, 170, 171; Celtidaceae as, 176; Chenopodiaceae as, 177; *Crataeva palmeri* as, 173–74; *Cyperus compressus* as, 183; *Diospyros sonorae* as, 184; *Eucnide hypomalaca* as, 222; Fabaceae as, 195, 214, 216, 218; *Forchhammeria watsonii* as, 106; *Guasuma ulmifolia* as, 261–62; *Hyptis suaveolens* as, 221; *Ipomoea bracteata* as, 179–80; *Jarilla chocola* as, 174; *Laguncularia racemosa* as, 178; Malpighiaceae as, 222, 224; *Marsdenia edulis* as, 141; Martyniaceae as, 226, 227; Moraceae as, 228, 229, 230, 231; *Pachycereus pectenaboriginum* as, 96–98, 167; *Passiflora arida* as, 235; *Portulaca* spp. as, 241; Rhamnaceae as, 243, 244; Rubiaceae as, 245, 246, 247; *Salpianthus arenarius* as, 233; *Sideroxylon* spp. as, 252; Solanaceae as, 255–56, 258, 259; *Stenocereus thurberi* as, 86–90; *Tabebuia impetiginosa* as, 152; *Tournefortia hartwegiana* as, 156; Verbenaceae as, 267–68, 269

Foothills thornscrub, 67–70, 131
*Forchhammeria watsonii* (jito; palo jito; lollipop tree), 65, 66, 105–6, 110, 113, 174
*Fouquieria* (murue; jaboncillo), 113, 115; *diguetii* (Baja tree ocotillo), 11–12, 65, 219; *macdougalii* (tree ocotillo), 12
Fouquieriaceae, 11, 219–20
Frangipani (*Plumeria rubra*), 74, 109–10, 138
Freezes, 10, 319–20n.9
French expeditionary force, 49, 326nn.35, 36
Frijol de codorniz (*Desmodium scorpiurus*), 199
Fuerte, Río, 7, 10, 13, 14, 15, 38, 50, 57, 127; dam on, 8, 15, 58; farming on, 12, 320n.10; Mayos on, 17–18, 46
Furniture, 114; Cactaceae used in, 85, 96, 167, 171; *Cedrella odorata*, 227; *Ceiba acuminata* as, 153; *Cordia parvifolia* used in, 81; Fabaceae used in, 191, 193, 201, 204, 214, 216; of *Guazuma ulmifolia*, 261, 262; *Montanoa rosei* used in, 146; Moraceae, 229; *Otatea acuminata* used in, 239; *Populus mexicana* used in, 250; *Taxodium distichum* used as, 183

Galloping cactus (*Stenocereus alamosensis*), 65, 115, 169–71
Gámez, Francisco, 80, 87, 277
Gámez Piña, Serapio, 87, 88, 201, 277
Gándara, Manuel, 48, 49
Garambullo (*Pisonia capitata*), 232–33
Garbanzo de coyote (*Bunchosia sonorensis*), 222
García, Maximiano, 176, 239, 244, 266, 277
Geckos: *Phyllodactylus homolepidurus*, 118; western banded (*Coleonyx variegatus*), 117–18
Geology, 13–15, 32–33
Gila monster (*Heloderma suspectum*), 77, 117
Gilbertson, Robert L., 182–183
Gloria (*Tecoma stans* var. *stans*), 109, 120, 152
*Gnaphalium leucocephalum*. See *Pseudognaphalium leucocephalum*
Goats, 6; forage and fodder for, 112, 134, 135, 141, 143, 144, 145, 148, 157, 163, 173, 174, 176, 177, 187, 195, 205, 209, 210, 212, 216, 218, 226, 233, 236, 239, 240, 241, 242, 243, 258
Gobernadora (*Larrea divaricata*), 116
Golondrina (*Euphorbia abramsiana*), 187
Golondrina del mar (*Palafoxia linearis* var. *linearis*), 147

Goma Sonorae, 121, 198
Gonzalez, María, 97, 260, 277
Goosefoot, fish (*Chenopodium neomexicana*), 177
Gordolobo (*Pseudognaphalium leucocephalum*), 149
*Gouinia virgata* (ba'aso síquiri), 239
Gourd, wild (*Cucurbita argyrosperma* subsp. *argyrosperma* var. *palmeri*), 180
Grama: six-weeks and Mayo (*Bouteloua barbata*), 238; six-weeks needle (*Bouteloua aristidoides*), 237–38
Granadilla (*Malpighia emarginata*), 65, 125, 223–24
Grasses (Poaceae), 130, 132, 236–40; Bermuda (*Cynodon dactylon*), 238; spider (*Aristida ternipes* var. *ternipes*), 236
Graythorn (*Ziziphus obtusifolia*), 120, 243–44
Grazing, 13, 58–59, 69
Grippe, 120–21
Guacaporao (*Parkinsonia aculeata*), 133
Guachapori (*Cenchrus* spp.), 238
*Guaiacum coulteri* (júyaguo; guayacán), 12, 65, 72, 112, 114, 124, 270–71
Guaje (*Leucaena lanceolata*), 206
Guamúchil (*Pithecellobium dulce*), 110, 115, 125, 133, 213–14
Guamuchilillo: *Calliandra emarginata*, 196; *Pithecellobium unguis-cati*, 214
Guarijíos, 16, 79, 138, 157, 174, 176, 193, 198, 218, 233, 243, 269, 271, 320–21n.6, 329n.3
Guasaves, 16
Guásima (*Guazuma ulmifolia*), 71, 72, 114, 261–62
Guasimilla (*Aphananthe monoica*), 175
Guayacán (*Guaiacum coulteri*), 12, 65, 72, 112, 114, 124, 270–71
Guayavilla (*Salpianthus arenarius*), 233
Guayavillo (*Acacia coulteri*), 67, 68, 191–92
Guaymas, 12, 17
Guayparín (*Diospyros sonorae*), 74, 125, 133, 183–84
Guazapares, 16
*Guazuma ulmifolia* (agia; guásima), 71, 72, 114, 261–62
Güerequi (*Ibervillea sonorae*), 65, 120, 124, 181–82
Güirocoba, Arroyo, 73
Güirote de culebra (*Serjania* spp.), 251
Gumbo limbo (*Bursera simaruba*), 161

Guzmán, Diego de, 33
Guzmán, Nuño de, 33

Hacendados, 53, 54, 57, 327–38n.44, 328n.51
Hackberry (*Celtis*), 71, 124, 125; desert (*pallida*), 176; netleaf (*reticulata*), 176; tropical (*iguanaea*), 175–76
*Haematoxylum* (tebcho; brasil chino), 203–4; *brasiletto* (júchajco; brasil), 67, 112, 113, 115, 120, 122–23, 126, 202–3
Hats, palm leaf, 139–40
*Havardia*: *mexicana* (chino; palo chino), 74, 204, 205; *sonorae* (jócona), 65, 112, 204–5
Hawks, 77; Harris (*Parabuteo unicinctus*), 77
Health system, 115–16
Heart problems, 152
Hechiceros (sorcerers), 118, 119, 243, 323n.15
*Hechtia montana* (huítbori; mescalito; false agave), 157
*Helicteres baruensis*, 74, 246
*Heliocarpus attenuatus* (sa'amo; samo baboso), 71, 115, 264
*Heliotropium*: *angiospermum* (laven co'oba; violin head), 156; *curasavicum* (mala mujer; alkali heliotrope), 124, 156
*Heloderma* spp., 77
Helodermatidae, 77
Hemorrhoids, 121, 172, 257
*Henrya insularis* (rama del toro), 130, 134
Hermosillo, 7
Heron, tiger (*Tigrisoma mexicanum*), 77
*Heteropteris palmeri* (compio), 223
Hidalgo Dam, Miguel, 15, 57
Hiedra (*Rhus radicans*), 183
Hierba de la víbora (*Stegnosperma halimifolium*), 236
Hierba de pasmo: *Baccharis sarothroides*, 144; *Hymonoclea monogyra*, 146
Hierba del indio (*Aristolochia watsonii*), 140–41
Hierba del venado (*Porphyllum gracile*), 123, 148
Hierba lanuda (*Tidestromia lanuginosa*), 138
Higuera (*Ficus trigonata*), 231
Higuerilla (*Ricinus communis*), 132, 190
*Hintonia latiflora* (tapichogua; copalquín), 67, 113, 119, 120, 122, 123, 124, 245
Holy Week ceremonies, 20–22, 81, 100, 125–26, 212, 216, 250, 321n.9

Honey, collection and preservation of, 102, 115
Horse-purslane (*Trianthema portulacastrum*), 137
Horses, 240, 324n.16
Houses, 113
Huatabampito, 63
Huatabampo, 6, 9
Huatabampo culture, 30
Huebampo, 118
Huevos de toro (*Stemmadenia tomentosa*), 138
Huichilame (*Ziziphus obtusifolia* var. *canescens*), 120, 243–44
Huichori (*Sarcostemma cynanchoides* subsp. *hartwegii*), 141–42
Huilanchi (*Digitaria bicornis*), 238
Huiloche (*Diphysa occidentalis*), 68, 200
Huisache (*Caesalpinia cacalaco*), 110, 194
Huites, 16, 17, 35
Huites Dam (Presa Luis Donaldo Colosio), 8, 15, 58
Huma, Norma, 20
Hunting, 42, 78, 101, 144, 176
Hurdaide, Diego Martínez de, 35, 36, 37, 38, 322n.9, 323n.11
Hurricanes, 11, 196
*Hybanthus mexicanus* (jépala; jarial), 269
Hydrophyllaceae, 220
*Hymenocallis sonorensis* (guoy cebolla; cebolla de coyote; spider lily), 138
*Hymenoclea monogyra* (jeco; hierba de pasmo, romerillo; cheesebush), 146
*Hyptis: albida* (bíbino; salvia; desert lavender [*H. emoryi*]), 110, 221; *suaveolens* (comba'ari; chani), 122, 125, 221

Ibarra, Domingo, 141, 169, 220, 239, 258, 261, 277
Ibarra, Francisco de, 34, 322n.6
*Ibervillea sonorae* (choya huani; güerequi), 65, 120, 124, 181–82
Igualama (*Vitex mollis*), 74, 125, 268–69
Iguana, spiny-tailed (*Ctenosaura hemilopha*), 77, 118, 218, 234
Incense, 160, 161
Incienso (*Encelia farinosa*), 145
*Indigofera suffruticosa* (chiju; añil; indigo), 115, 133, 205–6
Industrial Revolution, 33
Infectious disease, 123
Insects, 66, 180; bites and stings from, 116, 117, 138, 170, 178, 192; lac scale (*Tachardiella coursetiae*), 199
Instituto Nacional Indigenista, 56
Insurrections, 45, 48, 51
*Ipomoea*, 123; *arborescens* (jútuguo; palo santo; tree morning glory), 68, 71, 72, 112, 178–79; *bracteata* (tosa huíra; jícama), 179–80; *pedicellaris* (jícure; trompillo; morning glory), 180
Ironwood: *Chloroleucon mangense*, 67–68, 197; *Olneya tesota*, 62, 111, 210
Irrigation, 7–8, 50, 57
Ivy, poison (*Rhus radicans*), 183
Ixtle, 87; manufacture of, 92–94; weaving with, 94–95
Izábal, Rafael, 55

Jaboncillo (*Fouquieria diguetii*), 65, 113, 115
*Jacquinia macrocarpa* subsp. *pungens* (tásiro; sanjuanico), 65, 72, 120, 121, 123, 263–64
Jaguar's-claws (*Zanthoxylum mazatlanum*), 249
Jambiolobampo, 29
Jarial (*Hybanthus mexicanus*), 269
Jarilla chocola (chócola), 125, 174
*Jatropha*, 221; *cardiophylla* (sa'apo; sangrengrado azul; limber bush), 188; *cinerea* (sa'apo; sangrengado), 65, 68, 72, 115, 124, 188–89; *cordata* (sato'oro; torote panalero, papelío), 12, 65, 100–103, 115, 123, 189; *malacophylla* (sa'apo; sangrengado), 189
Javelina (*Tayassu tajacu*), 103
Jay: black-throated magpie (*Calocitta formosa collei*), 268; purplish-backed (*Cyanocorax beecheii*), 77
Jeco: *Hymenoclea monogyra*, 146; *Xylothamia diffusa*, 149
Jejeri (*Pereskiopsis porteri*), 11, 167
Jesuits, 38, 41, 46, 325n.23; accounts by, 17, 31, 32, 39–40; agriculture and, 43–44; missions, 41–42, 323n.15; and silver miners, 42, 43; and 1740 rebellion, 44–45
Jícama (*Ipomoea bracteata*), 179–80
Jícura (*Ipomoea* sp.), 180
Jímaro (*Euphorbia californica*), 187
Jócona (*Havardia sonorae*), 65, 112, 204–5
Jojoba (*Simmondsia chinensis*), 254
Jopopaco, 213
Joso (*Albizia sinaloensis*), 74, 115, 193
Joso de la sierra (*Conzattia multiflora*), 71, 198

Juárez, Benito, 49, 50, 326n.35
Júchica (*Sideroxylon* spp.), 76, 251–52
Jumping bean, Mexican (*Sebastiania pavoniana*), 110, 190
Jupachumi (*Petiveria alliacea*), 121
Jusacamea, Juan Ignacio, 48
Júsairo (*Croton* sp.), 68, 121, 186–87
Jusa'iro (*Cordia curassavica*), 154–55
*Justicia: californica* (semalucu; chuparosa), 134; *candicans* (ma'aso o'ota), 134
Jutuqui: *Opuntia* sp., 124, 167; *Ziziphus obtusifolia*, 120
Juvavena (*Capparis atamisquea*), 65, 124, 173
Juya jotoro (*Bernardia viridis*), 184

*Kallstroemia* spp. (jímuri; baiburín; summer poppy), 271
Kapok (*Ceiba acuminata*), 71, 153, 154
Karwinskia humboldtiana (aroyoguo; cacachila, tullidora), 67, 121, 122, 123, 242–43
Kidney ailments, 116, 117, 123, 140, 166, 189, 221, 232, 238, 244, 253, 267
Kidneywood (*Eysenhardtia orthocarpa*), 68, 201–2
Krameria: *erecta* (tajimsi; tajuí; range ratany [*K. parvifolia*]), 175, 201, 220; *sonorae* (cósahui; tajuí; Sonoran ratany), 65, 123, 124, 220–21
Krameriaceae, 65, 220–21

Labor, Mayos as, 42–43, 55, 57, 60–61
*Lagacea decipiens* (confiturilla amarilla), 130
*Laguncularia racemosa* (pacio tosa; mangle; white mangrove), 62, 130, 178
Lamiaceae, 110, 122, 221
Land tenure, 19, 325n.32; capitalist, 56, 57; commercial agriculture and, 57–58; under Mexico, 47–49, 53, 54, 56, 60, 326–27n.38, 327–28n.44, 328n.51; during Spanish colonial era, 44–45
*Lantana: achyranthifolia* (confiturilla blanca), 130; *camara* (ta'ampisa; confiturilla negra), 130, 266–67; *hispida* (confiturilla blanca), 130, 267
*Larrea divaricata* (gobernadora; creosotebush), 116
Las Bocas, 6, 19, 63, 93
*Lasiacis ruscifolia* (baca cupojume jicto; carrizito de huitlacoche, negrito), 239
Lavender, desert (*Hyptis albida*), 110, 221
Leadwort (*Plumbago scandens*), 124, 236

Leatherworking, 77; tanning, 115, 193, 204, 207, 214
Lechuguilla (*Agave aktites*), 135
Legumes, 12, 65. *See also* Fabaceae
Lenten ceremonies, 125–26
*Leptonycterus curasoae* (lesser long-nosed bat), 66
*Leptophis diplotropis* (parrot snake), 233
*Leucaena lanceolata* (huique; guaje, palo bofo), 206
Ley Almada, 47
Leyva, Francisco "Chico," 236, 277
Leyva, José María (Cajeme), 50
Lianas. *See* Vines
Lice, killing, 180, 254
Lily: spider (*Hymenocallis sonorensis*), 138; water (*Nymphaea elegans*), 233
Limber bush (*Jatropha cardiophylla*), 188
Limoncillo (*Zanthoxylum fagara*), 72, 249
*Linum* sp. (flax), 217
*Lippia: alba* (valeriana), 267; *graveolens* (bura mariola; orégano de burro), 120, 267; *palmeri* (orégano), 125, 267–68
Liso Jusaino, Genaro, 209, 277
Livestock, 13, 19, 43, 219, 233; forage and fodder for, 111–13, 133, 134, 135, 137, 141, 143, 144, 145, 148, 157, 163, 165, 166, 167, 173, 174, 176, 177, 179, 180, 187, 191, 194–95, 196, 197, 202, 205, 209, 210, 211, 212, 214, 216, 217, 218, 222, 225, 226, 233, 236, 238, 239, 240, 241, 243, 244, 258, 264, 269, 270
Lizards, 118; Clark's spiny (*Sceloporus clarkii*), 234; Mexican beaded (*Heloderma horridum*), 77, 117. *See also* Geckos
Loasaceae, 222
Lollipop tree (*Forchhammeria watsonii*), 105–6, 110, 174
*Lonchocarpus hermannii* (nesco; Venus tree), 12, 71, 72, 74, 124, 206–7
London rocket (*Sisymbrium irio*), 132
López, Fausto, 98, 196, 245, 277
López, José, 245, 277
López, Leobardo, 173, 244, 277
López Flores, Braulio, 164, 166, 180, 190, 223, 277
López García, Balbaneda "Nelo," 7, 149, 244, 266, 277
*Lophocereus schottii* var. *tenuis* (musue; sinita), 65, 120, 163–64
Loranthaceae, 222

Los Capomos, 13, 17, 19, *71*, 123, 153, 158, 320n.12, 321n.9
Luis of Sáric, 45
*Lycium* (wolfberry): *andersonii* (sigropo), 65, 125, 257–58; aff. *berlandieri* (jejiri), 258; *brevipes* (bahue mo'oco; chamisa del mar), 258; *fremontii* (to'oro bichu), 258
*Lysiloma: divaricatum* (mayo; mauto), 12, 67, *70*, *72*, 79, 112–13, 114, 115, 207–8, 322–23n.10; *watsonii* (macha'aguo; tepeguaje), 12, 71, 72, 113, 123, 209–10

*Machaeranthera tagetina* (mo'oso), 124, 146
Macoyahui (village), 6
Macoyahuis, 17
Maguey (*Agave vivipara*), *91;* uses of, 90, 92–95, 125
Mala mujer (*Heliotropium curassavicum*), 124
Mallow, Indian (*Abutilon* spp.), 224–25
*Malpighia* (sire): *emarginata* (granadilla, morita), 65, 125, 223–24; *glabra*, 224
Malpighiaceae, 65, 124, 222–24
Malva (Malvaceae), 130, 224, 225, 226
*Malva parviflora* (malva), 132
Malvaceae, 115, 130, 132, 224–26
*Malvastrum bicuspidatum* (to'oro jíchiquia; malva), 226
*Mammillaria* spp. (chicul ónore; biznaguita; fishhook pincushion), 11, 65, 164
Mamoa (*Erythroxylon mexicanum*), 72, 74, 184
Manglares, 131
Mangle dulce (*Maytenus phyllanthoides*), 174–75
Mangroves (mangle), 62, *63*, 130; black (negro; *Avicennia germinans*), 62, 130, 266; button (*Conocarpus erecta*), 130, 177–78; red (rojo; *Rhizophora mangle*), 62, 130, 244; white (*Laguncularia racemosa*), 62, 130, 178
Manteca, 118
Marijuana growers, 74
Mariola: *Aloysia sonorensis,* 120, 265–66; *Lippia graveolens,* 120, 267
*Marsdenia edulis* (mabe; tonchi), 141, *142*
*Martynia annua* (tan cócochi; aguaro), 125, 226
Martyniaceae, 125, 226–27
*Mascagnia macroptera*. See *Callaeum macropterum*

Masiaca, *6,* 9, 19, *21, 23,* 27, *28, 37,* 60, 62, 86, 207; land clearing around, 84–85; Mayos in, 17, 18. *See also* Mesa Masiaca
Masks, *80,* 158, 160, 201, 204, 250
*Masticophis flagellum* subsp. *cingulum* (Sonoran red racer), 233
Mataneni (*Callaeum macropterum*), 124, 223
*Matelea altatensis* (jo'osi pipi; teta del diablo; milkweed vine), 141
*Matricaria chamomilla* (chamomile), 269
Mats, 237, 239, 265
Matus, Francisco, *170*
Mauto, mayo (*Lysiloma divaricatum*), 12, 67, *70, 72,* 79, 322–23n.10; uses of, 112–13, 114, 115, 207–8
*Maximowiczia sonorae*. See *Ibervillea sonorae*
Mayo, Río, 3, 10, *14,* 15, 17, 50, 127, 207, 325n.24, 326n.37, 328n.46; agriculture on, 43–44, 53, 327n.42; delta of, 4, 5–6, 13, 43, 46, 57–58; farming on, 12–13; Mayos on, 17, 18, 46; Mexican settlers on, 48–49, 328n.45
Mayo language, 19–20, 26, 27, 320n.5
Mayos (Yoremem), 16, 17, 51, 320nn.5, 6, 322–23n.10; Catholicism and, 39, 55, 321n.11, 323nn.12, 13, 324n.20; dress of, 25–26; economic participation of, 59–60; epidemic diseases and, 40–41; fiestas of, 20–25; land tenure and, 47, 48, 49, 326–27n.38; language retention among, 19–20, 320–21n.6, 321n.8; linguistic differences among, 17–18; material culture of, 32, 33; Mexican government and, 47–48, 52–53, 55; and Mexican Revolution, 53–54; mission experience of, 41–42; names, 27, 29; rebellions by, 44–45, 48, 49–51, 52, 53, 56; repression of, 45–46; resistance by, 51–52; and silver mines, 42–43; in Sinaloa, 56–57; social organization of, 31–32, 52–53; Spanish and, 33–35, 38–39; subjugation of, 4–5, 35–36, 326n.37; and Yaquis, 34–35, 36–37
*Maytenus phyllanthoides* (pasio; mangle, mangle dulce), 63, 121, 130, 174–75
*Medicago sativa* (alfalfa), 132
Medicine, 115–24, 236; Acanthaceae use, 133, 134, 135; *Agave vivipara* as, 92, 137; *Agonandra racemosa* as, 235; *Aloe barbadensis* as, 142; *Alvaradoa amorphoides* as, 254; Apocynaceae as, 138; *Argemone ochroleuca* as, 235; *Aristolochia watsonii* as, 141; Asclepiadaceae as, 141–42; Asteraceae

350 / Index

Medicine *(continued)*
as, 143, 144, 145, 146, 147, 148, 149; Bignoniaceae as, 150; Bombacaceae as, 153, 154; *Bonplandia geminiflora* as, 240–41; Boraginaceae as, 156; Burseraceae as, 158, 159, 160, 161; Cactaceae as, 96, 163–64, 165, 166, 167, 170, 172; Capparaceae as, 173; Celtidaceae as, 176; Chenopodiaceae as, 177; *Cochlospermum vitifolium* as, 153; Convolvulaceae as, 178, 180; *Cordia parvifolia* as, 81; Cucurbitaceae as, 180, 182; *Descurainia pinnata* subsp. *halictorum* as, 156; Euphorbiaceae as, 185, 186, 187, 188–89, 190–91; Fabaceae as, 191, 192, 193, 194, 195–96, 198, 201, 202, 204, 205, 207, 208, 209, 210, 211, 212, 213, 214, 216, 217; *Fouquieria* spp. as, 219; *Guaiacum coulteri* as, 270–71; *Hymenocallis sonorensis* as, 138; *Jacquinia macrocarpa* as, 264; *Jatropha cordata* as, 102–3; Krameriaceae as, 220, 221; Lamiaceae as, 221; Malpighiaceae as, 223; Malvaceae as, 224; Martyniaceae as, 226; *Maytenus phyllanthoides* as, 174–75; Moraceae as, 228–29, 231; *Nama hispidum* as, 220; Nyctaginaceae as, 232, 233; *Petiveria alliacea* as, 235–36; *Phoradendron californicum* as, 270; Poaceae as, 239; *Populus mexicana* as, 250; *Pouzolzia guatemalana* as, 265; *Psidium sartorianum* as, 232; Rhamnaceae as, 243, 244; *Rhizophora mangle* as, 244; Rubiaceae as, 244, 245, 246; Rutaceae as, 248, 249; *Schoepfia schreveana* as, 234; *Serjania* as, 251; *Sideroxylon tepicense* as, 252; Solonaceae as, 257, 258, 259, 260–61; *Stenocereus thurberi* used for, 85–86; Sterculiaceae as, 261, 262–64; *Struthanthus palmeri* as, 222; *Tecoma stans* as, 109, 152; *Tidestromia lanuginosa* as, 138; *Turnera diffusa* as, 265; *Vallesia glabra* as, 99–100; Verbenaceae as, 266, 267, 269
Melchor Ocampo, 85
*Meleagris gallopavis* (guíjolos; turkeys), 219
Meliaceae, 227–28
*Melilotus indica* (trebolín; sour clover), 132
*Melochia* (té de lila), 262–63
Melons, 13; coyote (*Cucumis melo*), 180
Méndez, Pedro, 39, 323nn.12, 15
Mendibles Cota, Wencelao "Mono," 182, 277–78
Mendoza, Buenaventura, 86, 212, 264, 278
Mesa Masiaca, 9, 68, 70, 161, 162

Mescalito: *Hechtia montana*, 157; *Tillandsia* spp., 11, 157
Mexican Revolution, 5, 53–54, 327n.43
Mexican tea (*Chenopodium ambrosioides*), 121, 177
Mexico: independence of, 46–47; and Mayo and Yaqui lands, 4–5, 47–48, 57–58; and Mayo social organization, 52–53; resistance against, 48–52; rural development in, 58–59
Mezcal (*Agave* spp.), 91; uses of, 90, 92–95, 125, 136–37
Mezquital de Buiyacusi, 237
Mezquite (*Prosopis* spp.), 65, 72, 110, 114, 115; uses of, 111, 112, 113, 215
Mezquite hormiguillo (*Condalia globosa*), 242
*Micruroides euryxanthus* (coralillo; coral snake), 117
*Micrurus distans* (coralillo; coral snake), 117
Milkweeds (Asclepiadaceae), 141–42
Milpas, 12–13
Mimbro (*Cephalanthus occidentalis* var. *salicifolius*), 244
*Mimosa: asperata* (yete ogua; rama dormilona), 133, 209; *distachya* (nésuquera; uña de gato; catclaw [*M. laxiflora*]), 65, 209; *palmeri* (cho'opo; chopo), 74, 209
*Mimus polyglottos* (cenzontle; mockingbird), 135
Minas Nuevas, 9
Mining, 151, 207, 322n.6, 324–25n.21, 325n.23
Miona (*Commicarpus scandens*), 123, 232
Missionaries, missions, 323nn.12, 15; agricultural production of, 43–44; attitudes of, 39–40; disease and, 40–41; Franciscan, 39–40; Jesuit, 17, 31, 32, 38, 41–42
Mistletoe (*Struthanthus palmeri*), 222; desert (*Phoradendron californicum*), 269–70
Mochicahui, 36
Mochis, 123, 137, 138
Mockingbird (*Mimus polyglottos*), 135
Mocorito, Río, 17
Mocúzari, 6, 320n.1
Mocúzari Dam, 5, 9, 15, 57, 320n.1
Mojino, 12
Momotidae, 77
*Momotus mexicanus*, 77
*Monanthochloë littoralis* (bahue huitcha), 239
*Montanoa rosei* (batayaqui), 71, 113, 146–47
Monte, 131

Monte mojino, 131
Monte verde, 131
Montesclaros, 38
Mo'oso: *Bebbia juncea*, 124, 144; *Machaeranthera tagetina*), 124, 146
Mora (*Chlorophora tinctoria*), 71, 113, 228
Moraceae, 123, 228–32
Morales, Jesús María, 50
Morales, Ramón, 166, 238, 265, 278
Mordant, 115
Morita (*Malpighia emarginata*), 65, 125, 223–24
Morning glory (*Ipomoea*), 123, 180; tree (*arborescens*), 68, 178–79
Moroyoqui, Aurora, *114*, 278
Moroyoqui, Gregoria, 87, 93, 278
Moroyoqui, Jesús, 185
Moroyoqui, Miguel, 236, 278
Moroyoqui, Teresa, 221, 264
Moroyoqui family, 27
Moroyoqui Mendoza, María Soledad, 96, 227, 246, 278
Moroyoqui Zazueta, María Teresa, 98, 206, 259
Morrales, weaving of, 93–95
Mostaza (*Dryopetalon runcinatum*), 157
Motmot, russet-crowned (*Momotus mexicanus*), 77
Mountain Pima, 139
Mules, forage for, 143, 144, 240
Muni, 45
Muñoz Valdés, Jesús, 278
Musculoskeletal problems, 124, 158, 163–64, 212, 217. *See also* Rheumatism
Musical instruments, 100, *150*, 153, 159, 179, 204, 233, 237, 239
Musicapidae, 77
Mustard, tansy (*Descurainia pinnata* subsp. *halictorum*), 156
*Myadestes occidentalis* (brown-backed solitaire), 77
Myrtaceae, 231–32

Nacapuli (*Ficus* spp.), 110, 229
Nahuibampo, 6, 13, 19, 29, 127, 158
*Nama hispidum* (goy tabaco; tabaco de coyote; sand bells), 220
Names, naming, 27, 29
*Nasua narica* (cholugos; coatimundis), 118, 268
Navarete Valencia, Filemón, *82*, 278; songs sung by, 81, 90, 99, 103, 104, 260

Navojoa, 5, 50, 323n.14, 327n.43
Negrito: *Lasiacis ruscifolia*, 239; *Vitex pyramidata*, 74, 269
Nentvig, Juan, 39–40
*Neotoma* spp. (torim; packrats), 118
Nesco (*Lonchocarpus hermannii*), 12, 71, *72*, 74, 124, 206–7
Nettle, rock (*Eucnide hypomalaca*), 222
*Nicotiana: glauca* (tree tobacco), 132, 258; *obtusifolia* (tabaco de coyote; wild tobacco), 259
Nieblas, Cornelia, 96, 100, 278
Nieblas Moroyoqui, Jesús José, 218, 278
Nieblas Zazueta, Balbina, *225*, 278
Nieblas Zazueta, Julieta, 189, 278
Nightshade (*Solanum* spp.), 259–60
Nómom (*Peniocereus striatus*), 11
Nopal (*Opuntia* spp.), 65, 165, 166–67
Noseburn (*Tragia* sp.), 123; *T. jonesii*, 190–91
Nyctaginaceae, 123, 232–33
Nymphaceae, 233
*Nymphaea elegans* (capo segua; flor de campo; water lily), 233

Oak woodland, 73–74, 131
Oaks (*Quercus* spp.), 73, 218–19
Oaxaca, 55
Obregón, Alvaro, 53–54, 56, 328n.45
Ocoronis, 48
Ocotillo (*Fouquieria*): Baja tree (*diguetii*), 11–12, 65, 219; tree (*macdougalii*), 12, 219–20
*Odocoileus virginianus* subsp. *sinaloense* (ma'aso; Sinaloan whitetail deer), 175
Olacaceae, 76, 234
Oligarchies, and land control, 49, 327–28n.44
*Olneya tesota* (ejéa; palo fierro; ironwood), 62, 65, 111, 210
*Oncidium cebolleta* (orchid), 268
Ono jújugo: *Adelia virgata*, 65; *Colubrina viridis*, 242
Opatas, 48, 207, 325n.31
Opiliaceae, 234–35
*Opuntia*, 123–24, 125; *fulgida* (choya; chain-fruit cholla), 65, 164–65; *gosseliniana* (navo; nopal; Sonoran purple prickly pear), 165; *leptocaulis* var. *brittonii* (jijí'ica; choya; desert Christmas cactus), 11, 165; *leptocaulis* × *thurberi* (hybrid choya), 165–66; *pubescens* (to'otori huita; choya), 166; *thurberi* (sibiri; choya), 65, 124, 166; *wilcoxii* (navo; nopal; prickly pear), 65, 166–67

Orchids (*Oncidium cebolleta*), 268
Orégano (*Lippia palmeri*), 125, 267–68
Orégano de burro (*Lippia graveolens*), 267
Oreja de ratón (*Chiococca alba*), 244
Organpipe cactus (*Stenocereus thurberi*), 64, 82–90, 131, 172–73
Ornamental plants, 109–10
*Ortalis wagleri* (chacalacas), 77
Ortiguilla (*Urera baccifera*), 265
Ortija (*Tragia* sp.), 123
*Otatea acuminata* (ba'aca; otate; bamboo), 113, 239
Overgrazing, 3, 6
Owen, Albert, 57, 328nn.51, 52
Owls, 77, 106, 118
*Oxybelus aeneus* (brown vine snake), 233

*Pachycereus pecten-aboriginum* (etcho), 64, 66–67, 70; uses of, 95–99, 113, 124, 125, 167
Packrats (*Neotoma* spp.), 118
*Palafoxia linearis* var. *linearis* (bahue cuépari; golondrina del mar; Spanish needles), 147
Palm, fan (Arecaceae), 113, 139–40
Palma (*Brahea aculeata*), 113, 139
Palma real (*Sabal uresana*), 139–40
Palo amarillo (*Esenbeckia hartmanii*), 248
Palo barril, palo barriga (*Cochlospermum vitifolium*), 71, 152–53
Palo blanco: *Carlowrightia arizonica*, 133; *Piscidia mollis*, 68, 113, 213
Palo bofo (*Leucaena lanceolata*), 206
Palo cachora (*Schoepfia* spp.), 76, 234
Palo chino (*Havardia mexicana*), 74, 114, 115, 204, 205
Palo colorado (*Caesalpinia platyloba*), 67, 113, 195
Palo de asta (*Cordia sonorae*), 67, 114, 155–56
Palo dulce (*Eysenhardtia orthocarpa*), 68, 114, 115, 126, 201–2
Palo fierro: *Chloroleucon mangense*, 67–68, 197; *Olneya tesota*, 62, 65, 111, 210
Palo huaco (*Agonandra racemosa*), 234–35
Palo jito (*Forchhammeria watsonii*), 65, 66, 105–6, 110, 113, 174
Palo joso (*Acacia sinaloensis*), 74, 115, 193
Palo limón: *Amyris balsamifer*, 76, 110, 248; *Zanthoxylum* sp., 249
Palo mulato (*Bursera grandifolia*), 71, 123, 125, 158, 159
Palo pinto (*Chloroleucon mangense*), 67–68, 197

Palo piojo: *Brongniartia alamosana*, 12, 194; *Caesalpinia palmeri*, 65, 194–95
Palo piojo blanco (*Caesalpinia caladenia*), 68, 194
Palo pitillo (*Fouquieria* spp.), 11–12, 65, 219–20
Palo santo (*Ipomoea arborescens*), 68, 71, 72, 112, 178–79
Palo torsal (*Alvaradoa amorphoides*), 253–54
Palo verde: *Agonandra racemosa*, 234–35; blue (*Parkinsonia florida*), 65, 211; Mexican (*P. aculeata*), 133, 211
Palo zorillo (*Senna atomaria*), 71, 111, 112, 217
Paloma pitahayera (*Zenaida asiatica*), 87
Palomares, Artemisia, 139, 140, 278
Pamita (*Descurainia pinnata* subsp. *halictorum*), 156
Pamitón (*Sisymbrium irio*), 132
Papache (*Randia echinocarpa*), 72, 121, 125, 245–46
Papache borracho (*Randia* spp.), 68, 246–47
Papaveraceae, 235
Papelío (*Jatropha cordata*), 65; uses of, 100–103, 115
*Parabuteo unicinctus* (Harris' hawk), 77
Parasites, treatment of, 121–22, 143, 177
Parasitism, 119
*Parkinsonia*: *aculeata* (guacaporo; retama; Mexican palo verde), 133, 211; *florida* (caaro; palo verde; blue paloverde [*Cercidium floridum*]), 65, 211; *praecox* (choy; brea [*C. praecox*]), 65, 111, 112, 115, 126, 211–13, 212
Parrots: lilac-crowned (*Amazona finschi*), 77; white-fronted (*A. albifrons*), 77, 243
*Parthenium hysterophorus* (chíchibo; estafiate), 147–48
Pascola dancers, 6, 24–25, 80
*Passiflora arida* (ma'aso alócossim; talayote; passion flower), 235
Passifloraceae, 235
Pastures, 58–59, 69, 99
*Pectis coulteri* (goy sisi; chinchweed), 148
Pedaliaceae, 235
*Pedilanthus macrocarpus* (cantela oguo; candelilla), 65, 189–90
*Peniocereus*: *marianus* (nómom; sina), 167; *striatus* (nómom; sina), 11, 115, 167, 168
*Pennisetum ciliare* (zacate buffel; buffelgrass [*Cenchrus ciliare*]), 58–59, 69, 240
Peonia (*Erythrina flabelliformis*), 12, 113, 115, 200–201

Pepper, red bird (*Capsicum annuum*), 71, 125, 254–57
*Pereskiopsis porteri* (jejeri), 11, 167
*Perityle cordifolia* (oseu nácata; rock daisy), 148
Persimmon, Sonoran (*Diospyros sonorae*), 74, 183–84
Pesqueira, Ignacio, 48–50, 325n.37
*Petiveria alliacea* (jupachumi; cola de zorillo), 121, 235–36
Pfefferkorn, Ignaz, 39, 198, 255–56, 324n.16, 330n.6
*Phainopepla nitens* (phainopepla), 270
*Phaulothamnus spinescens* (ba'aco), 65, 135
*Phoradendron californicum* (pohótela; toji; desert mistletoe), 269–70
*Phragmites: australis*, 237, 239; *communis*, 132
*Phyllodactylus homolepidurus*, 118
*Physalis maxima* (tombrisi; tomatillo), 259
Phytolaccaceae, 235–36
*Piaya cayana* (squirrel cuckoo), 77
Pichicuate (*Agkistrodon bilineatus*), 117
Pigeon berry (*Rivina humilis*), 236
Pigs, 219, 241
Pigweed (*Amaranthus palmeri*), 125, 137–38
Piloncillo molds, 208, 215
*Pilosocereus alensis* (aaqui jímsera; pitahaya barbona), 168
Pimas, 45, 325n.29
Piña, Florentino "Queriño," 163, 278
Pinales (pine forests), 131
Pincushion cactus (*Mammillaria* spp.), 11, 65, 122, 164
Pintapán (*Abutilon abutiloides*), 224
Piocha (*Trichilia* spp.), 228
Pipevine (*Aristolochia watsonii*), 140–41
*Piscidia mollis* (joopo; palo blanco), 68, 113, 213
*Pisonia capitata* (baijuo; garambullo, vainoro), 232–33
Pitahaya (*Stenocereus thurberi*, *S. montanus*), 31, 64, 66, 69, 70, 71, 108, 131, 171–73; collecting, 87–88; and land clearing, 84–85; natural history of, 82–84; uses of, 85–87, 88–90, 113, 121, 125
Pitahaya barbona (*Pilosocereus alensis*), 168, 329n.1
Pitahayales, 60, 64, 66, 67, 131; clearing of, 84–85
*Pithecellobium: dulce* (maco otchini; guamúchil), 110, 115, 125, 133, 213–14; *undulatum*, 197; *unguis-cati* (jodimaco'otchini; guamuchilillo), 214. See also *Havardia*
Plan de Ayala, 54
Planting sticks, 184, 266
*Platymiscium trifoliolatum* (tampicerán), 214–15
Plumbaginaceae, 236
*Plumbago scandens* (toto'ora; cresta de gallo; leadwort), 124, 236
*Plumeria rubra* (cascalosúchil; frangipani), 74, 109–10, 138
Poaceae, 130, 132, 236–40
Pochote (*Ceiba acuminata*), 70, 71, 125, 153, 154
Poison, fish, 149. See also Toxins
Polemoniaceae, 240–41
Polygonaceae, 125, 241
Poppy: prickly (*Argemone ochroleuca* subsp. *ochroleuca*), 235; summer (*Kallstroemia* spp.), 271
*Populus mexicana* subsp. *dimorpha* (aba'aso; álamo; Mexican cottonwood), 74, 115, 126, 249–50
Porfiriato, 55, 327n.42, 328n.48
*Porophyllum gracile* (cuchu pusi; hierba de venado; deer weed), 123, 148
*Portulaca* spp. (huaro; verdolaga; purslane), 241
Portulacaceae, 241
Pottery making, 238
*Pouzolzia: guatemalana* var. *nivea* (chi'ini), 265; *occidentalis* [*P. palmeri*], 265
Prickly pear, 102; *Opuntia wilcoxii*, 65, 125, 166–67; Sonoran purple (*O. gosseliniana*), 165
Prieto, Cerro, 69
*Proboscidea* (aguaro): *altheaefolia* (desert unicorn plant), 125, 226–27; *parviflora* subsp. *parviflora* (devil's claw), 227
Promontorios, 42
Prophets, Mayo, 51–52, 53, 56
*Prosopis* (juupa; mezquite): *articulata* (sanéa), 215; *glandulosa* (juupa; mezquite), 65, 72, 110, 112, 113, 114, 115, 215–16
Prostatitis, 96, 124, 167, 191, 226, 235
Protestantism, evangelical, 56, 328n.49
*Pseudabutilon scabrum*, 130
*Pseudobombax palmeri* (cuajilote), 153–54, 155
*Pseudognaphalium leucocephalum* (talcampacate; gordolobo; white cudweed), 149

*Psidium sartorianum* (arrellán), 231–32
Psittacidae, 77
Ptilogonatidae, 270
Puebla, 54
Purslane (*Portulaca* spp.), 241

Quail, elegant (*Callipepla douglasii*), 77, 199
Queen's wreath (*Antigonon leptopus*), 125, 241
Quelite: *Amaranthus palmeri*, 137–38; *Chenopodium neomexicanum*, 177
*Quercus* (encino): *albocincta* (cusis), 73, 218; *chihuahensis*, 73, 218–19; *tuberculata* (encino roble), 74
Quijano, Damián, 51, 52
Quinta Chilla, 53

Rabbits, 78
Racer, Sonoran red (*Masticophis flagellum* subsp. *cingulum*), 233
Ragweed, canyon (*Ambrosia ambrosioides*), 119, 143
Rainfall, 7, 9; and milpas, 12–13; and seasons, 10–11
Rama amarga (*Brickellia brandegei*), 144
Rama blanca (*Encelia farinosa*), 145
Rama del toro (Acanthaceae), 130, 133, 134, 135
Rama dormilona (*Mimosa asperata*), 133, 209
Rama quemadora (*Tragia jonesii*), 190–91
Ramos, Alejandro, 233
*Randia*: *echinocarpa* (jósoina; papache), 72, 121, 125, 245–46; *laevigata* (sapuchi), 74, 246; *obcordata* (papache borracho), 246–47; *thurberi* (pisi; papache borracho), 68, 246–47
Ratany: range (*Krameria erecta*), 175, 201, 220; Sonoran (*Krameria sonorae*), 65, 124, 220–21
*Rathbunia alamosensis*. See *Stenocereus alamosensis*
Rattles, 150, 204
Rattlesnakes, 77, 117, 118
Rebellions: against Mexico, 48, 49–51, 52, 53; against Spanish, 44–45, 325nn.28, 29
Repression, 45–46, 323n.14
Reproductive problems, 123, 143, 148, 180
Reptiles, 117. See also by type
Resistance, Mayo, 51–52
Resurrection plant (*Selaginella* spp.), 123, 253

Retama (*Parkinsonia aculeata*), 133, 211
Rhamnaceae, 65, 67, 120, 121, 241–44
Rheumatism, treatment for, 143, 145, 156, 170, 182, 259
*Rhinocheilus lecontei* subsp. *antoni* (coralillo; long-nosed snake), 77
*Rhizophora mangle* (cánari; mangle rojo; red mangrove), 62, 130, 244
Rhizophoraceae, 62, 244
*Rhus radicans* (hiedra; poison ivy), 183
*Rhynchosia precatoria* (chanate pusi; rosary bean), 121, 216–17
*Ricinus communis* (higuerilla; castor bean), 132, 190
*Rivina humilis* (bacot mútica; coralillo cimarrón; pigeon berry), 236
Romerillo: *Baccharis sarothroides*, 144; *Hymenoclea monogyra*, 146
Ropes, vines as, 177, 180
Rosary bean (*Rhynchosia precatoria*), 121, 216–17
Rubiaceae, 67, 68, 72, 119, 244–47
*Ruellia nudiflora* (papachili; rama del toro), 134
Rural development, 58–59
Rutaceae, 72, 76, 110, 112, 248–49

*Sabal uresana* (ta'aco; palma real; fan palm), 139–40
Sábila (*Aloe barbadensis*), 142
Sabino (*Taxodium distichum*), 74, 75, 110, 114, 132, 182–83
Sacamanteca (*Solanum* spp.), 122, 260–61
Saddles, 211
Sahuaro (saguo; saguaro; *Carnegiea gigantea*), 62, 161–62
Saitilla (*Bouteloua aristidoides*), 237–38
Saituna (*Ziziphus amole*), 12, 65, 66, 125, 157, 243
Salamanquesa, 117–18
Salicaceae, 74, 249–50
*Salix* spp. (huata; sauce; willow), 74, 250
*Salpianthus arenarius* (cuey oguo; guayavilla [*S. macrodontus*]), 233
Saltbush (*Atriplex barclayana*), 177
Salt-tolerant plants, 63
Salvia (*Hyptis albida*), 110, 221
Samo (*Coursetia glandulosa*), 68, 198–99
Samo baboso (*Heliocarpus attenuatus*), 71, 115, 264
San Bernardo, 9

San Miguel, 17
Sand bells (*Nama hispidum*), 220
Sand mining, 76
Sandbur (*Cenchrus* spp.), 238
Sangrengado (*Jatropha* spp.), 65, 72, 122, 188–89, 221
Sangrengado azul (*Jatropha cardiophylla*), 188
Sanjuanico (*Jacquinia macrocarpa* subsp. *pungens*), 65, 72, 120, *121*, 123, 263–64
Sanmiguelito (*Antigonon leptopus*), 125, 241
Santa Bárbara, 43
Santa de Cabora, La (Teresa Urrea), 51–52, 327nn.40, 41
Sapindaceae, 74, 250–51
*Sapindus saponaria* (tupchi; abolillo, amolillo; soapberry), 74, 125, 251
Sapotaceae, 71, 251–52
Sapuchi (*Randia laevigata*), 74, 246
*Sarcostemma cynachoides* subsp. *hartwegii* (huichori), 141–42
Sauce (*Salix* spp.), 74, 250
Saúz (*Salix bonplandiana*), 74
Saya (*Amoreuxia* spp.), 103–4, 125, 152
Scaly-stem (*Elytraria imbricata*), 123
*Sceloporus clarkii* (Clark's spiny lizard), 234
*Schoepfia* (cuta béjori; palo cachora): *schreberi*, 234; *shreveana*, 76, 234
*Scirpus americanus* (tule; bulrush), 183
Seasons, 10–11
*Sebastiania pavoniana* (túbucti; brincador; Mexican jumping bean), 110, 190
Selaginaceae, 123, 253
*Selaginella* (teta segua; flor de piedra; resurrection plant): *novoleonensis*, 123, 253; *pallescens*, 253
*Selenicereus vagans* (sina cuenoji; sina volador), 168–69
Semana Santa. *See* Holy Week ceremonies
*Senna: atomaria* (jupachumi; palo zorillo), 71, 111, 112, 217; *obtusifolia, occidentalis* (ejotillo cafecillo), 217–18; *pallida* (huicori bísoro; flor de iguana), 71, 218
Seris, 45, 46, 160, 263, 321n.7, 324n.16, 325n.28
*Serjania* spp. (caamugia; güirote de culebra), 251
*Sesamum orientale* (ajonjolí; sesame), 13, 235
*Sesbania herbacea* (baiquillo; ejotillo), 218
*Setaria liebmannii* (hayas guasia; cola de la zorra), 240

Shade trees, 98–99, 110, 113, *197*, 210, 211, 213, 214, 216, 229, 230
Sheep, 240
Sibiricahui, Cerro, 136
*Sida* spp., 115, 130
*Sideroxylon* (júchica), 76, 130; *occidentale*, 251–52; *persimile* subsp. *subsessiflorum* (bebelama), 252; *tepicense* (ca'ja; tempisque), 71, 115, 125, 252
Sierra Madre Occidental, 9, 14–15
Silver mining, 33, 42–43, 207
Simaroubaceae, 253–54
*Simmondsia chinensis* (jojoba), 254
Simmondsiaceae, 254
Sina, 115; *Peniocereus marianus*, 167; *P. striatus*, 11, 167, *168*; *Stenocereus alamosenis*, 65, 169–71
Sina volador (*Selenicereus vagans*), 168–69
Sinaaqui (*Stenocereus alamosensis* × *S. thurberi*), *170*, 171
Sinaloa, 6, 7, 8, 9, 15, 17, 68, 321n.1; Mayos in, 19, 56–57; Spanish conquest and, 32–33; tropical deciduous forest in, 70–71
Sinaloans, 16, 34, 35
Sinita (*Lophocereus schottii* var. *tenuis*), 65, 120, 163–64
*Siphonoglossa mexicana* (cordoncillo), 134–35
Sirebampo, 29, 224
*Sisymbrion irio* (pamitón; London rocket), 132
Sitavaro (*Vallesia glabra*), 68, 99–100, 115, 122, 123, 124, 126, 139, 219
Skin conditions, 124; Asteraceae for, 144, 145, 147, 149, 170; Capparaceae for, 173; Celtidaceae for, 176; Euphorbiaceae for, 185, 187, 188–89, 190; Fabaceae used for, 207; *Guaiacum coulteri* for, 270–71; *Heliotropium curassavicum* as, 156; *Maytenus phyllanthoides* for, 175
Skunks, 78
Slavery, in silver mines, 42–43
Sleeping potion, 148
Snake bites, treatment of, 116, 192, 193, 235, 236, 258
Snakes, 76–77, 261; brown vine (*Oxybelis aeneus*), 233; coral (*Micruroides euryxanthus, Micrurus distans*), 117; indigo (*Drymarchon corais*), 77, 233; long-nosed (*Rhinocheilus lecontei* subsp. *antoni*), 77; parrot (*Leptophis diplotropis*), 233. *See also* Boa constrictor

Soap, shampoo, 115, 137, 144, 180, 212, 219, 264
Soapberry (*Sapindus saponaria*), 74, 125, 251
Social organization, 31–32, 52–53
Solanaceae, 110, 121, 122, 254–61
*Solanum: americanum* (mambia; chichiquelite; nightshade [*S. nodiflorum*]), 125, 259–60; *azureum* (paros pusi; sacamanteca), 122, 260–61; *nigrescens* (mambia; chichiquelite; nightshade [*S. douglasii*]), 125, 259–60; *seaforthianum* (bellísima), 110, 260; *tridynamum* (paros pusi; sacamanteca [*S. Amazonium*]), 122, 260–61
Solitaire, brown-backed (*Myadestes occidentalis*), 77
Solonaceae, 65, 71, 121
Songs: about *Amoreuxia*, 104; about *Cordia parvifolia*, 81–82; about *Desmanthus* spp., 199; about *Solanum* spp., 260; about *Stenocereus thurberi*, 90
Sonora, 7–8, 15, 325nn.26, 27, 68; Jesuit views of, 39–40; land tenure in, 47, 48–49, 54; Mayos in, 17, 19; porfiriato in, 55–56; strongmen from, 50, 54–55
Sonoran Desert, 62
Sorcerers (hechiceros), 118, 119, 323n.15
Spanish era, 324nn.16, 17, 18, 19, 20, 325n.31; contact, 32–35; Mayo reputation and, 38–39; missions, 41–42; rebellions during, 44–45; repression during, 45–46; silver mining and, 42–43; subjugation during, 35–36, 38; warfare in, 36–37, 323n.11, 325n.28
Spanish needles (*Palafoxia linearis* var. *linearis*), 147
Spells, curing of, 118, 119
Spiderling (*Boerhavia* spp.), 232
Springs, 76
Spurges (*Euphorbia* spp.), 124, 130, 187
Squash, 13
*Stegnosperma halimifolium* (jamyo olouama; hierba de la víbora), 236
*Stemmadenia tomentosa* var. *palmeri* (berraco; huevos de toro), 138
*Stenocereus: alamosensis* (sina; galloping cactus), 65, 115, 169–71; *alamosensis* × *thurberi* (sinaaqui), *170*, 171; *montanus* (sahuira; pitahaya), 71, 171–*72*; *thurberi* (aaqui; pitahaya; organpipe cactus), 64, 66, 69, 70, 82–90, *108*, 113, 121, 125, 131, 172–73
Sterculiaceae, 71, 74, 261–63

*Struthanthus palmeri* (chichel; toji; mistletoe), 222
*Suaeda moquinii* (bahue mo'oco; chamisa del mar), 177
Susto, 119
Sweetbush (*Bebbia juncea*), 124, 144

Tabaco de coyote: *Nama hispidum*, 220; *Nicotiana obtusifolia*, 259
*Tabebuia: chrysantha* (to'bo saguali; amapa amarillo), 150–51; *impetiginosa* (to'bo; amapa), 12, 71, 74, 110, 113, 151–52
Tábelojeca (*Capparis flexuosa*), 173
*Tachardiella coursetiae* (lac scale insect), 198
*Tagetes filifolia* (yerbanís), 149
Tajia Moroyoqui, Elvira, 269
Tajia Yocupicio, Vicente, 26, *60*, 77, 100, *108*, 110, 128, 132, 135, 141–42, 151, 157, 160, 173, 183, 219, 220, 221, 222, 233, 235, 236, 238, 240, 241, 243, 246, 251, 253, 255, 261, 278; on Asteraceae, 144, 145, 146, 148; on Cactaceae, *83*, 86, 90, 95, 96, 163, 164, 165, 167, 170, 171; on Convolvulaceae, 178–79; on *Cordia parvifolia*, 80–81; on Cucurbitaceae, 180, 182; on Euphorbiaceae, 185, 187, 188–89, 190; on Fabaceae, 193, 196, 199, 205, 206, 210, 212, 214; on jito, 105–6; on Moraceae, 229, 231; on Solanaceae, 257, 259; on Verbenaceae, 266, 267
Tajuí (*Krameria* spp.), 65, 123, 124, 175, 201, 220
Talayote (*Passiflora arida*), 235
*Talinum paniculatum* (cochizayam; pink baby's breath), 241
Tampicerán (*Platymiscium trifoliolatum*), 214–15
Tatachinole (*Tournefortia hartwegiana*), 72, 156
Tavachín (*Caesalpinia pulcherrima*), 72, 110, 133
Taxodiaceae, 74
*Taxodium distichum* (ahuehuete; sabino; Mexican bald cypress [*T. mucronatum*]), 74, *75*, 110, 114, 132, 182–83
Tayassidae, 103
*Tayassu tajacu* (tayaso; jabalí; javelina), 103
Tchuna (*Ficus* spp.), 110
Té de lila (*Melochia* spp.), 262–63
Teachive, 10, 32, 77, 86–87, 95, 158, 110
Teas. *See* Medicine
*Tecoma stans* (gloria, lluvia de oro; yellow

trumpet bush): var. *angustata*, 152; var.
  *stans*, 109, 120, 152
Tehuecos, 16, 35, 36, 45, 48
Temperatures, 10
Tempisque (*Sideroxylon tepicense*), 71, 115,
  125, 252
Temporales, 7, 12
Tepahuis, 17
Tepeguaje (*Lysiloma watsonii*), 12, 71, 72, 74,
  113, 123, 208
Tepehuanes, 36
Tepozana (*Eupatorium quadrangulare*), 145
Terúcuchi, Cerro, 68, 77
Tescalama (*Ficus petiolaris*), 72, 124, 230–31
Tesia, 6
Teso (*Acacia occidentalis*), 114, 192
Teta del diablo (*Matelea altatensis*), 141
Teta segua, 123
Tetas del venado (*Asclepias subulata*), 141
*Tetramerium abditum* (rama del toro), 130, 135
Theophrastaceae, 65, 263–64
Thorn apple, desert (*Datura discolor*), 257
Thornscrub, 3, 4, 8, 9, 329n.4. *See also*
  Coastal thornscrub; Foothills thornscrub
Thrasher, curve-billed (*Toxostoma curvirostre*), 157
*Tidestromia lanuginosa* (mochis; hierba
  lanuda), 123, 138
*Tigrisoma mexicanum* (tiger heron), 77
Tilapia, 149
Tiliaceae, 71, 264
*Tillandsia* (huírivis cu'u; mescalito; ballmoss):
  *exserta*, 11, 157; *recurvata*, 11, 157
Tobacco: tree (*Nicotiana glauca*), 132, 258;
  wild (*Nicotiana obtusifolia*), 259
Tóbari, Estero, 62, 328n.1
Toji: *Phoradendron californicum*, 269;
  *Struthanthus palmeri*, 222
Toloache (*Datura* spp.), 121, 257
Tomatillo (*Physalis maxima*), 259
Tonchi (*Marsdenia edulis*), 141, 142
Tonics, tónicos, 119
Tonsillitis, 124, 144, 146
Tools, tool handles, 114, 156, 176, 201, 202,
  209, 250, 252
Topolobampo, 6, 9, 57
Tornillo (*Helicteres baruensis*), 74, 246
Torote acensio (de incienso) (*Bursera
  penicillata*), 71, 160
Torote copal: *Bursera lancifolia*, 158–59; *B.
  stenophylla*, 71, 113, 161

Torote de vaca (*Bursera fagaroides*), 158
Torote lechosa (*Euphorbia gentryi*), 68
Torote panalero (*Jatropha cordata*), 12; uses
  of, 100–103, 123, 189
Torote prieto (*Bursera laxiflora*), 12, 65–66,
  67, 68, 69, 120, 159–60
Torotes (to'oro): *Bursera* spp., 67, 113, 114,
  130, 158–61; *Jatropha* sp., 189
Torres, Luis, 50
Tortillas, etcho seed, 97–98
Tortoises, 77
Totoliguoqui, 53
*Tournefortia hartwegiana* (tatachinole),
  72, 156
Toxins: in *Cephalanthus occidentalis* var.
  *salicifolius*, 244; in *Karwinskia humboldtiana*, 242–43
*Toxostoma curvirostre* (huiviris; huitlacoche;
  curve-billed thrasher), 157
*Tragia* sp. (nata'ari; ortiga; noseburn), 123;
  *jonesii*, 190–91
Trebolín (*Melilotus indica*), 132
*Trianthema portulacastrum* (mochis; verdolagas; horse-purslane), 137
*Trichilia* spp. (síquiri tájcara; bola colorada,
  piocha), 228
Trompillo (*Ipomoea* sp.), 123, 180
Tropical deciduous forest, 6, 8, 9, 12, 70–73,
  74, 75, 76, 131, 329n.4
Trucha (*Croton ciliatoglandulifer*), 185
Trumpet bush, yellow (*Tecoma stans*), 152
Tule: *Scirpus americanus*, 183; *Typha
  domingensis*, 265
Tullidora (*Karwinskia humboldtiana*), 67,
  121, 122, 123, 242–43
Turkeys (*Melagris gallopavo*), 219
*Turnera diffusa* (damiana), 265
Turneraceae, 265
*Typha domingensis* (tule; cattail), 265
Typhaceae, 265

Uña de gato (*Mimosa distachya*), 65, 209
Unicorn plant, desert (*Proboscidea altheaefolia*), 125, 226–27
Upper respiratory ailments, 120–21, 217, 235,
  236
*Urera baccifera* (ortiguilla), 265
Urinary ailments, 96, 116, 117, 123, 143, 189,
  191, 201
Urrea, Teresa (La Santa de Cabora), 51–52,
  327nn.40, 41

Urticaceae, 265
Utopian communities, 57
Uvalama (*Vitex mollis*), 74, 125, 268–69

Vainoro (*Pisonia capitata*), 232–33
Valencia Moroyoqui, Seferino, 10, 144, 190, 234, 242, 248, 278
Valenzuela, Alvaro, 172, 279
Valenzuela, Benjamín, 126, 239, 258, 279
Valenzuela, Berta, 268
Valenzuela, Felicita, 219–20
Valenzuela, Fernando, 3
Valenzuela, Luciano "Chano," 97, 239, 248, 279
Valenzuela, Luis, 144, 157, 228, 234, 279
Valenzuela, Marcelino, 174, 190–91, 217, 222, 279
Valenzuela, Marcos, 149, 168, 279
Valenzuela, Rosario "Chalo," 138, 145, 161, 165, 172, 189, 250–51, 254, 279
Valenzuela, Sara, 177, 279
Valenzuela, Vicenta, 204
Valenzuela, Wencelao "Mono," 250
Valenzuela Nolasco, Francisco "Pancho," 74, 86, 100, 118, 128, 153, 156, 157, 174, 180, 189, 206, 218, 219, 221, 223, 226, 239, 240, 246, 253–54, 264, 265, 270; on Acanthaceae, 133, 134, 135; on Asteraceae, 143, 144, 146, 149; on Fabaceae, 191, 193, 199, 200, 201, 202, 206, 214
Valenzuela Yocupicio, Santiago, 166, 187, 222, 225, 236, 241, 257, 279
Valeriana (*Lippia alba*), 267
*Vallesia glabra* (sitavaro), 68, 99–100, 115, 123, 124, 126, 139
Vara blanca (*Croton fantzianus*), 71, 113, 185–86
Vara prieta: *Brongniarti alamosana*, 194; *Cordia parvifolia*, 65, 79–83, 114, 126, 155; *Croton* spp., 68, 185, 186–87
Venereal disease, 123–24, 167, 244
Venus tree (*Lonchocarpus hermannii*), 12, 71, 72, 74, 124, 206–7
Verbenaceae, 74, 120, 125, 130, 265–69
Verdolaga: *Portulaca* spp., 241; *Trianthema portulacastrum*, 137
Verdugo Rojo, Amelia, 25, 144, 232, 269, 279
Villages, 8; distribution of, 6–7; social organization of, 52–53
Vines, 65, 141, 177, 179–80, 181, 183, 216–17, 223, 241, 251, 270

Vinorama (*Acacia farnesiana*), 65, 191, 192
Violaceae, 269
Violin head (*Heliotropium angiospermum*), 156
Viperidae, 77
Viscaceae, 269–70
Vitaceae, 270
*Vitex: mollis* (júvare; igualama, uvalama), 74, 125, 268–69; *pyramidata* (jupa'are; negrito), 74, 269

Water rights, 7–8
Weaving industry, 114; *Adela virgata* in, 184; *Agave vivipara* fiber in, 94–95; Cactaceae used in, 96, 170–71; *Citharexylum scabrum* used in, 266; *Condalia globosa* used in, 242; *Cordia parvifolia* used in, 81, 155; with grasses, 237, 239; *Haematoxylum brasiletto* used in, 203; *Havardia mexicana* used in, 204; and morrales, 93–95; with palms, 139–40; tule used in, 183, 265; Verbenaceae used in, 266
Willow: Bonpland (*Salix bonplandiana*), 74, 250; Goodding (*S. gooddingii*), 74, 250; seep (*Baccharis salicifolia*), 144
*Wimmeria mexicana* (chi'ini; algodoncillo), 71, 112, 113, 175
*Wissadula hernandioides*, 130
Witches (brujos), 119
Wolfberry (*Lycium* spp.), 65, 125, 257–58

*Xylothamia diffusa* (jeco), 149

Yaqui, Río, 8, 13; development of, 50, 325n.24, 326n.37, 327n.42
Yaquis, 16, 17, 18, 19, 26, 51, 160, 180, 324n.16, 325n.36; land tenure and, 47, 326–27n.38; and Mexican Revolution, 54, 328n.46; Mexican subjugation of, 50, 326n.37; rebellions by, 44–45, 48, 50–51; resistance of, 55–56; Spanish and, 4–5, 33, 34, 38, 44, 46, 323n.11; warfare with, 34–35, 36–37
Yavaros, 63
Yerbanís (*Tagetes filifolia*), 149
Yocogigua, 29
Yocupicio, Román, 56
Yocupicio Anguamea, Felipe, 83, 97, 133, 173, 184, 196, 262–63, 264, 266, 279–80
Yópori, 58
Yucatán, 55
Yucohuira (*Cissus* spp.), 270

Zacate (Poaceae), 130, 132, 236–40
Zacate de lana (*Cynodon dactylon*), 238
*Zanthoxylum* (palo limón), 249; *fagara* (o'ouse suctu; limoncillo), 72, 249; *mazatlanum* (yocutsuctu; jaguar's-claws), 249
Zapata, Emiliano, 54
Zapote (*Casimiroa edulis*), 112, 248
Zazueta, Felipe, 194
Zazueta, Lidia, 100, 187, 280
Zazueta, María Teresa, 92, 280
*Zenaida asiatica* (paloma pitahayera; white-winged dove), 87
*Ziziphus: amole* (báis cápora; saituna), 12, 65, 66, 125, 157, 243; *obtusifolia* var. *canescens* (jutuqui; ciruela del monte, huichilame; graythorn), 120, 243–44
Zuaques, 16, 17, 34, 35, 36, 45, 48
Zygophyllaceae, 12, 116, 270–71

| | |
|---:|:---|
| Compositor: | Integrated Composition Systems |
| Text: | 10/13 Minion |
| Display: | Minion |
| Printer/Binder: | Edwards Brothers, Inc. |
| Indexer: | Linda Gregonis |